The **Political Economy** of South-East Asia

AN INTRODUCTION

EDITED BY

GARRY RODAN

KEVIN HEWISON

RICHARD ROBISON

Melbourne

OXFORD UNIVERSITY PRESS

Oxford   Auckland   New York

OXFORD UNIVERSITY PRESS AUSTRALIA

Oxford   New York
Athens   Auckland   Bangkok   Bombay
Calcutta   Cape Town   Dar es Salaam   Delhi
Florence   Hong Kong   Istanbul   Karachi
Kuala Lumpur   Madras   Madrid   Melbourne
Mexico City   Nairobi   Paris   Port Moresby
Singapore   Taipei   Tokyo   Toronto

and associated companies in
Berlin Ibadan

OXFORD is a trade mark of Oxford University Press

National Library of Australia
Cataloguing-in-Publication data

Rodan, Garry, 1955– .
The political economy of South-East Asia: an introduction

Bibliography.
Includes index.
ISBN 0 19 553736 X. (pbk)
ISBN 0 19 550654 5.

1. Political science — Asia, Southeastern. 2. Asia,
Southeastern — Economic conditions. 3. Asia,
Southeastern — Economic policy. 4. Asia,
Southeastern — Foreign economic relations. 5.
Asia, Southeastern — Politics and government. 6.
Asia, Southeastern — Commercial policy. 7. Asia,
Southeastern — Foreign relations. I. Hewison,
Kevin J. II. Robison, Richard, 1943– . III. Title.

330.959053

Edited by Lee White
Cover design by Anitra Blackford
Typeset by Desktop Concepts P/L, Melbourne
Printed by Kyodo Printing Co. Pty Ltd, Singapore
Published by Oxford University Press,
253 Normanby Road, South Melbourne, Australia

Contents

Contributors

Melanie Beresford is a Senior Lecturer in Economics at Macquarie University.

Frederic Deyo is Professor of Development Studies in the Faculty of Arts at the University of Auckland.

Kevin Hewison holds the Foundation Chair in Asian Languages and Societies at the University of New England.

Jane Hutchison is a Lecturer in the Politics Programme at Murdoch University.

Andrew MacIntyre is Associate Professor and Associate Dean in the Graduate School of International Relations and Pacific Studies at the University of California, San Diego.

James Parsonage is a PhD candidate in the Politics Programme, Murdoch University, researching regional strategies for capital accumulation from Singapore.

Rajah Rasiah is an Associate Professor in Economics at Universiti Kebangsaan Malaysia.

Richard Robison is Director of the Asia Research Centre and Professor of Southeast Asian Studies at Murdoch University.

Garry Rodan is a Senior Research Fellow of the Asia Research Centre and a Senior Lecturer in the Politics Programme at Murdoch University.

Abbreviations

ABRI	Angkatan Bersenjata Republik Indonesia
ADB	Asian Development Bank
AFTA	ASEAN Free Trade Agreement
APEC	Asia-Pacific Economic Cooperation
APKINDO	Asosiasi Panel Kayu Indonesia
ASB	Amanah Saham Bumiputera
ASEAN	Association of South-East Asian Nations
ASEM	Asia–Europe Meeting
ASN	Amanah Saham Nasional
BBD	Bank Bumi Daya
BCIC	Bumiputera Commercial and Industrial Community
BKPM	Badan Koordinasi Penanaman Modal
BoI	Board of Investment
BPIS	Badan Pengelola Industri Strategis
BPPC	Badan Penyangga dan Pemasaran Cengkeh
BPPT	Badan Pengkajian dan Penevapan Teknologi
BRI	Bank Rakyat Indonesia
BS	Barisan Sosialis
CEDA	Committee for the Economic Development of Australia
CER	Australia–New Zealand Closer Economic Relationship
CGI	Consultative Group on Indonesia
CIDES	Center for International Development Studies
COLT	Commercial Offshore Loan Team
CPF	Central Provident Fund
CPP	Communist Party of the Philippines
DBS	Development Bank of Singapore
DRV	Democratic Republic of Vietnam

EAEC	East Asian Economic Caucus
EAEG	East Asian Economic Group
EAGA	East ASEAN Growth Area
EDB	Economic Development Board
ELIPS	Economic Law and Improved Procurement System
EOI	export-oriented industrialisation
EU	European Union
FDI	foreign direct investment
FELDA	Federal Land Development Authority
FTZ	Free Trade Zone
FYP	five year plan
GAPPRI	Gabungan Pengusaha Pabrik Rokok Indonesia
GATT	General Agreement on Tariffs and Trade
GIC	Government of Singapore Investment Corporation
GLC	government-linked company
GMS	Greater Mekong Subregion
HDB	Housing Development Board
HICOM	Heavy Industry Corporation of Malaysia
HPH	Hak Pengusahaan Hutan
HRDC	Human Resource Development Council
HRDF	Human Resource Development Fund
HTI	Hutan Tanaman Industri
IBRD	International Bank for Reconstruction and Development
IAC	International Advisory Council
ICA	Industrial Coordination Act
ICMI	Ikatan Cendekiawan Muslim Indonesia
IGGI	Inter Governmental Group on Indonesia
ILO	International Labour Organization
IMF	International Monetary Fund
IMS-GZ	Indonesia–Malaysia–Singapore Growth Zone
IPTN	Industri Pesawat Terbang Nusantara
ISI	import substitution industrialisation
JSEDC	Johor State Economic Development Corporation
JSX	Jakarta Stock Exchange
JTC	Jurong Town Corporation

MARA	Majelis Amanah Rakyat
MCA	Malaysian Chinese Association
MERCOSUR	Mercado Comun del Cono Sur
MFA	Multi Fiber Agreement
MFN	Most Favoured Nation
MIGHT	Malaysia Industry-Government Group of High Technology
MITI	Ministry of International Trade and Industry
MNC	multinational corporation
MOSTE	Ministry of Science, Technology and Environment
MTDC	Malaysian Technology Development Corporation
NAFTA	North American Free Trade Agreement
NEP	New Economic Policy
NESDB	National Economic and Social Development Board
NGO	non-government organisation
NIC	newly industrialised country
NLF	National Front for the Liberation of Southern Vietnam
NOL	Neptune Orient Lines
NPA	New People's Army
NST	*New Straits Times*
NTUC	National Trades Union Congress
NWC	National Wages Council
OCBC	Overseas Chinese Banking Corporation
ODA	Overseas Development Aid
OECD	Organization for Economic Cooperation and Development
OHQ	operational headquarters
PAFTAD	Pacific Trade and Development
PAP	People's Action Party
PAS	Parti Islam SeMalaysia
PBEC	Pacific Basin Economic Council
PBS	Parti Bersatu Sabah
PECC	Pacific Economic Cooperation Council
PERNAS	Perbadanan Nasional
PIA	Promotion of Investments Act
PIEU	Pioneer Industries Employees' Union
PKI	Partai Komunis Indonesia
PLN	Perusahaan Listrik Negara
PSDC	Penang Skills Development Centre

R&D	Research and Development
RBF	Regional Business Forum
RISDA	Rubber Industry Smallholders Development Authority
RVN	Republic of Vietnam
SDP	Singapore Democratic Party
SGZ	subregional growth zone
SET	Securities Exchange of Thailand
SEZ	special economic zone
SILO	Singapore Industrial Labour Organisation
SLORC	State Law and Order Restoration Council
SME	small and medium enterprises
SOBSI	Sentral Organisasi Buruh Seluruh Indonesia
SOE	state-owned enterprise
SPSI	Serikat Pekerja Seluruh Indonesia
STC	Singapore Technologies Corporation
TCC	transnational capitalist class
TFP	total factor productivity
TNC	transnational corporation
TVRI	Televisi Republik Indonesia
UMNO	United Malays National Organisation
USAID	United States Agency for International Development
USTR	United States Trade Representative
VAT	value-added tax
VDP	Vendor Development Programme
WTO	World Trade Organisation

Tables, Figures and Maps

Maps

Acknowledgments

This project has been possible because of the support from the Asia Research Centre at Murdoch University, which is a Special Research Centre of the Australian Research Council. Academic colleagues and support staff of the Centre and University have made valued contributions to the realisation of this volume. These people include Del Blakeway, Geoff Paton, Lesley-Anne Philps, Cisca Spencer, Andrew Brown, David Brown, Rob Lambert, Ian Scott and Helen Bradbury. The editors are also grateful to Jomo K. S. for his constructive criticisms at a two-day workshop assisting us to refine the manuscript. The editors are especially appreciative of the considerable assistance of publications officer Amanda Miller and research assistant Robert Roche. Finally, we thank the contributors themselves for their commitment, cooperation and good work.

Garry Rodan, Kevin Hewison and Richard Robison

1

Introduction

Richard Robison, Garry Rodan and Kevin Hewison

Despite the widespread adoption of policies aimed at securing economic growth and development, the economies of South-East Asia were to struggle in the immediate post-Second World War period (Tarling 1966: 238–45). Except for Singapore, South-East Asian countries continued to be characterised by primary commodity production—the vast majority of people engaged in agricultural production, much of it for subsistence. Industry was generally rudimentary, small-scale and often craft-oriented. However, over the past three decades the economies of the region have undergone remarkable transformations. Rates of economic growth have increased exponentially, and the region is now a significant exporter of industrial goods. Such developments have been accompanied by urban con-glomeration and the rise of significant middle and working classes. South-East Asia is no longer an economic backwater but an influential eco-nomic region, with increasing internal cooperation in trade and production.

Again with the exception of Singapore, this shift to industrialism in South-East Asia took place two decades after the dramatic industrialisation of the newly industrialising countries (NICs) of North-East Asia, namely Taiwan, South Korea and Hong Kong. Consequently, there is a discernible difference in emphasis in much of the literature examining North-East and South-East Asia, the former having a greater preoccupation with the ques-tion of how industrialisation was achieved. In North-East Asia powerful states instigated national development strategies based on export-oriented industrialisation (EOI), involving various collective arrangements with cor-porate capital. Where successful economic growth had been so firmly associated with market systems in orthodox liberal thought, it is not sur-prising that the focus of research was upon the role of the state; either to propose state-led markets as an alternative to the liberal model, or to explain the significance of the state in these cases.

In explanations of economic growth in South-East Asia, the role of the state has not been given the same prominence. To some extent this reflects an ideological bias in favour of the market among writers who tend to see the North-East Asian experience as an aberration and wish to propose South-East Asia as a different route to growth, one without substantial state involvement. But it also reflects the extraordinary diversity and complexity of state involvement in economic life in South-East Asia which often masks the extent of this role, frequently embedding it in a diverse range of relationships with social forces. For example, in the Philippines, political and economic power is often seen to revolve around powerful oligarchies, while in Indonesia the power of the state resembles a form of command capitalism, even though the market is less constrained than it was in past decades. Vietnam, of course, has the added challenge of moving from an economy dominated by state enterprises to one where the private sector takes a leading role. In Singapore, state officials have had a crucial role in directing development in a society where the domestic capitalist class has embarked on increasing cooperation, rather than competition, with state capital. In Thailand, the state has facilitated the emergence of one of the most vigorous capitalist classes in the region.

In all South-East Asian economies there has been a continued opening to the world economy and a retreat from the nationalist and state-led economic regimes of the immediate postwar period. In recent years, privatisation has accompanied programs of market deregulation, while domestic capital has become increasingly internationalised, entering international financial markets, investing overseas and extending links with foreign corporate groups. Private banking systems and capital markets have developed and expanded significantly. These moves have been encouraged by a range of international institutions, including the World Bank, International Monetary Fund, Asian Development Bank, Asia-Pacific Economic Cooperation, General Agreement on Tariffs and Trade and World Trade Organisation.

However, while the institutions of liberal capitalism are apparently being reproduced in the region, their emergence is not without problems. Despite all the changes, states continue to play a critical economic role. To varying degrees, they continue to maintain systems of trade protection, develop and implement strategic industry policies, and public sector investment remains significant. In addition, legal-rational systems of authority have not yet established a clear distinction between public authority and private power.

Such observations have seen a number of questions emerge, both in academic circles and the policy arena. First: what are the dynamics at work? Is there an inexorable advance of the market system, driven by a functional superiority of such systems in generating growth and prosperity? Is change being driven by a rational process of institution building in which new collective rules, legal systems, property rights and regulatory frameworks are incrementally established as a functional response to new problems in the organisation of political and economic life? Is change driven by the interests of new social forces through political contests for power and control of the policy agenda?

Second: where are the economies of the region heading? Towards various liberal models of economic and social organisation characterised by deregulated markets and legal frameworks protecting the individual from the state? Or towards systems of 'collective capitalism' in which the state plays a leading and coordinating role, and strategic industry and trade policies? Or perhaps towards systems of unconstrained capitalism in which public power becomes the possession of contending oligarchies or groups of state bureaucrats, and markets are dominated by the exercise of political authority rather than formal and enforceable rules. Can such systems coexist with industrial capitalism in the long term? Or are these economies reproducing many of the same problems and issues which arose in the earlier industrialisations in Western Europe, North America and Japan, but in a 'pressure-cooker' environment of exceptionally swift social and economic change? Or, more simply, is there an 'Asian model' of capitalism?

Third: what are the social and political consequences of the rapid emergence of capitalism in South-East Asia? What friction and conflicts are associated with the transformation? What are the likely political responses to these? How do patterns of social and political domination change? Does rapid economic development mean that parliamentary political forms are inevitable? What role or roles can we expect rapidly expanding middle classes to play in political development in the region? Will this mirror the experience of earlier industrialisations? Will rapid industrialisation open the way for sizeable working classes to emerge, capable of organised, independent social and political action? Is there to be a development of a welfare state? Or will South-East Asian states seek other means of dealing with the inequalities often resulting from capitalism?

In this chapter, we set out some of the theoretical divisions shaping the political economy literature that attempts to answer such questions. We will also introduce the chapters that comprise this volume, highlighting the central political economy arguments embodied in each.

Explaining Development

The first concerted attempts to explain the problems of transformation in South-East Asia drew heavily on a body of theory, conveniently categorised as the 'modernisation approach'. Many believed that it was possible for the former colonies to replicate the 'original transition' of Western Europe, from traditional to modern society (see Roxborough 1979: chs 1–2). Modernisation theories begin by establishing a dichotomy between tradition and modernity, ' and seeing an evolutionary movement from the former to the latter. A variety of modernisation approaches emerge from the different emphases placed on sociological, psychological and economic factors in the transition (see Larrain, 1989: 87–98; Hoogvelt 1982: chs 3–4). Traditional societies were seen as 'pre-state, pre-rational and pre-industrial' (Higgott et al. 1985: 17–18). In order to modernise, traditional societies needed to adopt the same organisational structures and social and political values of the West. Significantly, this included the adoption of liberal-democratic political arrangements.

However, this optimism had begun to wane by the mid-1960s as economic growth languished. It became clear that many of the assumptions about the conditions required for development were not emerging, and young democracies were being replaced by authoritarian regimes. While others preceded him (for example, Olson 1963), it was Samuel Huntington (1968) who made the strongest case that modernisation was threatened by political instability. For him, strong government and political order were required if growth and development were to succeed.

Many of the governments of South-East Asia struggled against political instability and Huntington's ideas provided a rationale for the suppression of opposition. This was complemented by policies for economic development, fostered by international agencies, which emphasised the need to maximise growth so that its benefits would eventually 'trickle down' to all levels of society. Growth was to be enhanced through policies attractive to foreign investment, thereby alleviating any shortage of domestic capital (Higgott et al. 1985: 24–7).

At the same time that modernisation theory was being revised to account for the apparent failure of development, what became known as 'dependency theory' challenged modernisation approaches in a quite fundamental way. Dependency theory had its origins in Latin America, where economists contested the notion that modernisation could be diffused to poor countries (see Larrain 1989: 102–10). This was developed into a radical critique of modernisation theory by A. G. Frank (1967, 1969).

Frank's approach was radical in that he referred to the 'development of underdevelopment', turning modernisation theory on its head, arguing that the development of the already developed countries depended on the *under*development of poor countries. Caught in a web of international political and economic relationships, Frank argued that poor countries stayed poor because of the exploitative relationship established between them and the already developed countries.

In explaining how this occurred, Frank argued that the economic surplus of poor countries is lost precisely because the structures considered important by modernisation theorists—foreign investment through transnational corporations, foreign aid, international loans, and international trade regimes—are the structures which extract the surplus from the underdeveloped countries. Local capitalists and the state act as compradors, providing the links between the developed and underdeveloped economies necessary for surplus extraction.

Frank's position was considered by some—including theorists sympathetic to his position—to be too simplistic. Samir Amin (1974, 1976), for example, argued that the mechanism of surplus extraction is far more complex. For him, exploitation was embedded in unequal trade, where the value of labour congealed in commodities from the First World and Third World is quite different. Others, such as Fernando Cardoso and Enzo Faletto (1979) and Peter Evans (1979), attributed a greater degree of importance to local class developments than did Frank, even if they ultimately concluded that the political capacity of the local state was substantially compromised by the power of international capital. Writers, such as Gary Gereffi (1982) and Douglas Bennett and Kenneth Sharpe (1985), also examined specific cases of local state intervention to promote certain industries which achieved some success. In each of these studies, though, multinational corporations were seen as having considerable bargaining power to the detriment of Third World states. Despite the greater sophistication of these works they maintained the primary theoretical proposition of dependency theory—that stunted or incomplete development in the Third World resulted from decisions taken at the headquarters of multinationals in the advanced capitalist centres.

Asian Economic 'Miracles'

Dependency theory made an important contribution in turning critical attention to the international factors complicating development prospects in the Third World. Notwithstanding this, it was unable to answer a host of theoretical challenges posed by critics, emphasising the importance of social and economic developments within countries rather than international patterns of exploitation as the factors of development and underdevelopment (Brenner 1977; Kay 1975; Warren 1973). However, it was the emergence of countries like South Korea, Taiwan, Hong Kong and Singapore as important industrial exporters in the late 1970s that delivered a mortal blow to the claims of dependency theorists. Rather than being consigned to perpetual underdevelopment in a global system of exploitation dominated by the traditional industrial centres, these so-called peripheral economies were to become the victors in savage global contests of trade in manufactured goods. Their move into higher technology production and sophisticated services sectors, together with the development of substantial, well-educated middle classes, suggested a development pattern similar to that in the industrialised countries. Did the East Asian experience introduce a new model of managed capitalism to challenge the ideal of liberal capitalism in which growth was produced by the free operation of individual self-interest?

In explaining this, neo-classical economists were quick to claim victory. The success of the East Asian industrial superstars, they argued, reflected the adoption of policies that embraced global market forces. Export orientation was seen as crucial in rapid industrialisation, regarded as a force for greater competitive discipline and, in turn, a more efficient manufacturing sector. This was contrasted with production primarily for domestic markets, which was often heavily protected and seen as inefficient (see Krueger 1981). Importantly, the basis of export success was considered to be the exploitation of 'comparative advantage'—concentrating production in areas of relative endowment abundance in either land, labour or capital. Initially, in those economies that were labour rich and capital scarce, this meant low-cost, labour-intensive production (Little 1981). Seen in this way, policy-makers in East Asia had simply chosen technically correct policies and the lesson was there for others to emulate. This perspective exerted considerable influence on the World Bank which, in its 1981 *World Development Report*, called on developing countries to follow the South Korean example. Subsequently, loans to countries were often conditional on the acceptance of an EOI strategy, trade liberalisation and similar free market policies.

However, some writers, conveniently labeled as 'statist' theorists, took a different view. They argued that the remarkable feature of industrialisation in these countries was not the 'freeing' of the market, but the role played by the state in orchestrating public and private capital in achieving strategic national economic goals. Development in East Asia was achieved by the 'management' of markets and by 'getting the prices wrong' within the framework of complex trade and industry regimes, export strategies and systems of state-business cooperation (Wade 1990; Amsden 1989; Rodan 1989; Weiss and Hobson 1995; Matthews and Ravenhill 1996). The importance of developmental elites building powerful institutions to coordinate markets was added to this explanation by Chalmers Johnson, Laura Tyson and John Zysman (1989). On the back foot, neo-classical economists were to initially deny the importance of the state's role but were then to redefine it as 'market facilitating'—assisting decisions that would have been made by the market in any case. State intervention in itself was no longer the critical point, but rather, whether such intervention impeded decision-making in the private sector.

This was essentially the argument adopted by the World Bank (1993) in an influential study, *The East Asian Miracle*. The miracle is defined as an achievement of high growth with equity. Significantly, this serious consideration of the role of the state followed Japanese pressure regarding the Bank's consistent free market prescriptions for all developing countries (Awanohara 1993: 79). In summary, the *East Asian Miracle* team concluded that the East Asian economies were successful because they got the basics right: (i) private domestic investment and growing human capital were the engines of growth; (ii) agriculture declined in relative importance, but continued to grow; and (iii) population growth was reduced (World Bank 1993: 5). The team also acknowledged the significance of 'sound development policy'.

It was argued that the governments of these countries have been important in establishing: (i) good macroeconomic management and stable economic performance, providing a framework for private investment; (ii) policies for the banking system which enhanced its integrity and raised national savings; (iii) emphasis on primary and secondary education; (iv) policies which enhanced agricultural productivity; (v) fiscal regimes where price distortions were kept to a minimum; and (vi) a society open to foreign ideas and technology (World Bank 1993: 5). The conclusion was that, in most of the economies studied, the government intervened, systemically and through multiple channels, to foster development, including development of specific industries. At least some of these interventions were seen to violate the dictum of establishing a neutral incentives regime for private investment (World Bank 1993: 5–6). Interestingly, the report argued that South-East Asian governments had played a much less prominent and often less constructive role in economic success than had their East Asian counterparts.

While it is significant that the World Bank's theoretical inclinations have been scrutinised, this has essentially been an exercise in revising the growth theory which had always been influential within the Bank, rather than challenging it. The 'new' growth theory sees government as important, keeping inflation low, maintaining political and macroeconomic stability, establishing the rule of law, encouraging education, keeping taxes low, promoting trade, and encouraging investment, both foreign and local (see *Economist* 25 May 1996: 23–5). The policy lesson is that governments should ensure that they have measures in place which support the growth of private investment.

A significant problem with this approach is the tendency for government to be conceived in neutral terms. The theoretical outcome is that 'good' public policy is relatively insulated from political influences. This is an interesting perspective, for it suggests that democratic or representative politics may be an impediment to good policy development. The Thailand study associated with the World Bank's *Miracles* document makes this clear when the legislature is described as having been relatively unproductive 'in making laws, especially when members of parliament are elected ...' Hence, the military coup is seen to 'perform an important function'. The junta 'assumes broad legislative powers, and ... break[s] the legislative logjam developed in previous elected parliaments' (Christensen et al. 1993: 19–20). Almost as an afterthought it is added that bureaucrats are not immune to extra-bureaucratic demands.

There were a number of partial endorsements of an economic role for the state from a somewhat different angle, although still within the mainstream of neo-classical economics. Alwyn Young (1995), for example, argued that state intervention may be useful at certain early stages in the industrialisation process where the capacity to apply vast amounts of capital and labour is decisive. However, when innovation and technology become critical factors then it is argued that a transition away from state intervention to a freer operation of the market is essential (Krugman 1994).

While the conclusion by the World Bank that the state in South-East Asia had been generally less significant to industrialisation than its counterpart in East Asia is contestable, the *Miracles* report did draw attention to the fact that the state in South-East Asia has been less cohesive in its control and marshaling of economic organisation and strategy than in South Korea, Taiwan or Japan. Protective trade and investment policies and state ownership were seldom tied into strategic, highly disciplined national programs for building export competitiveness. They were often devices for giving preferential access or bolstering the interests of the state and its officials. Unlike the states of East Asia, they were, in neo-classical political economy, more likely to be predatory states with rent-seeking bureaucrats, rather than the far-sighted bureaucrats idealised in the East Asian models.

Consequently, the neo-classical debate over South-East Asia has tended to be concerned with the question of transition from predatory and statist economic systems to markets. A central question has been to understand how to construct markets and regulatory frameworks in environments of chronic government failure, predatory states, rent-seeking and economic nationalist ideology. As was the case in the East Asian debate, however, an important theoretical divide has emerged between neo-classical political economy and varieties of political economy based broadly in theories of society. While the high public policy ground has been taken by variations on the neo-classical theme, there is a long tradition of alternative political economy approaches which place a greater emphasis on societal or class influences on the state. In other words, the state is not seen as independent of the social formation in which it arises. These political economy approaches explain state policy in the context of how power is organised in society as systems of social, political and economic domination.

Essentially there are three major political economy approaches relevant to South-East Asia. As with all attempts to categorise, these cannot be rigidly applied as there will always be writers who draw selectively from one or more approach. But the categories do serve as a useful guide to the most significant points of theoretical departure between the majority of authors. As will be demonstrated, these theoretical divides have their resonances in major international policy circles, where the sorts of ideas which have been influential have changed over time.

1 *Neo-classical economics and political economy*: emphasises the importance of market forces and prescribes further deregulation to ensure current economic development is consolidated and extended.
2 *New institutional political economy*: maintains that market development is contingent upon the establishment of institutional structures. These must provide the neutral regulatory framework necessary for the operation of free markets.
3 *Pluralist and Marxist-derived political economy*: employs social and political theories of markets in which markets and market institutions are seen as the products of social and political interest and conflict. This conflict inevitably shapes the course of development, not necessarily to the detriment of a market economy.

Neo-Classical Economics and Political Economy

In the neo-classical view, it is the individual pursuit of self-interest that results in the most efficient allocation of resources and, in the long run, the greatest wealth for society as a whole. Hence, social welfare and wealth are best achieved by allowing the free operation of the market as a natural and neutral, self-correcting and regulating mechanism separate from the dynamics of social and political power. Viewing government failure produced by intervention in markets as the primary cause of poor economic performance, neo-classical economists urge the withdrawal of the state from economic activity (Little, Scitovsky and Scott 1970; Bauer 1970; Balassa

1971; Krueger 1974). Given that the choice, in the neo-classical view, is a rational one between the efficiency and prosperity markets can deliver and the inefficiency and poverty produced by state intervention, it is no surprise that many of these theorists have been perplexed by the refusal of governments to seize with enthusiasm policies that allow markets to determine prices by deregulating trade, investment and financial regimes.

An explanation for this failure was to be provided by neo-classical political economy, which emerged in the 1970s as 'public choice theory'. The pluralist conception of the benign state committed to the common good was replaced with the neo-classical idea of the state as predator and its officials and elected representatives as self-seeking rentiers. Politics was conceived as a marketplace of transactions between individual politicians, officials and contending interest lobbies, its product logically being the virtual appropriation and sale of public policy and public goods in return for political support. At another level, the state itself was seen to have its own interests in accumulating ever-increasing revenues (Lal 1983; Krueger 1974; Buchanan and Tullock 1962).

Mancur Olson (1982) saw economic growth in democratic political systems strangled by vested interests seeking to share the spoils of growth (distributional coalitions), preventing good policy in the public interest and diverting scarce resources from productive investment. The predatory interests of political leaders frequently locked them into a pattern of rule aimed at keeping contending interests satisfied, ruining the economy in the process (see Bates 1981). Such perspectives reinforced the neo-classical theme that the essential struggle was between economics and politics, between the assumed rationality in markets and that embodied in politics. Approaches influenced by neo-classical political economy, juxtaposing rational economic technocrats seeking market reforms with self-seeking officials and politicians dealing in rents, corruption and grandiose schemes, were applied widely to government in South-East Asia (Liddle 1992; Soesastro 1989; Christensen and Ammar 1993).

What, then, are the neo-classical/public choice answers? For the industrial economies, they advocated the restriction of the state's economic role by imposing deregulation, removing the structural opportunities for senior bureaucrats to profit from their positions (rent-seeking) and for others to derive benefits from the market without having to pay for them (free-riding). In addition, the incentive and ability to raise taxes and run deficit budgets was reduced (Olson 1982). This proposition, however, presumes a substantial and politically organised middle class and business sector ideologically committed to freedom from state control. Ironically, the absence of this in much of Asia and Latin America led some neo-classical economists to propose that only benign authoritarian leadership, able to enforce free markets in the face of vested interests, could provide an answer in developing countries (Lal 1983; Srinivasan 1985).

New Institutional Political Economy

Neo-classical political economy was to come under attack from what has been termed the new institutional political economy. Emerging from within the ranks of liberal reformers, critics were to argue that the focus on the individual as the explanatory factor in analysis of state, market and economy was misguided. In the process of development and economic growth it was argued that the solution of collective problems through general systems of rules, regulation and property rights was the central issue. This required the building of a complex institutional capacity. Contrary to the position of public choice theory, these theorists were not convinced of the inherent rationality of the market and argued that unconstrained markets could lead to chaos, collusion, gangsterism and widespread market failure. Rather than dismantling states, it was only the state which could do that which individuals could not: by avoiding the free-rider problem and addressing long-term collective interests. The efficient operation of markets, in this view, required highly developed institutional frameworks for the regulation of economic activity, including rules for the use of common property and the private ownership of property, regulations for banking and exchange, mechanisms for the effective collection of taxes and the operation of welfare programs (Doner 1992; Haggard and Kaufman 1992).

In policy terms, the new institutionalist propositions forced those on the neo-classical side to argue that deregulation had to be complemented by the creation of 'institutional capacity' if markets were to be successful. In practice, the new institutional political economy was to be applied to the problem of uncovering impediments to reform and creating institutional capacity to facilitate deregulation in the marketplace.

Part of the policy impact of the new institutionalist approach was seen in the World Bank's (1991a) *World Development Report*, which added a concern for institution building, public sector reform, and regulatory frameworks for ensuring competition and legal and property rights to its market-friendly agenda (World Bank 1991a: 7, 9, 128–48). The Bank (1991b: 3) also noted in its report, *Managing Development: The Governance Dimension*, that: 'without the institutions and supportive framework of the state to create and enforce the rules [to make markets work more effectively], to establish law and order, and to ensure property rights, production and investment will be deterred and development hindered'.

The impact of its approach is also evident in the *Miracles* report, and especially in some of the background papers to the final report. The emphasis is on 'getting policies right'—this is the state's critical role, and these concerns are seen in a country's policies. In Indonesia and Vietnam, the Bank has become interested in questions of governance. For example, the Bank of Indonesia has complained of the 'lack of transparent, predictable and enforceable rules for business', and has cited malpractices in state procurement, the allocation of licenses and tendering processes for infrastructure contracts (World Bank 1993: 135–64; World Bank 1995: xv–xvii). It has initiated a range of projects to improve economic legislation and government procedures and systems.

There is a substantial element of voluntarism in the World Bank's approach and a belief among many of its staff and consultants that institutional building is essentially a technical matter. However, there is another view where the process is seen to have a social and political component. In one report, the Bank insists that: 'While donors and outsiders can contribute resources and ideas to improve governance, for change to be effective, it must be rooted firmly in the societies concerned, and cannot be imposed from outside'.

It recognises the social underpinnings of corrupt or weak institutional frameworks, including the capture of governments by vested interests, distributional coalitions and military groups. It is, however, less sure about how to go about the political processes of reform, observing, rather lamely, that citizens 'need to demand good governance' and, governments 'need to prove responsive to these demands' (World Bank, 1991b: 6–7).

How, then, were factors of social power and interest to be incorporated into what was a technical concept of markets and market institutions? In explaining why some countries proved better at constructing efficient institutions it is not surprising to observe a return to the cultural explanations of modernisation theory. Douglass North (1994: 366–7) relied heavily on culture as the factor seen to be driving both resistance and innovation, arguing that it was the 'mental models of the actors that will shape choices'. For new institutionalists who emerged out of the statist school of political economy, the choices made by developmental elites are seen as taking place in a context of constraints imposed by various coalitions of social and political interest. Building effective institutions depends on restraining self-interested behaviour and by establishing institutional designs intended to provide greater insulation from distributional pressures as well as increasing the efficiency of revenue collection and regulatory frameworks (see Haggard and Chung-in Moon 1990).

Rational policy choices emerge, in this view, when elites are autonomous of the squabbling associated with distributional coalitions. The examples cited are usually of bureaucratic developmental elites in Japan, South Korea and Taiwan, or the People's Action Party apparatchiks-cum-bureaucrats in Singapore. There has been some revision to this approach, recognising that appropriate market institutions can be successfully developed by policy-makers and others. This occurs where the social power of a capitalist class has advanced to the stage where profit-making is secured through general rules rather than in the pursuit of political favour and monopolies. This is best achieved through the formation of a dominant coalition of social inter-ests, of which the state is a part. In the case of North-East Asia, state autonomy from vested interest must give way to support from, and coop-eration with, the private sector if these economies are to successfully negotiate a new era of deregulation and market reform (Evans 1992). Markets, in this view, are regarded more as deliberate political constructs involving the conflict of contending interests than rational and technical, self-regulating mechanisms (see Chaudrey 1994). Interestingly, the institu-tional approach has evolved from a neo-classical emphasis on individual rational choice and the consequent importance of institution building as a technical problem. There is now greater emphasis on the structural con-straints on individual actions.

Pluralist and Marxist-Derived Political Economy

Because neo-liberal theories are based on a concept of markets as tech-nically superior mechanisms of economic organisation, they necessarily see the political struggles over their introduction as contests between rational-ity and vested interest, between selfishness and the common good of the system as a whole. Opposition to markets, in this view, came from those who sought to achieve an advantage by using political power to establish monopoly or privileged positions.

This contrasts with the position taken by various pluralist writers who have focused on the importance of social groups, civil societies and politi-cal cultures in the emergence of appropriate institutions (see Diamond, Linz and Lipset 1989, especially the Introduction; and Diamond 1993). Their point is that there are definite social and political preconditions for institu-tions to emerge and work effectively. These pluralists see the possibility of a range of competing social interests exerting influence over the form of institutions—the exact importance of each being primarily an empirical question. Nevertheless, they require the prior emergence of professional and entrepreneurial middle classes for whom systems of political tutelage and control are obstacles to opportunities and careers based on expertise, credentials or investment. Such approaches give theoretical priority to the emergence of organised interest groups in society. These groups are seen to provide a challenge to state power and entrenched interests. Such extra-bureaucratic influences force the state to establish systems of interest mediation and consultation (see Anek 1992). Exactly how the state makes policy decisions remains a weakness of pluralist theory. It tends to draw on earlier approaches—which see a neutral state working for a society's best interest—or views interests as being organised within systems of mediation operating inside the apparatus of the state, privileging particular interests.

Like the liberal theorists he criticised, Karl Marx regarded market institutions as the natural institutional framework for the organisation of economic life and production in capitalist society. He argued that the capitalist state must emerge to provide 'substance to the will of private property', guaranteeing the common interests of all capital and managing the conflict between classes that is the inevitable product of capitalist society. In this process, the state and its officials are subordinated to the requirements of business relations. This involved establishing the accountability of the public officials and public expenditure, transparency of procedures within both the public and private spheres and the guarantee of private property. Rule of law and equality before the law ensured, in Marx's view: 'a public system that is not faced with any privileged exclusivity' (cited in Miliband 1989: 89).

Interestingly, and in contrast to the previously mentioned approaches, Marxist-derived political economy emphasises the historical specificity of capitalist markets and underlines that markets *per se* are not the creation of the capitalist system. Thus, the focus is on the emergence of new and unique systems of production and circulation, and the specific social or class relations inherent in them. These generate particular institutions in property, government, legal, business and cultural affairs. The capitalist market is seen to be but one part of these arrangements and therefore has a different theoretical meaning and significance compared with other approaches.

Marxist political economists argue that class relations are the significant factor in determining the distribution and use of power in society. The nature of domination is seen to be structured by these relations and the relationship between elements in the economy, state and society. From this perspective, state policy cannot be neutral, nor can it be the outcome of a process of professional decision-making based on an analysis of interest group inputs. Policy is a reflection of the nature of domination in society.

The issue is not to identify 'good' and 'bad' policy choices, but to understand why it is that particular policy agendas emerge and hold sway under particular political and economic regimes. Thus, the questions and issues Marxist political economists find important relate to the nature of economic power and control, the character and power of the state and its political regime, and the development of civil society. Central also are the struggles and competition which these involve between capitalists and labour, between capitalists and the peasantry, within the capitalist class and between state and society.

Even though state bureaucracies may achieve well-defined interests of their own and relative autonomy from direct control by capitalists—who are, in any case, regarded by Marx as congenitally unable to organise themselves—in the long run, they must provide the conditions for capital accumulation and mediate the conflicts within capitalist society. This task is in the collective interest of the whole capitalist class, even though it may involve a contradiction of the short-term interests of individual capitalists and firms (Draper 1977: 189–93, 484–9).

Arguments of the Book

In this volume the issues, concepts and debates that separate the contending political economy approaches will be considered in the context of six country studies and three chapters which deal separately with labour in South-East Asia, regional economic institution-building and the emergence of subregional economic growth zones. The positions taken by the various authors here lie outside the neo-classical political economy approach but instead draw on approaches which emphasise the social and political dimensions of economic development. Thus, elements of the new institutionalist, pluralist and Marxist-derived approaches are to be found in this collection. However, the conflictual nature of economic development and the importance of the state as an arena for this are central analytical features of all chapters.

In chapter 2, Richard Robison analyses Indonesia where, since the mid-1980s, a range of significant deregulatory reforms has been introduced in trade, investment and financial sectors. However, for Robison, this does not represent the shift towards a more rational policy process and the beginnings of an inevitable institutionalisation of neutral market-supportive institutions. What is unfolding is a selective reorganisation of the economic role of the state which involves a change in the complexion and alliances of interests, rather than a repudiation of political influence in the market. Four economic agendas, and attached political and social interests, that have emerged in post-colonial history are observed: economic nationalism, economic populism, predatory bureaucratism and liberalism.

As Robison sees it: 'the complex coalition of bureaucratic, family and corporate interests nurtured within the economic nationalist and rent-seeking regimes' is confronted with the need to accommodate other interests— notably the more internationalised and competitive interests supporting a liberal economic agenda. Structural factors behind this include the demise of oil prices, which has meant fiscal pressures on the state and the need for new export sources of revenue. The response has been effective to date, with a significant increase in non-oil exports and revenues, increases in private sector investment and greater integration with the international economy.

However, according to Robison, it still remains that: 'In strategic instances, political power is drafted to override decisions based in law and formal procedure by state managers'. Furthermore, and against the predictions of neo-classical and new institutional political economy, the absence of transparent and predictable rules for business have not prevented a continued influx of foreign investment. Investors have either adjusted to these conditions or, in the case of Taiwanese and Korean capitalists, have adeptly exploited the politico-business networks within Indonesia. Neither do frictions between different elements of domestic capital appear likely to be resolved through the agitation for market institutions and regulatory frameworks based in law.

It is not structural economic pressure which Robison sees as the most compelling force behind new institutional frameworks. Rather, in the post-Soeharto era there is a strong likelihood of an authority vacuum and a jockeying of powerful social and economic interests attempting to fill this. Yet, even in this scenario Robison does not necessarily envisage a move towards liberal democracy or some sort of 'highly structured legal-rational state apparatus'. He sees the continued development of the market being compatible with non-liberal political and social structures.

The Philippines stands in contrast with most of its South-East Asian neighbours that have embarked on, or achieved, dramatic economic growth over recent decades. As Jane Hutchison points out in chapter 3, after a period of comparatively positive import-substitution industrialisation in the 1950s, the last thirty years have brought low levels of economic growth and declining living standards for the bulk of the population. The structural dependence on the export of primary products, such as sugar and coconuts, has yielded diminished returns. Although a shift towards export-oriented, labour-intensive manufacturing production, notably of garments and electronics products, has taken place and economic growth and investment levels have improved since the early 1990s, fundamental problems remain.

In particular, Hutchison contends that political reforms devolving authority to local administrative levels in the post-Marcos period only further weaken the Philippines state and militate against the harnessing of economic growth to broad social gains. Fiscal problems associated with foreign debt continue to exert the major force on the state for structural economic change. Meanwhile, privatisation and other market-oriented economic reforms are seen by Hutchison to open up new opportunities for capital accumulation in manufacturing, construction, telecommunications, real estate and other services. Consequently, while the traditional oligarchies may continue to enjoy prominence, often in conjunction with surviving business cronies of former President Marcos, the bourgeoisie has certainly become more diversified. It includes sections geared to global manufacturing production that are largely removed from political patronage. In all of the changes, though, the political and economic positions of peasants and rural and urban workers have not altered.

The likelihood of any redress of this situation is contingent not just on political will, but a strengthening of the capacity of the state to effect social reform and economic redistribution. However, the prevailing orthodoxy, which derives in large part from neo-classical economic thinking, emphasises the importance of economic liberalisation and the curbing of rent-seeking. Lingering resentment of patrimonialism also fuels domestic support for the dismantling of the state. Yet Hutchison argues 'government processes are still widely subject to particularism and patronage' and without institutional changes to strengthen the state this cannot be arrested. The force behind pressure to deregulate the economy is essentially external and fiscal-related, she maintains, and is not indicative of any systematic assault on rent-seeking. In the absence of a strong state, Hutchison concludes, there is no guarantee that 'the market' will redistribute wealth and power to the Philippine majority.

In chapter 4, Kevin Hewison takes issue with work for the World Bank which depicts Thailand as a case where 'the market has compensated for government failures'. This tends to portray economic development in technical terms, and reduces it to questions of 'good' versus 'bad' policy choices. Instead, Hewison argues that the state has been centrally important to development, but that its roles have varied according to the particular accumulation regime which prevails. This, in turn, is depicted by Hewison as the product of such factors as class relations and the changing international context of development.

Hewison identifies the period from 1932 to 1957 as one characterised by an accumulation regime with an uncertain investment climate generated by 'haphazard and increasingly personalised arrangements'. The state and its enterprises dominated, with heavy involvement by military leaders and those with political connections. Policy was populist and ostensibly meant to promote the non-agricultural employment and investment of ethnic Thais. Following the twin coups of 1957–58, the incoming authoritarian government brought with it a commitment to import substitution industrialisation (ISI). While this was encouraged by international agencies, the role for the state in infrastructure development that was being prescribed happened also to create positions for 'a rising group of young civilian officials … in newly-created economic agencies'. According to Hewison, though, the 'big ISI winners were industrial conglomerates and their financial partners, and most notably the fifteen to twenty families who dominated the highly protected domestic commercial banks'.

As in the case of Indonesia, structural economic pressures contributed to a shift towards export-oriented industrialisation, in this case beginning in the late 1970s: the steady rise of the *baht* which undermined the export competitiveness of primary commodities; the changing pattern of international investment, notably the relocation of East Asian firms offshore; and the fall in international prices for agricultural commodities. This gave rise to a radical devaluation of the *baht* and a concerted push into EOI. An investment boom followed, particularly from the mid-1980s, and this involved increasing investment from abroad as well as from the domestic capitalist class. The surge of foreign investment included the search for local partners, extending well beyond the established bank-dominated clique. The position of local banks was also challenged by the development of the Securities Exchange of Thailand (SET), which enabled various new companies and groups to emerge and broadened the economic base to include telecommunications, real estate, tourism and various services. This new accumulation regime necessarily upset the cosy relationships between politicians and technocrats of the ISI phase.

While the EOI strategy has instigated economic growth, Hewison draws attention to income and wealth disparities as an increasing political problem—especially as it involves striking disparities between rural and urban incomes. The government response to date has centred around further expansion of EOI in provincial areas, rather than embarking on targeted welfare. Although this has reaped little so far, Hewison contends that the current EOI regime 'is unlikely to allow movement beyond the adoption of increased emphasis on increasing the productive capacity of the population'.

Rajah Rasiah's chapter 5 on Malaysia examines the changing role of the state in economic and social development and the interests attached to this. Attention is paid to the way in which ethnicity and class have intersected to shape the activities and rationale of the state's role. Rasiah begins with an

historical analysis of this intersection, pointing out how colonialism produced a division of labour which separated the ethnic Chinese, Indians and
Malays in socio-economic terms. This was essentially reinforced in the
1960s after Malaysia's independence. The economic marginalisation of the
Malays in particular remained striking. The failure of the economic
strategy, due in no small part to the ill-conceived and inadequately supported ISI program, culminated in acute inter-ethnic rivalry towards the end
of the 1960s. Class antagonisms and inter-ethnic tensions fed off each other
to produce a social and political crisis in Malaysia.

The political response to this was the New Economic Policy (NEP),
launched with the Second Malaysia Plan (1971–75). This was characterised
by affirmative action to promote the social and economic development of
Malays, including a target of 30 per cent Malay corporate equity by 1990,
and vigorous state promotion of EOI to generate employment. The public
sector became especially important, both for general employment and managerial opportunities for Malays. As manufacturing gathered momentum
the Malay peasant class declined in favour of an enlarged proletariat.
Indians and Chinese business people were also drawn to the cities. State
attempts to encourage a broader industrial base in the 1980s were linked to
the objective of creating a strong Malay bourgeoisie, as was privatisation.
The NEP, Rasiah contends, produced a number of achievements, including
greater inter-ethnic collaboration, but the industrial sector was beset with
structural limitations.

According to Rasiah, the advent of Prime Minister Mahathir Mohamad's
political supremacy within the ruling coalition, together with a recession-
induced reliance on greater foreign investment as the state-sponsored heavy
industries program faltered, precipitated a major redirection. Such circumstances strengthened the hand of policy-makers and capitalists geared
towards competitive, higher technology production. This meant a more
inclusive set of investment incentives and state supports for business, with
the promotion of an ethnic Malay, or *Bumiputera*, class a less critical consideration. This has been accompanied by wider and more institutionalised
state-capital consultations. The more general question of the social distribution of wealth has also lessened in priority in the 1990s. Rasiah argues that
the political feasibility of this new direction in public policy rests, in part, on
the extent to which the more powerful elements of recently emerged
Bumiputera interests are serviced by the current economic strategy.

Garry Rodan's chapter 6 on Singapore deals with South-East Asia's most
developed economy, and where policy-makers' attention is focused on the
promotion of economic diversification in the wake of successful EOI. As
Rodan explains, the state continues to play a pervasive role in Singapore's

development, even if the forms and emphases are changing in this new phase of the city state's political economy. In particular, Rodan details the centrality of the state in effecting regional economic integration opening up opportunities for a range of service industries involving Singapore-based companies. Indeed, much of this economic activity is shaped by new international accumulation strategies by state-owned or state-linked companies. At the same time, new forms of cooperation and commercial alliance are being formed with sections of the domestic bourgeoisie capable of contributing to the development of an 'external economy'; that is, offshore investment which more fully integrates Singapore-based enterprises with the region's expanding economies.

Alongside changing relations between the state and business, Rodan also sees Singapore's latest phase of economic development involving widening domestic material disparities. Among other factors, the growth in demand and remuneration for the highly skilled sections of the workforce has been accompanied by a relative decline in the opportunities and rewards for other sections of the workforce. The political challenge the ruling People's Action Party (PAP) has set itself is to address mounting public concerns about inequality without compromising institutionalised elitist values and structures, or embarking on redistributive policies akin to those adopted in industrialised liberal democracies. This has involved substantial general outlays on social development. However, as Rodan demonstrates, so far this approach is proving somewhat problematic.

In chapter 7 on Vietnam, Melanie Beresford examines a society that has undergone a major social, political and economic transformation since the unification of North and South in 1976 leading to the demise of the planned economy. The roots of conflict and struggle embodied in this transformation, argues Beresford, can actually be traced to the structures that pre-dated unification. Before 1976, production incentive problems and conflicts of interest inherent to the planned economy had been manifested in attempts to operate on the black market. After 1976 though, the interests of relatively independent capital accumulation began to gather momentum—a process aided by serious economic crises in 1979–80 and the mid-1980s and the essentially pragmatic nature of Communist Party leadership. However, the transition is not simply driven out of technical imperative, as moves to a market system are often depicted by neo-classical economists and some writers from the new institutionalist school, but rather a shift in the balance and make-up of interests involved.

Indeed, Beresford emphasises that the planned economy in Vietnam was not simply a 'top-down' system but one 'in which the plan emerged as the outcome of a process of negotiation between groups with differing, often conflicting, interests'. These included central authorities, local authorities and enterprise managers within the state system and collectives, and those who did not derive benefits from the planning system owing to limited access to established networks. In the ensuing contention between competing factions and interests in the late 1970s and early 1980s, Vietnamese leaders formed coalitions with interests that were seen to be most facilitative of the consolidation of their political positions. This was the background to the 1986 Congress which committed the Communist Party to the building of a market economy through a process of *Doi Moi* (Renovation).

Thereafter, the embryonic market economy has been actively promoted by the state. Sweeping structural and institutional change has taken place, including decollectivisation and the welcoming of foreign investment and aid. Foreign companies and aid donors now constitute a significant interest group agitating for further market-oriented reforms. Meanwhile, high growth rates, the arrest of inflation and significant improvements in the efficiency of resource utilisation have occurred. Attention now focuses on the limitations of the financial sector on the sustained development of the market economy. Accompanying this is a range of social issues arising out of the market economy, notably environmental and equity considerations affecting poor peasants and displaced and underemployed workers. As Beresford sees it, however, the trajectory of the Vietnamese political economy is not likely to see a wholesale shift towards a conventional capitalist economy. Rather, state business interests will remain a powerful brake against the establishment of an economy comprehensively based on private property; with equitisation (involving cooperating with other interests) rather than privatisation a likely model of ownership reform of the state sector. Moreover, there remains sufficient political sensitivity within the Party and bureaucracy to the interests of the poor 'to block the rise of opposition parties based on interests associated with private capital'.

Frederic Deyo's focus in chapter 8 is on the implications of recent industrial development in South-East Asia for labour. As he sees it, two global transformations are heavily shaping the political economy of South-East Asia: the growing influence of neo-liberalism which emphasises market forces, and the adoption of more flexible post-Fordist production systems. The net effect of these has been to limit the capacity of enlarged working classes in South-East Asia to significantly enhance their political and organisational strength. Instead, elite strategies of industrial management are being consolidated. No more clearly is this evidenced than in the inability of workers to take advantage of opportunities arising from democratic reforms, notably in Thailand and the Philippines. Labour's political influence has generally been restricted to small-scale, spontaneous action, often via 'community groups protesting government policies or inaction in matters of local concern'.

Deyo points out that economic strategies adopted in South-East Asia include light EOI programs, economic liberalisation and structural adjustment, industrial deepening under 'second-stage' EOI, and, most recently, post-Fordist flexible production systems in manufacturing. However, none of these have favoured increased bargaining power or organisational strength for labour. Indeed, the predominant response to intensified international economic pressure by managers has been cost-cutting measures directed at reducing labour costs, reflecting and compounding the political weakness of labour. Yet, in the few instances of dynamic flexibility strategies involving the pursuit of product niches based on quality, innovation and improvements in process and product technologies, as in Singapore, the position of labour has also remained constrained, even if the coopting of labour has become more extensive. According to Deyo: 'In general, flexibility-enhancing organisational reforms are overwhelmingly attentive to managerial agendas driven by competitive economic pressures, to the exclusion of the social agendas of workers and unions'.

This is not to suggest there are not significant changes afoot in relations between state and labour in South-East Asia. On the contrary, 'as employers have increasingly gained the upper hand in their dealings with workers, labour market "deregulation", the counterpart of economic liberalisation and marketisation, has proven a more effective policy, than continued repression'. Moreover, with militant, independent labour structurally excluded from the political process, Deyo suggests dominant elites may have a greater measure of tolerance for parliamentary institutions. However, the continued exclusion of the popular sector from politics is likely to promote 'inequities and alienation among large sections of the population, thus eroding this essential base and in the long term bring increased resistance even to the democratic institutions increasingly favoured by the elites'.

In chapter 9, Andrew MacIntyre examines the reasons behind the establishment of the Asia-Pacific Economic Cooperation (APEC) and the interests embodied in this institution. MacIntyre points out that even before APEC's initiation in 1989 there had been regional economic cooperation under the aegis of the Pacific Economic Cooperation Council (PECC) but this was informal and involved tripartite teams of academic economists, business people and government officials participating in an unofficial capacity. The advent of APEC formalised and elevated this as an exclusively inter-governmental institution. Unlike other economic groupings such as the European Union (EU), the North American Free Trade Agreement (NAFTA) or the ASEAN Free Trade Agreement (AFTA), APEC has not involved the negotiation of industry-specific trade deals. In fact, as a political process it has been largely remote from all but government ministers, officials and an elite of policy-oriented economists.

MacIntyre maintains that, as a group, the economic bargaining power of South-East Asian governments appears to have been enhanced by this subregional association, but differences within the region are nevertheless real and a sense of community yet to be firmly established. The most conspicuous expression of this has been Malaysian Prime Minister Mahathir's attempt to foster an alternative Asian-only institution while this has been flatly rejected by the governments of Indonesia, Singapore and South Korea. Mahathir's position, argues MacIntyre, derives not just from anti-Western sentiment but domestic political considerations. In any case, MacIntyre points out that the preconditions for the sort of vision Mahathir holds are lacking, notably the preparedness for Japan to assume the leadership role that would be required. Nevertheless, Mahathir's rhetoric does tap an important emerging sentiment of pan-Asian pride and self-confidence ensuing from the region's remarkable economic achievements. Furthermore, to the extent that Mahathir's objective has been to limit Western domination, the 1996 ASEAN-sponsored Asia–Europe Meeting (ASEM) in Bangkok did, by excluding the USA, signal a growing awareness of the potential political capacity of South-East Asian governments as a collective force.

On close inspection of the institution itself, MacIntyre emphasises that APEC is held together by a characteristically loose, voluntary and consensual set of principles, rather then being tightly rule-bound and involving reciprocal benefits and obligations for members. In evaluating the different accounts of what explains APEC, MacIntyre draws attention to the domestic political ramifications of anything more than the current arrangement. In particular, the South-East Asian economists and policy-makers who have dominated the process 'are not oblivious to the fact that their countries would bear some of the highest adjustment burdens if APEC were to have a rules-based regime featuring genuine reciprocity'. APEC's most vital function is seen by MacIntyre to be as a 'confidence-building' institution in the Pacific. It is, first and foremost, a foreign policy exercise which 'brings together nearly all of the countries in what is widely recognised to be a diverse, volatile, and fast-changing region'. The problem over the medium to long term is that, without the benefits normally associated with a rules-based regime emphasising reciprocity, it may be difficult to maintain the interest of the USA.

The final chapter by James Parsonage addresses an interesting and increasingly-prominent phenomenon in the region—attempts to creatively organise economic activity that break from the concept of the national economy or the integration of national economies. Instead, 'subregional growth zones' (SGZs) are intended to integrate distinct geographic areas of different nation states into a coherent economic unit based on complementary production specialisations. The most advanced of these to date is the so-called 'Growth Triangle' involving Singapore, the neighbouring Malaysian state of Johor and the Indonesian province of Riau. However, whereas neo-classical economists herald such developments as testimony to the inexorable diminution of the state in the face of economic globalisation, Parsonage emphasises the pivotal economic, social and political roles played by the state in providing the pre-conditions for SGZs and managing their consolidation.

In substantiating his claim that the state is paramount to the development of these SGZs, Parsonage not only emphasises the infrastructural needs of capital, but the political insurance state involvement provides for ventures, and the need for effective management of legal, social and ethnic frictions deriving from new economic relationships across state borders. His message is, paradoxically: no trans-state development without the state.

In accounting for the current spate of SGZ initiatives, Parsonage makes the point that throughout South-East Asia there is a new, more positive mood among policy-makers favouring economic regionalism. This he attributes to the end of the Cold War, which has ensured political legitimacy is even more closely tied to economic performance, and the salutary effects of the mid-1980s recession which has generated greater commitment to the attraction of foreign investment. This has, according to Parsonage, become a joint exercise involving new forms of cooperation between the different states of South-East Asia. The absence of a 'globally competitive domestic manufacturing class in South-East Asia and a reliance upon external capital for technology and global market access' has thus induced efforts to create new, more attractive investment environments. Much of this is directed at East Asian-based capital, not just from Japan but the Asian newly industrialised countries such as Taiwan and South Korea.

However, rather than threatening the positions of domestic-based capital in the region, Parsonage argues that the formation of SGZs is opening up investment opportunities for both state-linked companies and powerful elements of the private sectors in South-East Asia to constitute broad state-capital alliances. Indeed, he maintains that SGZs give expression to converging interests between states and local and external elements of capital which might be understood as a variant of Leslie Sklair's (1991) concept of a 'transnational capitalist class'. Mutually reinforcing economic interests among regional political and economic elites are seen by Parsonage as a major force behind these SGZs.

References

Amin, Samir (1974) *Accumulation on a World Scale*, New York: Monthly Review Press.

Amin, Samir (1976) *Unequal Development*, New York: Monthly Review Press.

Amsden, Alice (1989) *Asia's Next Giant: South Korea and Late Industrialization*, New York: Oxford University Press.

Anek Laothamatas (1992) *Business Associations and the New Political Economy of Thailand: From Bureaucratic Polity to Liberal Corporatism*, Boulder: Westview Press.

Awanohara, Susumu (1993) 'The magnificent eight', *Far Eastern Economic Review*, 22 July: 79–80.

Balassa, Bela (1971) 'Trade policies in developing countries', *American Economic Review; Papers and Proceedings*, 61:168–87.

Bates, Robert (1981) *Markets and States in Tropical Africa*, Berkeley: University of California Press.

Bauer, Peter T. (1970) *Dissent on Development*, London: Weidenfeld and Nicholson.

Bennett, Douglas C. and Kenneth E. Sharp (1985) 'The worldwide automobile industry and its implications', in Richard S. Newfarmer (ed.) *Profits, Progress and Poverty: Case Studies of International Industries in Latin America*, Notre Dame: Notre Dame University Press.

Brenner, Robert (1977) 'The origins of capitalist development: a critique of neo-Smithian Marxism', *New Left Review*, 104: 25–93.

Buchanan, James M. and Tullock, Gordon (1962) *The Calculus of Consent*, Anne Arbor: University of Michigan Press.

Cardoso, Fernando Henrique and Faletto, Enzo (1979) *Dependency and Development in Latin America*, Berkeley: University of California Press.

Chaudhry, Kiren Aziz (1994) 'Economic liberalization and the lineages of the rentier state', *Comparative Politics* 27 (1): 1–25.

Christensen, Scott and Ammar Siamwalla (1993) 'Beyond patronage: tasks for the Thai state', Paper, Thailand Development Research Institute 1993 Year-End Conference, 10–11 December, Jomtien.

Christensen, Scott, et al. (1993) *The Lessons of East Asia. Thailand: The Institutional and Political Underpinnings of Growth*, Washington DC: World Bank.

Diamond, Larry, Linz, Juan J. and Lipset, Seymour Martin (eds) (1989) *Democracy in Developing Countries, Volume Three, Asia*, Boulder: Lynne Rienner Publishers.

Diamond, Larry (ed.) (1993) *Political Culture and Democracy in Developing Countries*, Boulder: Lynne Rienner Publishers.

di Palma, Gabriel (1978) 'Dependency: a formal theory of underdevelopment or a methodology for the analysis of concrete situations of underdevelopment?', *World Development*, 6 (7/8): 881–924.

Draper, Hal (1977) *Karl Marx's Theory of Revolution: State and Bureaucracy*, New York: Monthly Review Press.

Doner, Richard (1992) 'Limits of state strength: toward an institutional view of economic development', *World Politics*, 44 (3): 398–431.

Evans, Peter (1979) *Dependent Development: The Alliance of Multinational, State, and Local Capital in Brazil*, Princeton: Princeton University Press.

Evans, Peter (1992) 'The state as problem and solution: predation, embedded autonomy and structural change', in Stephan R. Haggard and Robert Kaufman (eds) *The Politicsof Economic Adjustment*, Princeton: Princeton University Press, pp. 139–81.

Frank, André Gunder (1967) *Capitalism and Underdevelopment in Latin America*, New York: Monthly Review Press.

Frank, André Gunder (1969) *Latin America: Underdevelopment or Revolution*, New York: Monthly Review Press.

Gereffi, Gary (1982) *The Pharmaceutical Industry and Dependency in the Third World*, Princeton: Princeton University Press.

Haggard, Stephan and Chung-in Moon (1990) 'Institutions and economic policy', *World Politics*, 42 (2): 210–37.

Haggard, Stephan and Kaufman, Robert B. (1992) 'Introduction: institutions and economic adjustment', in Stephan Haggard and Robert R. Kaufman (eds) *The Politics of Economic Adjustment: International Constraints, Distributive Conflicts, and the State*, Princeton: Princeton University Press.

Higgott, Richard A. (1983) *Political Development Theory*, London: Croom Helm.

Higgott, Richard, et al. (1985) 'Theories of development and underdevelopment: implications for the study of Southeast Asia', in Richard Higgott and Richard Robison (eds) *Southeast Asia: Essays in the Political Economy of Structural Change*, London: Routledge & Kegan Paul.

Hoogvelt, Ankie M. M. (1982) *The Third World in Global Development*, London: Macmillan.

Huntington, Samuel P. (1968) *Political Order in Changing Societies*, New Haven: Yale University Press.

Johnson Chalmers, Tyson, Laura and Zysman, John (eds) (1989) *Politics and Productivity: The Real Story of Why Japan Works*, Cambridge: Ballinger.

Kay, Geoffrey (1975) *Development of Underdevelopment: A Marxist Analysis*, New York: St Martin's.

Krueger, Anne O. (1974) 'The political economy of the rent-seeking society', *American Economic Review*; *Papers and Proceedings*, 64: 291–303.

Krueger, Anne O. (1981) 'Export-led industrial growth reconsidered', in W. Hong and L. B. Krause (eds) *Trade and Growth of the Advanced Developed Countries in the Pacific Basin*, Seoul: Korea Development Institute.

Krugman, Paul (1994) 'The myth of Asia's miracle', *Foreign Affairs*, 73 (6): 62–78.

Lal, Deepak (1983) *The Poverty of 'Development Economics'*, London: Institute of Economic Affairs.

Larrain, Jorge (1989) *Theories of Development: Capitalism, Colonialism and Dependency*, Cambridge: Polity Press.

Liddle, William R. (1992) 'The politics of development policy', *World Development*, 20 (6): 793–807.

Little, Ian, Scitovsky, Tibor and Scott, Maurice (1970) *Industry and Trade in Some Developing Countries: A Comparative Study*, London: Oxford University Press.

Little, Ian (1981) 'The experience and causes of rapid labour-intensive development in Korea, Taiwan Province, Hong Kong and Singapore; and the possibilities of emulation', in Eddy Lee (ed) *Export-Led Industrialisation and Development*, Geneva: International Labour Organization.

Matthews, Trevor and Ravenhill, John (1996) 'The neo-classical ascendancy: the Australian economic policy community and Northeast Asian economic growth', in Richard Robison (ed.) *Pathways to Asia: the Politics of Engagement*, St. Leonards: Allen & Unwin.

Miliband, Ralph (1989) 'Marx and the state', in Graeme Duncan (ed.) *Democracy and the Capitalist State*, Cambridge: Cambridge University Press.

North, Douglass (1994) 'Economic performance through time', *American Economic Review*, 84 (3): 359–368.

Olson, Mancur (1963) 'Rapid growth as a destabilizing force', *Journal of Economic History*, 23 (4): 529–58.

Olson, Mancur (1982) *The Rise and Decline of Nations: Economic Growth, Stagflation and Social Rigidities*, New Haven: Yale University Press.

Rodan, Garry (1989) *The Political Economy of Singapore's Industrialization: National State and International Capital*, Basingstoke: Macmillan.

Roxborough, Ian (1979) *Theories of Underdevelopment*, London: Macmillan.

Sklair, Lesley (1991) *Sociology of the Global System*, Hertfordshire: Harvester.

Soesastro, Hadi M (1989) 'The political economy of deregulation in Indonesia', *Asia Survey*, 29 (9): 853–69.

Srinivasan, T. N. (1985) *Neoclassical Political Economy: the State and Economic Development*, New Haven: Economic Growth Center, Yale University.

Tarling, Nicholas (1966) *Southeast Asia: Past and Present*, Melbourne: Cheshire.

Wade, Robert (1990) *Governing the Market: Economic Theory and the Role of Government in East Asian Industrialization*, Princeton: Princeton University Press.

Warren, Bill (1973) 'Imperialism and capitalist development', *New Left Review*, 81: 2–44.

Weiss, Linda and Hobson, John M. (1995) *States and Economic Development: A Comparative Historical Analysis*, Cambridge: Polity Press.

World Bank (1981) *World Development Report 1981*, New York: Oxford University Press.

World Bank (1991a) *World Development Report 1991*, New York: Oxford University Press.

World Bank (1991b) *Managing Development: The Governance Dimension. A Discussion Paper*, Washington DC: World Bank.

World Bank (1993) *The East Asian Miracle: Economic Growth and Public Policy*, New York: Oxford University Press.

World Bank (1995) *Indonesia: Improving Efficiency and Equity—Changes in the Public Sector's Role*, Jakarta: Country Department III, East Asia and Pacific Region, World Bank.

Young, Alwyn (1995) 'The tyranny of numbers: confronting the statistical realities of the East Asian growth experience', *Quarterly Journal of Economics*, 110 (3): 641–80.

2

Politics and Markets in Indonesia's Post-oil Era

Richard Robison

In the half century of Indonesia's post-colonial economic history its economic structures and policy frameworks have been shaped by four major agendas, each with specific sets of political and social interests. These are: economic nationalism, economic populism, predatory bureaucratism and liberalism.

Economic Nationalism. A central theme in Indonesian policy-making has been the desire to transform the economy from one focused on low value-added commodity or industrial production to a technologically advanced industrial economy with a capacity for the production of capital and inter-mediate goods and with a sophisticated services sector. Hence, much of Indonesia's economic history has been characterised by state-led strategic industry policies, protective trade regimes and high levels of state invest-ment. This has not been solely an ideological vision but a policy that has enhanced the authority and power of the state and its officials over access and allocation. Intensive regulation and control of economic activity has been necessarily exercised through various state institutions, including the strategic ministries of Trade, Forestry and Industry, the Capital Investment Board (BKPM), the state oil company (Pertamina), the state electricity authority (PLN), state banks and the strategic Industries Board (BPIS).

Economic Populism. A strong theme in the anti-colonial struggle was a form of economic nationalism which focused on agricultural and small business cooperatives heavily influenced by anti-Chinese xenophobia. This tradition has continued as the basis of opposition to economic conglomer-ation as small-scale traditional economic activity is progressively swallowed by industrial growth. Ironically, populism has also reinforced the priorities placed on political stability and social order by the political and military hierarchies. Hence, contrary to liberal market agendas, Indonesian policy has contained populist objectives, including subsidised pricing of basic commodities, provision of targeted development funding and credit for regional and local projects and programs. These have not only pre-empted social unrest by alleviating rising prices but generated considerable popular legitimacy for the regime.

Predatory Bureacratism. Appropriation of public authority by political and bureaucratic interests and, more recently, by powerful families, has shaped the operation of both state and market capitalism. Because of its capture by these interests, the economic nationalist agenda did not produce powerful industrial economies like Korea or Japan. Vigorous market systems have emerged, defined to an important degree by politically conferred licences, concessions, trade monopolies, bank credit, state contracts and joint ventures rather than by sets of regulations based in law that provide common rules for business operation. Such a system is essential to maintaining the networks of patronage and control which sustain the politico-bureaucratic strata that effectively is in possession of the state. An increasingly important aspect of this agenda has been the exercise of state power to establish a base of social power in capital for families drawn from the politico-bureaucratic strata. As these business groups have grown so the function of the state, as the possession of powerful politico-bureaucrat elements, has shifted from serving itself to facilitating these new interests.

Liberalism. This has been a fairly slender reed in Indonesia's economic history. Its implications for the systems of economic nationalism, populism and predatory bureaucratism are obvious. Relying on state protection from foreign capital or, in the case of indigenous business, from larger Chinese business groups, Indonesia's weak bourgeoisie was never a champion of open and free markets. Emerging on the basis of political access to markets and within specific networks of patronage, it became reliant on the political gatekeepers. Only with the collapse of the oil revenues that sustained such a system, and with important structural changes in the world economy which forced Indonesia into global niches of competitive advantage, have pressures for deregulation and the construction of regulatory institutions become substantial. The government has been forced to pay increasing attention to appropriate macro-economic management of inflation, debt, and fiscal and monetary stability.

Since the mid-1980s a series of reform packages has brought substantial deregulation in trade, investment and financial sectors, threatening to unravel the economic and policy institutions that underpin the dominant political coalition. In the view of neo-liberal economists, these developments represent a rational and technical process in which market institutions emerge in a process of natural functional rationality, 'getting the prices right' in the quest for allocative efficiency. Such a process according to them, confronts the impediments of politics and ideology which are regarded as naturally rent-seeking and by definition irrational, dysfunctional and inefficient (Bhattacharya and Pangestu 1992; Soesastro 1989; Sjahrir 1988: 38).

Whether or not the transition from *dirigisme* and rents to more competitive markets results in greater productivity and general prosperity, the reconstruction of Indonesia's markets has been a political rather than a technical process, in which contending interests, working within a range of structural constraints and pressures, seek to shape the nature of market institutions and policy regimes in post-oil Indonesia.

The complex coalition of bureaucratic, family and corporate interests nurtured within the economic nationalist and rent-seeking regimes now faces the certainty that these frameworks cannot be sustained in their present form. the collapse of oil prices has required policy shifts to address balance of payments problems, maintain investment levels and secure state revenues. Policy choices are constrained by increasing pressures on fiscal resources, the necessary shift to an internationally competitive export economy based on non-oil products with all its current account difficulties, the growth of the private sector and the increasing importance of foreign capital with its requirement for market institutions and regulatory frameworks.

At the same time, as the balance within the ruling coalition shifts towards corporate conglomerates and politico-business families, and as their capacity for large scale investment expands, the state becomes, in certain areas, a constraint on the expansion and internationalisation of their enterprises. Political uncertainties of the coming transition from Soeharto's rule has also stimulated apprehension that corporate power based on political patronage brings its own vulnerabilities. Consequently, elements among the conglomerates and politico-business families now seek to reorganise the economic role of the state and their own place in the economy, selectively retaining *dirigiste* and rent-seeking frameworks that guarantee protection and privileged access while opening opportunities for international business alliances and entry into potentially lucrative economic sectors held as state monopolies.

Out of the very changing economy and its class structures, new contradictions are produced. Pressures to distribute wealth, jobs and income contradict their increasing concentration in the big cities and in the hands of large conglomerates and politico-business families. Predatory and rent-seeking activities of the state and its politico-bureaucratic strata increasingly constrain the more internationalised elements of capital and the emerging middle class, introducing pressures for the introduction of transparent regulatory frameworks for economic activity based on law and regulation. Export manufacturers competing in a global economy confront those whose interests are vested in subsidised upstream industries providing such intermediate goods as cement, steel, chemicals and paper or in various market monopolies.

What, then, is the nature of the Indonesian transition? Are we witnessing a critical transitionary phase in which powerful capitalist families reconstruct themselves to survive the era of state/renter capitalism into a more pervasive form of market capitalism? Does the process replicate the historical development of Western bourgeoisie? Is Indonesia now in the 'robber baron' period of USA capitalism or the so-called laissez-faire period of English capitalism as mercantilism becomes industrial capitalism? Or is the Indonesian situation quite different? Will Indonesian capitalism and its market institutions be characterised by a greater role for the state and for various forms of state-capital cooperation and integration, a greater fusion of public and private, a weaker civil society and a poorly developed public sphere?

Indonesian Capitalism: 1949–86

The pattern of state pre-eminence in ownership and in the planning of economic life was to begin almost as soon as the Republic was established. No powerful domestic bourgeoisie existed capable of replacing the Dutch banking, trading and estate houses or of forging a new industrial economy to supersede the declining agrarian-based colonial economy. Even Chinese-Indonesian business was confined primarily within medium and small-scale trade and, in any case, the prospect of Chinese business formally assuming a central role was politically unacceptable. It was almost as a surrogate that the state stepped in to assume the burden of investment and to construct the industrial base of Indonesia's new economy.

The pre-eminence of the state was to be dramatically extended in 1957 when Dutch enterprises were nationalised and converted into state-owned enterprises, a large number of them becoming military-run companies. With the rise of Soekarno's Guided Economy in 1960 a more intensive form of state capitalism was to involve centralised planning of a new industrial economy based on extensive state ownership through a vast array of state corporations and in which state control was to be exercised, not only through policy instruments but also through allocation of trading licences, import and production quotas, bank credit, state contracts and forestry concessions (Thomas and Panglaykim 1973: 56–100; Tan 1967).

Coming to power in 1965, the Soeharto government was to introduce two major changes. By destroying the Indonesian Communist Party (PKI), the main obstacle to a private propertied class, capitalist markets and foreign investment were removed. Rehabilitation of the economy through an ongoing debt-rescheduling package and agreements on loans and aid tied Indonesia closely to the requirements of Western creditors and international financial institutions in relation to macro-economic management. It was as a result of this development that liberal economic technocrats were to secure an important role in the policy arena as macro-economic managers and mediators with the World Bank and the major international creditors' consortia (IGGI). Selective opening of Indonesia to foreign investment in the resources sector and import substitution manufacture began to draw Indonesian business into the international arena through expanding corporate alliances.

However, with the surge in oil prices and production in the early 1970s, a new era of economic nationalism was to emerge. Presided over by Ibnu Sutowo, chief of Indonesia's state-owned oil company, Pertamina, and then by Ministers Soehoed and Hartarto within the Department of Industry, the state undertook a series of huge investments in large upstream industrial projects, including petrochemicals, steel and steel products, cement, fertilisers and forestry products. The State Secretariat (Sekneg) was also to secure a powerful and strategic position in the economy through control over allocation of supply and construction contracts for government funded projects under Presidential Decisions 10 and 14 of 1979 and 1980. As a focal point for sourcing of off-budget revenue for the regime and the most important channel for distribution of state patronage, Sekneg enabled a surge in the influence of economic nationalists, notably Ginandjar Kartasasmita. Under its patronage an important group of indigenous (*pribumi*) capitalists (the so-called contractors) was nurtured (Pangaribuan 1995: 51–67; Winters 1996: 123–41).

In the private sector Chinese corporate conglomerates were to emerge as the predominant element. Using domestic and regional credit and distribution networks, they were to benefit from protective policies and preferential access to monopolies and licences. By the mid-1980s such capitalists as Liem Sioe Liong, William Soerjadjaja and Eka Tjipta Widjaja presided over corporate empires that ranged from manufacture to banking, trade and construction as well as spreading through the region (Robison 1986: 271–322). Centres of politico-bureaucratic power were to engage in the market initially through the ubiquitous *Yayasan* (foundation), which collected 'contributions' from private sector beneficiaries and operated as non-accountable instruments of investment and political expenditure.[1] However, the families of powerful power holders were also to begin constructing extensive corporate alliances and discrete corporate empires. By far the most important was the Soeharto family, which began with equity holdings in the companies of large Chinese conglomerates, including the cement giant, Indocement, the Bank Central Asia and the monopoly flour mill. By the mid-1980s the Soeharto children were building their own companies with Pertamina distributorships, import monopolies and property investments (Jones and Pura 1986; Jones 1986; Pura 1986; Robison 1986: 345–50).

Structural Adjustment in the 1980s and 1990s

The catalyst for structural adjustment in the 1980s was a collapse in oil prices in 1981/82 and again in 1985/86. Whereas oil taxes constituted Rp. 8628 billion, or 70.6 per cent of total government revenues in 1981/82, this had declined to Rp. 6338 billion, or 57.1 per cent in 1985/86 (World Bank 1993: 185). Net oil exports peaked at US$6.016 billion in 1983/84, falling to US$1.426 billion in 1986/87 (World Bank 1994: 205). Clearly, policy shifts were required to stimulate non-oil exports, to generate new revenue sources and to replace state investment as the engine of economic growth. The economic role of the state was to be concerned increasingly with facilitating the development of the private sector.

In a series of reform packages beginning in the mid-1980s, the government was to substantially deregulate the financial and trade sectors, to relax foreign investment requirements and to open a range of government monopolies to private sector investment. In trade, reforms were driven by the absolute requirement that Indonesia develop international competitiveness in a range of non-oil sectors, particularly manufacture. Where they affected inputs essential to export manufacturing industries, import monopolies were dismantled, including important monopolies in plastics, tin plate and steel products held by well-connected business groups. By 1995, not only had import controls through quotas and sole import status been replaced substantially with tariff controls but also tariff barriers declined from the 37 per cent average tariff plus tariff surcharge which had applied in the pre-1985 period to 22 per cent in 1990 and to 15 per cent in 1995, placing Indonesia well in line with the pace of trade reform in the region (World Bank 1995: 40–2).

A series of reform packages intended to attract foreign investment significantly reduced the number of closed sectors and relaxed requirements that foreign firms progressively divest ownership to local partners. Such areas as power generation, telecommunications, ports and roads, long regarded as strategically sensitive were opened to foreign participation as Indonesia's need for infrastructure investment became increasingly urgent (World Bank 1995: 40–2). Foreign investment approvals surged from around US$9 billion per year in 1990–93 to US$23 billion in 1994 and US$39.9 billion in 1995 (keeping in mind that realisation has historically stood at around 40 per cent). Korean, Taiwanese and Hong Kong investment in low-wage export production in textiles and garments, footwear, plastic products and sporting goods has been superseded by investment in large upstream industrial projects in chemicals, paper, pulp and metal goods, and in infrastructure projects in power generation and construction in which Japanese, British and USA companies have played a larger role (World Bank 1996: 12–14).

Extensive reforms in the financial sector have enabled a freer mobilisation of savings and investment domestically, and from international sources as the exchange rate was deregulated and an open capital account operated. Domestically, the number of private banks proliferated and by 1995 they numbered around 200, holding 53 per cent of outstanding bank funds and 47.7 per cent of outstanding bank credit compared to the state bank's 37 per cent and 41.9 per cent (World Bank 1995: 17; Econit 1996a: 4–7).

A start has also been made on dismantling the state sector which, for so long, has dominated the Indonesian economy. While sale of public companies to private groups has been limited, public offering of shares in high profile companies such as Indosat, Telekom, Bank Negara Indonesia 1946 and the state tin miner, Timah, has proceeded with others planned, including PT Bank Exim and Krakatau Steel. Most important, a range of public monopolies in infrastructure, public utilities, television, banking and airlines has been opened to the private sector. Public fixed investment had declined dramatically in the 1980s, from 69 per cent of the total in 1979/80 to 27 per cent in 1993/94. However, the state corporate sector continued to be important. In 1995, 180 public corporations had a book value of assets totalling US$140 billion and produced 15 per cent of GDP (World Bank 1995: 29, 51).

These reforms were not entirely imposed on an unwilling government by inexorable structural pressures. In many areas, domestic business groups incubated within the New Order had developed to the stage where the economic nationalist policies of the state and the activities of its corporations had become a constraint in certain areas. Burgeoning conglomerates increasingly possessed the capacity to move into such lucrative sectors as banking, infrastructure, television and transportation currently held as government monopolies. They found themselves increasingly able to benefit from access to international finance and the corporate and technical resources offered by partnerships with foreign capital.

Markets and the Concentration of Economic and Political Power

Paradoxically, deregulation has appeared to reinforce rather than undermine the importance of state power in determining markets and the concentration of corporate power. This unexpected outcome is explained by liberals as the consequence of a technical error in deregulating finance and trade regimes before addressing deregulation in the real sector and the construction of regulatory frameworks (Soesastro 1989; Bhattarcharya and Pangestu 1992). However, the choice of sequences in themselves reflect the possibilities offered by prevailing configurations of social and political power and interest rather than technical mistakes. Structural pressures for development of non-oil exports may have forced state planners, corporate conglomerates and politico-business families to give way in the export-competitive sector but not in resource-related sectors or in sectors involved in production for domestic markets like automobiles or petrochemicals. A deregulated finance sector offered lucrative new opportunities in banking, while privatisation of public monopolies in infrastructure offered opportunities as long as access could be politically controlled.

To some degree it is logical that large conglomerates and politico-business families have been best placed to take advantage of deregulation because of the capital resources and organisational structures they had accumulated within the incubator of state tutelage. At the same time, their capacity to politically control many of the levers of access to the new deregulated markets ensured that further concentration of corporate power was enhanced. The progress of the newly deregulated banking industry illustrates some of the dynamics at work. By 1996 private banks had made major inroads into a sector previously dominated by the state. The top twenty banks by assets value include twelve private domestic banks, four of them in the top ten (*Infobank* May 1996: 22–3). Almost all of the larger, new, private domestic banks were elements within the corporate empires of the major corporate conglomerates, politico-business families and Yayasans (*Economic and Business Review Indonesia* 2 December 1995: 21). Operating in many cases as lenders to other companies within the same group, these new private banks were to play an important role in further concentrating the grip of the major conglomerates (*Infobank* October 1995: 14–34).

The new Jakarta Stock Exchange was also to facilitate the concentration of capital by providing a new source of cheap funding which was most effectively accessed by the conglomerates and politico-business families (*Eksekutif* August 1995: 12, 13). Once again, deregulation was compromised by lack of effective regulation. Inadequate rules and enforcement capacity allowed companies to go public without adequate disclosure, insider trading was rife and fake share scandals occurred frequently (Kwik 1993; *FEER* 2 April 1992: 46; *Tempo* 10 April 1993: 14–16). Often heavily exposed and highly geared, companies with privileged access to lucrative monopolies, particularly in the cement, foodstuffs or forestry and forest products industry, or with strategic positions in the infrastructure sector, found the stock exchange a valuable source of cheap equity funds as international portfolio investors scrambled to seize a share of these burgeoning industries.

Conglomerates and politico-business families were also to concentrate and reconstruct their ascendancy in sectors not susceptible to urgent structural pressures for international competitiveness: in domestic trade, the real sector or gatekeeping in non-export sectors. Five types of restriction were identified by the World Bank: cartels (cement, paper plywood and fertiliser); price controls (cement, sugar and rice); entry and exit controls (plywood, retail marketing); exclusive licensing (clove marketing, wheat flour milling, soymeal); public sector dominance (steel, fertiliser, refined oil products).[2]

In particular, the State Procurement Agency, Bulog, was to continue as a pivotal institution in the determination of market access in the domestic food industry. Within its network of sole distributorships and the protection of its pricing controls and minimum domestic content requirements, strategic enterprises were to develop in flour milling, flour-based food products, milk products, soymeal and sugar. Owned jointly (through Indocement) by Indonesia's largest capitalist, Liem Sioe Liong, the President's step-brother, Sudwikatmono, and the government, the flour miller, Bogasari, has long been a glittering prize. As the sole importer of wheat, Bulog had appointed Bogasari as sole flour miller in 1969, providing it with a secure and lucrative 30 per cent margin on milling and, perhaps more important in the long term, a base for a strategic position in the domestic food industry. By 1994 the same group controlled 75 per cent of the noodles market, 33 per cent of milk and 20 per cent of baby food (World Bank 1995: 43–4; Schwarz 1994: 110–12).

Such monopolies are not merely relics of an earlier era. In 1988 Bulog awarded a monopoly to mill soymeal to P. T. Sarpindo, a company owned by Hutomo Mandala Putera and Bob Hasan. Sustained by a price differential of 23 per cent above imported parity, Sarpindo was to earn US$12 per tonne in a milling fee from Bulog as well as US$72 per tonne from the soybean oil byproduct which it was able to retain (World Bank 1994: 94; Schwarz 1994: 132). A controversial monopoly on fertiliser pellets was allocated to P. T. Ariyo Seto Wijoyo, owned by Soeharto grandson, Ari Sigit (Econit 1996b), and a Ministry of Trade decision in 1990 awarded monopoly rights to purchase and sell Indonesia's clove crop to BPPC, a Board controlled by Hutomo Mandala Putera (Tommy Soeharto). This was facilitated by US$350 million in low interest liquidity loans from Bank Indonesia normally reserved for agricultural producers and by government requirements that the cigarette producers purchase their cloves from BPPC (*Prospek* 7 March 1992: 70–81, 10 July 1993: 67; *Tempo* 12 January 1992: 79–86, 10 July 1993: 74, 25 September 1993: 88–89; Schwarz 1994: 153–7).

Another refuge was to be found in the automobile industry. Initially in the hands of various military groups and senior officials, ownership was to gravitate into the hands of the larger Chinese conglomerates and the Soeharto family as the New Order progressed and as Japanese companies gained ascendancy. By the mid-1990s the Astra group (Toyota, Daihatsu), a consortia of Toyota, domestic Chinese business interests and government banks and companies accounted for 54 per cent of production while Liem Sioe Liong's Indomobil group (Suzuki, Mazda, Volvo, Nissan), accounted for 21 per cent of production (*Warta Ekonomi* 11 March 1996: 19).

Flourishing as sole agencies for the major auto producers within highly protective trade regimes in which tariffs of between 100 per cent to 300 per cent were applied to imports of completely built up (CBU) passenger vehicles, pressures to increase local content and to develop a capacity to produce engines and components as the basis for a national industry had largely failed. This was particularly true in the sedan sector where by 1995 few domestic sedans contained more than 10 per cent local components (*Jakarta-Jakarta* 16–22 March 1991). As we shall see later in this chapter, it was not deregulation that was to transform this situation but a decision to develop a national car based on a new business alliance between Soeharto family interests and a Korean auto major.

However, the most important new areas of investment were to be found in the so-called mega-projects concentrated in power generation and public infrastructure, petrochemicals, fertiliser plants, plantations, forestry, pulp and paper, industrial and residential estates, transport and construction. By the early 1990s investment in mega-projects committed or planned was estimated at US$80 billion.[3] For potential local partners in these new, lucrative sectors, cooperative arrangements with state corporations, access to licenses, state contracts for supply and construction, and state bank credit, rather than trade protection, became the critical currencies.

The country's massive forest resources provided obvious opportunities for the development of vertically and horizontally integrated industries in plywood, pulp, paper and rayon. With the banning of raw log exports in 1980, forestry production in Indonesia was channelled into the plywood industry and more recently into pulp and paper, forcing undercapitalised concession-holders, including most of the old military ventures, to sell to the bigger conglomerates or to enter joint ventures or management agreements.[4] The major groups could be assured of a high degree of control over the forestry resource base because the allocation of concessions (HPH), continued to be a closed process without public and transparent mechanisms of tender and relatively free of environmental constraints. Forestry companies were able to ignore most attempts by the Environment Ministry to control illegal and damaging logging practices (*FEER* 12 March 1992: 45). Most important, the forestry industry attracted remarkably low government economic rents. Compared to the 85 per cent rents imposed on the oil and gas industry, forestry was liable to only 17 per cent (World Bank 1993: 44–9; Ramli 1992; *Tempo* 26 October 1991: 26–32).

Because of its potential to contribute to foreign earnings through exports, the government made special loans available to investors willing to establish forestry plantations (HTI), to create a long-term sustainable supply of logs and ease pressure on the diminishing natural forests (*Prospek* 17 August 1991: 91). So attractive were the potential rewards in the industry that companies had little difficulty attracting international investors eager to share in the action through syndicated banking loans, portfolio investment and equity shareholding.[5]

Driven by the increasing incapacity of Indonesia's arteries of roads, ports, telephones and power grids to keep pace with demand generated by new industrial growth, the government was to open the infrastructure sector to private investors. Although public investment in infrastructure was projected by the World Bank to remain the main funding source, rising from Rp.49.1 trillion in 1989/90–1993/94 to Rp.72.8 trillion in 1994/95–1998/99, private investment was expected to play a greater role, rising from Rp.15.3 trillion to Rp.37.7 trillion in the same period (World Bank 1992: 114). Well-connected conglomerates and politico-business families stood as gatekeepers to the industry, necessary elements in any bids by foreign investors or domestic consortia seeking to gain access to these lucrative new areas. As virtual components of the licence, Soeharto family companies were able to secure, without experience or substantial equity resources, an important share of the industry where access was decided by a tendering process that was, again, neither public nor transparent.

Contracts for supply, construction and local manufacture in the telephone sector worth over US$2 billion were let to Japanese and USA consortia in 1990 but with the companies Bimantara, Humpuss and Citra Telekomunikasi Indonesia, owned by the Soeharto children, playing a major role as local sole agents, local suppliers and manufacturers (*FEER* 8 March 1990: 54, 9 August 1990: 54, 24 January 1991: 40–1; *AWSJ* 15 January 1991: 1; Schwarz 1994: 144–5). Bambang's Bimantara group was also successful in securing operation and management contracts for state-owned satellites, again a puzzling move given that the company brought nothing tangible to the existing state operation in terms of equity, experience or technology (*Prospek* 16 January 1993: 32; *Tempo* 6 March 1993: 90–1; *AWSJ* 15 April 1994: 1, 9–10, 11 February 1995: 1, 2; Schwarz 1994: 149).

Construction and operation of ports, roads and airports, formerly a state monopoly, were also to prove bonanzas for Soeharto-related companies. Allocated to pre-selected bidders without transparent processes, Soeharto family companies were to secure the bulk of licences, often as partners of state companies (*AWSJ* 21-22 October 1994: 4; World Bank 1995: xiv–xvii). The major player was to be Siti Hardiyanti Rukmana's Citra

Lamtorogung group which secured domestic and international toll road projects worth Rp.6.9 trillion in partnership with the state road builder, Jasa Marga (*Infobank* May 1996: 80–1).

In the electricity generation and distribution industry it was planned that the private sector would contribute US$9.1 billion of investment for 1994–98 while the public sector would contribute US$7.9 billion plus US$7.1 billion in transport and distribution (World Bank 1995: 71). Foreign companies were to construct the plants, provide the technology, expertise and organise the funding which was to come mainly from international banks. Because they were expected to generate or sell electricity to PLN on a 'take or pay' contract negotiated with pre-selected bidders, they were effectively required to include in their consortia a local partner, usually with around 20 per cent equity. The main local players were to be Liem Sioe Liong, Sudwikatmono, Bambang Trihatmojo, Hutomo Mandala Putera, Sigit Hardjojujanto, rising star, Hasjim Djojohadikusumo, a Soeharto in-law, the Bakrie group and Poo Djie Gwan, a long-time supplier to PLN (*Prospek* 27 June 1992: 70–3; *Editor* 1, 8 February 1992 and 27 June 1992: 68–9; *Warta Ekonomi* 22 June 1992: 24; *AWSJ* 9, 14 February 1994: 1, 13 September 1994: 1, 20 October 1994: 3, 3 November 1994: 1, 2, 24 April 1995: 4, 5).

Coinciding with the economic nationalist agenda of adding value to natural resources and reducing the cost of imports, the petrochemical industry has witnessed a surge of massive investments by international companies. As gatekeepers to the industry licences and subsidised inputs from state corporations, including electricity, cement, steel and Pertamina feedstock, local partners, sole agents, and sub-contractors for supply and construction have come predominantly from within the Soeharto family and the conglomerates of Liem and Prajogo.[6]

Ironically, then, state power has been essential to the dramatic expansion of the conglomerates and politico-business families in the period of deregulation. Substantial flows of state bank credit were provided to leading conglomerates and politico-business families. Lists circulated in 1993 and alleged to come from officials within the Bank of Indonesia nominated as leading creditors a virtual who's who of the large conglomerates as well as three of the Soeharto family groups. They also revealed extraordinary levels of chronic bad debt. The companies Golden Key, Kanindo, Bentoel and Mantrust, all soon to collapse, were high on the lists of outstanding debt, as well as Prajogo Pangestu, the forestry and petrochemical tycoon, who was listed as having loans of around US$2 billion, of which 24 per cent was in the bad or doubtful loan category.[7]

Distribution monopolies, contracts and subsidised inputs allocated by state corporations have also been important elements in the growth of politico-family business groups. Bimantara and Humpuss emerged in the mid-1980s as suppliers and distributors to Pertamina and as insurers of Pertamina tankers, the state airlines, Garuda and Merpati, and the Palapa satellites, and as leasers of aircraft to Garuda.[8] Several controversial investments by the state superannuation fund, Taspen, have also been important in supporting a range of private companies, in the case of Prajogo Pangestu's Barito group, boosting equity requirements in preparation for public listing (*AWSJ* 8 July 1993: 1 and 7, 17 July 1993: 1, 4; *Jakarta Post* 16 July 1993: 1; *Detik* 14-20 July 1993: 19; *Tempo* 24 July 1993: 88–9). Subsidised feedstock from Pertamina has assisted private petrochemical ventures (Nasution 1995: 16), while state companies, TVRI and Jasa Marga, played important roles in Siti Hardiyanti Rukmana's entry into television and road building (*Jakarta Post* 7 August 1992: 1; *Matra* August 1992; *Tempo* 25 August 1990: 75–8).

Contrary to expectations, the reforms did not create a generalised system of open markets and free competition, nor did they shift the larger domestic corporations into export competitive manufacturing sectors. The World Bank continued to report high average rates of concentration of ownership in the manufacturing sector in areas of low export orientation such as food, paper, chemicals, basic metals and machinery, cement and glass (World Bank 1993: 91–2; World Bank 1995: 49, 50). While the World Bank saw failure to create the 'level playing field' as a problem of an unfinished deregulation agenda (World Bank 1996: xxvii), the uneven playing field might best be seen as a necessary part of a market structured by a particular alliance of politico-bureaucratic and corporate interest.

However, such a regime was not to be an untroubled or unchallenged one. Contradictions were to emerge or to be rekindled: between ideals of equity and social justice and the increasing concentration of economic and political power; between Chinese and *pribumi*, between economic nationalism and increasing integration with the global economy; between downstream export manufacturers and upstream producers or import monopoly-holders; between state managers committed to fiscal discipline and predatory capitalists seeking state protection or credit.

Response by Vested Interests to New Pressures

The systems of economic and corporate power that were to emerge during the decade of deregulation were contested at every step. Populist notions of social justice re-emerged in the countryside and the world of peasant and rural small business was unravelled by the encroachment of industrial estates, residential development and commercial investment in agribusiness. While the populist opposition to concentration of wealth and the new enclosures were primarily articulated by urban middle-class students, it was the prospect of rural reaction under an umbrella of Islam and the examples of Egypt, Iran and Algeria that were to concern the government. In this context it was the denial of liberalism and individualism that became important policy objectives for the government.

Resentment of the concentration of wealth and economic power in the corporate conglomerates also reactivated anti-Chinese sentiment. Not only did this take traditionally xenophobic forms among declining elements of *pribumi* business, but also it was a sentiment manipulated by larger *pribumi* business groups on the margins of power in the contest for patronage and access. Both of these issues are important and will be dealt with elsewhere. For the purposes of this chapter the focus will be on the particular challenge from liberal reformers market institutions.

Liberal pressure for continuing market reforms has been intense. Continuing restrictive trade practices, the persistence of state-supported monopolies and cartels in trade, special pricing and licensing arrangements and economic nationalist agendas in policy were targeted by successive World Bank Reports. Since the early 1990s the Bank has also become interested in questions of governance and the weakness of regulatory legal and administrative frameworks. Complaining of the 'present lack of transparent, predictable and enforceable rules for business' (World Bank 1995: xv), it has cited malpractices in state procurement, the allocation of licences and tendering processes for infrastructure contracts (World Bank 1993: 135–64; World Bank 1995: xv–xvii), while USAID has initiated a range of projects to improve economic legislation and government procurement systems through its ELIPS program.

World Bank assaults on rents and protective trade policies have been supported by sections of an influential liberal intelligentsia located within the burgeoning professional, managerial and corporate middle class. Based primarily within the mushrooming consultancy sector and in professional firms supplying business and capital market intelligence, economic analysis and more traditional services in law and financial/tax management services, these critics possess a degree of independence from the state that did not exist even a decade ago. Their clients are to be found primarily in the private sector; in international banking and financial services and the media industry.[9]

Public criticism of costs imposed on business by practices of corruption, collusion and protection was initiated by former Finance Minister and now consultant and businessman, Professor Sumitro (*Kompas* 23, 24 August 1985). Entering the conglomerates' debate of the early 1990s, other liberal reformist critics drew a distinction between firms whose position was achieved in competition in open markets and those founded in political manipulation, monopolies, cartels, trusts and distorted markets, collusion in banking and on the stock exchange, high levels of debt and collaboration between business and rulers (*Tempo* 4 December 1993: 83–4, 21 August 1993: 29, 28 August 1993: 79, 6 January 1990: 102; *Jakarta Post* 15 April 1993; *Prospek* 6 March 1993; *AWSJ* 8 July 1993). It was the liberal intellectuals who were to dissect in the public media the recurring instances of corporate collapse and financial scandal, bad debts and the state banks, financial irregularity and ineptitude in the private banks and on the stock exchange, collusion and debt in the collapse of private companies, protection and favour in the automobile and petrochemical industry.

Although frustration with corruption, incompetence and arbitrary authority has resulted in increased criticism from Indonesia's liberal reformers, there are no political mechanisms through which a new policy regime might be introduced. Caught within the institutional channels of the organic state and its sponsored parties and organisations, no possibility exists for a reformist party or for change in government through electoral competition. In any case, reformers are not dominant even within Indonesia's middle class. It is the desire for 'good government' within existing ideological paradigms rather than democratic reform that appears to be the prevailing agenda of a middle class apprehensive of economic and political disintegration.

Prospects for the liberal cause lie, therefore, in the increasing structural contradictions that accompany the evolution of Indonesia's capitalist society and its position in the global economy. Indonesia's *dirigiste* policies face a count-down as regional and international trade agreements, including WTO/GATT, AFTA and APEC, which require progressive dismantling of trade protection, take effect. Rent-seeking institutions and protective frameworks are also heavily dependent on continuing inflows of grant and loan funds into the state budget through the Consultative Group on Indonesia (CGI), and on the Japanese, giving outside forces potential leverage to enforce policy change.[10] Another growing concern is Indonesia's foreign debt, now standing at around US$100 million and under pressure as non-oil exports begin to flag.[11]

Contradictions are emerging within domestic capital, between downstream producers of export goods and subsidised and protected importers and upstream producers, between holders of sole licences, cartels and other manufactures and producers. At the same time there are pressures from foreign capital and portfolio investors for transparent and predictable regulatory frameworks for business activity. Paralleling these are conflicts caused when state managers seeking to implement regulation and fiscal and monetary policies are confronted with demands from politico-business families for licences, protection and state bank funds.

Cartels and Licences

Continued state support for various cartels and systems of exclusive licensing were to provoke resistance not only from the World Bank and liberal critics but also from state managers seeking to impose general rules for business activity or to defend the fiscal integrity of their departments, as well as from business interests disadvantaged by the monopolies.

Nowhere have these tensions been more evident than in the plywood producers' association (APKINDO), established by the Chinese business tycoon, Bob Hasan. With the authority to require the membership of all Indonesia's plywood producers it imposed an iron discipline, allocating prices, quotas and markets to members in a strategy to seize control of world prices and markets from Japan and Korea. These heavy-handed actions increased resentment of the cartel's control among its members, particularly as plywood prices plummeted in 1995 and Japan was able to secure alternative sources of supply (Pura 1995).

Domestic trading monopolies, too, were to create conflict within business. Bambang Trihatmojo's attempt to corner the citrus monopoly in Kalimantan ran into unexpected and effective opposition from traders entrenched in the industry, eventually forcing him to withdraw (*Prospek* 20 June 1992: 70–80). However, monopolies in flour, cloves and soymeal were to continue despite the financial burden they clearly placed on the state and the opposition of competing industry interests and state managers (Schwarz 1994: 132–3, 154–5). It was Tommy Soeharto's clove monopoly that was to provoke the most intense confrontation and test the resolve of the contending forces. Claiming to act in the interests of clove producers against the big cigarette companies Tommy proposed to purchase cloves at Rp.7000–8000 per kilo and sell at Rp.13 000 per kilo, doubling the farmers' income and, in the process, generating an expected profit of US$100 million per year for his consortia (BPPC). With government approval for the cartel in hand and low interest loans from Bank Indonesia secured to set up the clove stock, the outcome looked uncomplicated.

Encouraged by the promise of higher prices, growers had flooded the BPPC warehouses. As BPPC proved unable to absorb the burgeoning stocks, prices fell to levels lower than they were before and Tommy Soeharto advised growers to burn half their crop (*Prospek* 12 August 1991; *Warta Ekonomi* 12 August 1991; *Matra* May 1992). As the disaster intensified, it was announced that interest payments on the Bank Indonesia loan could not be met by BPPC. The government was forced to come to the rescue. In 1993 new loans were provided from state banks, BRI and BBD at subsidised rates of credit. In 1992 the government arranged for the buying monopoly, the costly and troublesome part of the arrangement, to be handed over to the Federation of Cooperatives while BPPC retained the selling monopoly (*Prospek* 7 March 1992: 70–81, 10 July 1993: 67; *Tempo* 12 January 1992: 79–86, 10 July 1993: 74, 25 September 1993: 88–9; Schwarz 1994: 153–7).

At the other end of the operation, BPPC found the cigarette manufacturers association (GARPRI), surprisingly resistant and critical (*Editor* 7 July 1990). Initially, they refused to purchase BPPC cloves, thus reducing the cash flow to a cartel with huge debt servicing obligations (Schwarz 1994: 155). At this stage, the government was once again mobilised. The Minister of Trade, Siregar, reminded cigarette manufacturers that they required certification that their cloves came from BPPC. Substantial fines and confiscation were threatened for attempts to break the cartel (*Kompas* 30 April 1991; *Suara Pembaruan* 25 May 1991). In the end, the manufacturers capitulated.

Trade monopolies have not been the only form of exclusive licence. Soeharto family groups and their associates have, at various times, been given highly unpopular monopolies, including the state lottery and collection rights for auto registration and licensing (*Forum Keadilan* 11 November 1993: 9–16; *Editor* 24 April 1993: 17–29). Violent opposition from Muslim groups to the lottery forced its cancellation. An attempt in 1995 by the Governor of Bali to give P. T. Arbamass, a company owned by Soeharto grandson, Ari Sigit, a licence to collect a Rp.200/bottle tax on beer (to which Arbamass to add a Rp.400 fee), was thwarted by opposition from Bali's hotel owners who stated their intention to halt purchase of beer. This action constituted a threat to Bali's huge tourist industry, a source of substantial foreign earnings for Indonesia. It also confronted Indonesia's beer industry. The hapless Ari Sigit had managed to galvanise into united opposition a raft of interests, including, ironically, Soeharto family interests in beer production and Bali hotels. The President was forced to order the Department of Trade to rescind the levy. This was Ari Sigit's second unsuccessful attempt to secure a source of income through tax farming (*Independen* 31 January 1995; *AWSJ* 27 January 1996: 1, 4).

Privatisation and the Role of the State Sector

One of the central objectives of liberal reformers has been the privatisation of Indonesia's huge public sector. Private investment as a percentage of GDP stood at 12 per cent in 1993 and was expected to rise to 15 per cent by the year 2000. However, public investment was also expected to rise from 8 per cent to 10 per cent in the same period meaning that the private sector share of total investment is around 60 per cent, compared with a 75 per cent average in the rest of East Asia (World Bank 1993: 103). The continuing relative importance of public sector investment is attributed by the Bank to the heavy demand for investment in physical and human infrastructure in the coming decade which cannot be met by the private sector.

Nevertheless, there are powerful incentives for privatisation. A large proportion of state companies are poorly managed and inefficient, classified by the government itself as unsound and constituting a financial burden (World Bank 1995: 51; *Tempo* 6 January 1990: 100–3). Other well-performed companies offer the government potential fiscal windfalls on the stock exchanges and, indeed, a range of these, including Indosat, Telkom, Timah, were placed on local and international exchanges and a portion of the proceeds, US$760 million in the case of Telkom and Timah, used to prepay high interest debt to the Asian Development Bank and the World Bank (World Bank 1996: 10).

Despite the initial enthusiasm of reformers, privatisation by direct sale has proceeded slowly. The most commercially attractive state companies are usually strategically important in achieving national policy objectives and as focal points in the networks of political patronage and therefore not available for direct purchase. On the other hand, as noted above, because a large proportion of state companies are commercially unsound they are hardly attractive commercial propositions for private purchasers. Many are unaudited and unprepared to compete in the open market and will not carry with them the monopoly status that had sustained them in the public sector.

The enthusiasm of policy reformers for the sale of state companies to private purchasers was also dampened as it became clear that where state companies were being sold they went to well connected conglomerates and politico-business families at low prices and without open and transparent divestiture procedures, reinforcing the system of concentration and monopoly (*Prospek* 18 May 1991: 1; *Warta Ekonomi* 2 August 1993: 44–5; *Editor* 13 June 1992: 75; Habir 1995: 87; World Bank 1995: 49). Indeed, Bambang Trihatmojo was to argue that the transfer of state assets backed with state bank credit be used as a deliberate tool for shifting the balance of power towards *pribumis*, an interesting legitimation of the Soeharto family dominance of the transfer process (*Media Indonesia* 24 May 1991). Privatisation, therefore, meant strengthening the major politico-business families and Chinese conglomerates without necessarily satisfying the liberal reformers' objectives of a shift to free markets. This was recognised by reformist ministers, including Finance Minister, Mar'ie Muhammad, who began to talk of the state sector as a balance for conglomeration (*Warta Ekonomi* 2 August 1993: 44–5).

Where privatisation proceeded most rapidly it was to mainly involve rescinding state monopoly in potentially lucrative areas of economic activity, such as television, telecommunications, the generation and supply of electricity, construction and operation of roads and ports and air transportation. Again, without appropriate regulatory institutions in place and given the configuration of political power, licences were to pass without transparent and public tender into the hands of major conglomerates and politico-business families (World Bank 1995: 79; *Editor* 13 June 1992: 75). Ironically, the importance of state companies as providers of subsidised inputs, credit and distribution monopolies required that these public firms at least remained public so that the state could bear the cost.

Establishing Fiscal and Monetary Discipline

Attempts by state managers to impose regulations and fiscal and monetary discipline were to clash with the interests of conglomerates and politico-bureaucrat families. Increasing opportunities for investment in mega projects substantially raised the demand for borrowing from state banks and international institutions. It was estimated in 1991 that there were thirty-seven mega projects worth over Rp.1 trillion each (US$500 approximately), including projects like Chandra Asri and the pulp and plantation complex, Enim Lestari which each involved over US$2 billion (*Warta Ekonomi* 29 April 1991). It was calculated that in 1991, at the height of the debt blowout, mega projects potentially involved US$70 million in foreign loans. With a debt service ratio of 31 per cent and rapidly growing foreign borrowings and debt, the pressures on inflation and the currency were increased (*Tempo* 27 July 1991).

The government was therefore faced with the tasks of reigning back the domestic money supply, curtailing overseas loans for mega projects and imposing systems for prudential control in the banking sector. In the so-called Sumarlin shocks of 1987 and 1991, the government drastically reduced the liquidity of state banks, and in 1991 Rp.8 trillion was withdrawn by state companies from eight state banks sending interest rates to levels above 30 per cent, slowing investment and economic activity. Later that year, the government announced the formation of a ten-person team (Team 39 or COLT) to regulate foreign loans for commercial projects, resulting in the postponement of four projects worth US$9.8 billion, including the giant Chandra Asri Olefins project and others owned by conglomerates and the Soeharto business groups (*FEER* 19 September 1991: 80–1; 24 October 1991: 76–7; *Kompas* 14 October 1991).

This exercise in financial discipline dramatically raised interest rates and restricted access to credit at a time when the big conglomerates and politico-business families carried heavy debt exposure, often in the form of short-term loans, exacerbated in some cases by the strengthening of the yen. In several high profile cases, large companies went to the wall, notably Bank Summa, Bentoel and Mantrust. While state managers attempted to impose monetary discipline by containing foreign loans, the state banks, as noted earlier, continued to provide access to funds for selected borrowers, often at concessionary rates. Attempts by Team 39 to halt the massive borrowings that were required for the Chandra Asri project were sidestepped when the Indonesian owners reorganised their investment through Hong Kong subsidiaries, thereby making ownership technically fully foreign. In the end, a finance package was constructed involving six state and fifty foreign banks in which the foreign partners raised US$600 in loans and US$100 in equity, while Prajogo and Bambang raised US$100 in loans and US$300 in equity (*Tempo* 24 August 1991: 91, and 20 June 1992: 94; *FEER* 12 March 1992: 45, and 7 June 1993: 52; *Warta Ekonomi* 29 April 1991: 30–9; *AWSJ* 12-13 August 1991: 4, 9).

The tight money policy also led to a so-called capital flight in which a large number of companies simply took their funds offshore (including those obtained from state banks), to speculate on currency changes or to invest in more conducive financial environments (*Bisnis Indonesia* 3 August 1991: 1; *Tempo* 17 August 1991: 86).

The Politics of Regulation

As deregulation failed to produce liberal markets, attention began to be directed to the absence of strong market institutions and regulatory frameworks. Regarded by neo-classical political economists as primarily a matter of supplying a technical framework of rules and procedures or providing training courses, the struggle to establish a regulatory framework was to confront powerful interests within the bureaucracy and among conglomerates and the politico-business families.

The Indonesian bureaucracy has a long history of appropriating development funds and demanding extra-legal payments from business for cooperation and licences. In 1993 Professor Sumitro brought these issues into the public arena by alleging that up to 30 per cent of development funds were leaked in the process of allocation (including over-invoicing), in operations and as a consequence of a general culture of corruption. His claims were backed by a range of observers, including Sri Bintang Pamungkas and Ginandjar Kartasasmita, who noted that Indonesia had a very low incremental capital output ratio, requiring relatively high levels of investment capital to achieve GDP growth (*Editor* 9 December 1993: 60–1). As well as leakages, illegal levies by bureaucrats has been another persistent area of concern (*Kompas* 13 May 1996; *Forum Keadilan* 25 March 1996: 95–108).

A further focus for regulatory action has been in the area of tendering and procurement. Assuming that institutional capacity is the core of the problem, the World Bank supported a range of programs aimed at increasing skills and improving procedural regimes for the stock exchange, government departments, banks and the legal apparatus (Trobek et al. 1993; World Bank 1993: 135–64). However, difficulties of implementation rather than the absence of procedures and rules appear to be the critical factor in several cases. Continuing dismissals of state managers following disputes over tendering and use of state facilities, including prominent examples in Garuda, TVRI, PLN and Telkom, have illustrated the weakness of their position and the political nature of the exercise.[12]

Not all the outcomes have favoured the predators. A significant development has been the 1995 tendering exercise for telephone contracts worth US$2 billion, which was reportedly conducted according to international standards of transparency, excluding bids by several well-connected companies, among them Rajawali, Maharani, Elektrindo Nusantara, Sinar Mas and the Army's Pension Fund. Last-minute efforts by the Minister for Telecommunications, Joop Ave, to convince the successful consortia to include selected unsuccessful bidders were not productive (*AWSJ* 15 June 1995: 1 and 2, 20 June 1995: 1, 24; *Jakarta Post* 20 June 1995: 1).

Two factors appear to explain this apparent victory of procedure over the power of rent-seekers. First, the imminent listing of Telkom shares on the New York stock exchange which was expected to raise US$2 billion and to contribute to a more rapid disbursement of Indonesia's debt required confidence among potential investors that Indonesian Telkom operated on the basis of international standards. Second, the need for private investment in the infrastructure sector also required some indication that the tendering process was governed by predictable rules if reasonable numbers of bidders were to be attracted.

Problems of prudential management and intra-group lending in the private banking sector and insider trading and compliance with disclosure requirements in the JSX were also to be the subject of attempts to improve regulatory frameworks. However, it has been the imposition of regulatory control over the state banks that has proven most difficult and politically contentious. State banks have long been subject to 'command credit' allocated to support national economic strategies or as rents to individual firms (MacIntyre 1993: 151). Liberal reformers argued that state bank credit intended for industrial projects was often obtained without collateral and on the basis of inflated and unsubstantiated cost projections, freeing such credit for speculative use such as deposit in high interest bank accounts or sent overseas to speculate on currency fluctuations. They also noted that in time of tight money, well-connected borrowers were still able to obtain large amounts of credit at favourable rates (Kwik 1993; *Tempo* 9 March 1990; *Tempo* 9 March 1991: 86–90, 11 May 1991).

It was the spectre of bad debts that was to drag state banks into the political arena. As early as May and June 1993, Kwik Kian Gie had claimed in *Kompas* (4 May and 24 June), that bad debts stood at around Rp.10 trillion at least, or 7 per cent of all debt. The issue was brought to a head in July 1993, when documents alleged to come from officials within the Bank Indonesia, were publicly circulated with details of the levels of bad debt and the names of debtors. It was claimed that the state bank for industrial development, Bapindo, had bad or doubtful debts of Rp.2453 billion (28.7 per cent of total outstanding loans), and that the figures for Bank Bumi Daya, Bank Dagang Negara and Bank Rakyat Indonesia were higher.

The Jakarta press took up the story with a vengeance, detailing the huge extent of state bank loans to the textile, petrochemical and forestry industries. A picture was drawn of widespread disregard for banking regulations, collusion between officials, powerful patrons and borrowers, and the dubious nature of many of the projects receiving state bank funds. While debates between officials and commentators over the exact size of bad and doubtful debts were inconclusive, one thing was agreed: the state banks had a serious problem of non-performing loans and the processes of making loans were questionable, to say the least. It was also clear that the government was not prepared to take firm action (*Tempo* 3 July 1993; *Prospek* 3 July 1993; *AWSJ* 3 July 1993; *Warta Ekonomi* 5 July 1993; *Jakarta Post* 5 July 1993).

However, in a dramatic development in February 1994, it was revealed that Bapindo had loaned US$430 million to Eddy Tansil of the Golden Key group, that the loan had violated banking laws, and that the total debt to the state was US$614 million with interest. A 'red clause' facility to draw a letter of credit for US$241 million had been illegally issued for the import of equipment for a proposed petrochemical plant whose asset value had been grossly overvalued. It appeared that no attempt had been made to import the equipment. In an orgy of intense media scrutiny, the Bapindo–Golden Key scandal and the court proceedings were dissected in great detail (*Tempo* 19 February 1994, 26 February 1994, 5 March 1994,

12 March 1994, 2 April 1994; *FEER* 3 March 1994, 23 June 1994; *AWSJ* 9, 11 May 1994).

A series of letters from former Manpower Minister and State Security Chief, Sudomo, to Bapindo supporting the Tansil credit application, and subsequent revelations from defendants in the case that former Finance Minister Sumarlin had also lent his support to the application suggested an embarrassing trail into the heart of power. More interesting, it appeared that Tommy Soeharto had been a partner in the project when the red-clause letter of credit had been issued and remained a partner of Tansil in other ventures (*Tempo* 26 February 1994: 21–30; *Bisnis Indonesia* 27 February 1994: 7).

In the years following the Bapindo affair, the replacement of all state bank directors, including the influential Surasa of Bank Bumi Daya, the appointment of Standard Chartered Bank to provide management advice to Bapindo and the arrest and imprisonment of Eddy Tansil and several bank officials (despite the highly controversial 'escape' of Tansil in May 1996 (*Forum Keadilan* 17 June 1996: 12–20), suggests that the Department of Finance may be having some success in cleaning up the banks. However, as recently as December 1994, the World Bank reported that state bank 'classified assets' (loans classified as 'substandard', 'doubtful', or 'loss'), stood at 18.6 per cent, (World Bank 1995: 19), suggesting that the problems still remain.

The Struggle Over Policy

In what has become popularly known as the contest between the engineers and the economists, the policy environment remains one of intense conflict. Leadership of the 'engineers' was to be assumed by B. J. Habibie, a German-trained engineer who was to draw heavily on the historical experiences of Japan and Germany in developing a concept of industry policy intended to leapfrog Indonesia from the stage of reliance on natural resources and cheap labour directly to that of an advanced economy by means of technological and human resource development. By investing in long-term technological projects such as aircraft and ship manufacture, Habibie and his researchers in the think-tank, CIDES, argued that a new competitive advantage could be created through the by-products of technology transfer and a highly trained workforce (Habibie 1983; *Tempo* 10 October 1992: 21–33; *Tiara* 23 May 1993).

With the support of Soeharto, who was clearly attracted by the vision of Indonesia as an advanced economy of real substance, Habibie developed an extensive institutional base, initially in BPPT (the Board of Research and Technology Application), and then as Minister for Research and Technology. In 1989, Habibie was appointed to manage a new strategic industries board (BPIS), which included ten state enterprises to be isolated from the privatisation process, among them the state-owned aircraft factory (IPTN), a shipbuilding enterprise (PAL), Krakatau Steel and various engineering, transport and munitions firms with a total asset value in 1992 of US$4.6 billion. From this base, Habibie was able to develop considerable authority over state procurement and tendering processes and has reportedly been influential in determining key appointments in PLN and Telkom (*Prospek* 31 October 1992: 16–27; *Warta Ekonomi* 8 October 1989).

Politically he was also to become a central figure as Head of ICMI, an organisation designed to mobilise the support of middle-class Muslims and create a political channel of recruitment additional to the military and the state political party, Golkar. Not surprisingly, fears were raised within liberal circles that an abrupt change in policy was about to occur when several ICMI figures were appointed to the new cabinet of 1993, coinciding with the departure of the last of the 'economists' Sumarlin, Wardhana and Mooy (*FEER* 1 April 1993: 72–5). However, the new economic Ministers, Saleh Afiff (Coordinating Minister for Economy and Finance), Mar'ie Muhammad (Finance), and Soodradjad Djiwandono (Governor of Bank Indonesia), have continued to defend the conservative fiscal and monetary policies that tend to reflect the universal interests of financial ministries and central banks.

Opposition to Habibie has come from several sources. Liberal critics, including the World Bank, technocrats and others have argued that Habibie's ventures have resulted in heavy losses, noting that sales have been confined largely to captive markets among Indonesian state corporations, and in politically organised swap deals with other countries such as the planes-for-cars deal with Malaysia and the planes-for-rice deal with Thailand (*FEER* 11 June 1987: 110–16; *Prospek* 1 May 1993: 16–25; *AWSJ* 11, 12 August 1995: 1, 5). Yet, Habibie continues to attract support from Soeharto and it has been estimated that US$1.6 billion has been injected by the government into IPTN, including a recent and controversial injection of Rp.400 million from funds set aside for reafforestation (*Kompas* 5 July 1994: 2). Plans to build a US$2 billion jet aircraft in the USA have also been supported by Soeharto. Suggested sources of funding include an offering of public shares in the project and further selling of the assets of state companies (*AWSJ* 11–12 August 1995: 1).

In the struggle for funds and for authority over state procurement and the allocation of tenders, Habibie was to come into conflict with the military and with state managers in the economic departments. A vivid illustration of these struggles was revealed in the public dispute over the purchase of thirty-nine former East German warships by Habibie on behalf of the Indonesian government. The initial proposal by Habibie involved expenditure of US$1.1 billion, largely for repairs and refurbishment in Indonesia involving procurement and contracts favouring BPPT and PAL. The ships were to cost only US$12.7 million. In the outcome to public and often bitter debates between Habibie, the Finance Ministry and armed forces (ABRI) commander, Faisal Tanjung, Finance Minister, Mar'ie Muhammad was prepared to allocate only US$319 million which included cuts to the cost of repairs in P. T. PAL from US$64 to US$9.5 million (*Tempo* 11 June 1994: 21–7). However, the state managers were not always so successful. An attempt by the head of the state airline, Merpati, to refuse leasing IPTN's C235 on the terms offered led to his dismissal (*Forum Keadilan* 20 November 1995: 36–40; *Gatra* 4 November 1995: 21–8). With so little success selling the planes on the international market, Habibie could not permit captive domestic markets to demur.

In general, however, the position of the 'engineers' that deregulation and industry policy could be strategically integrated was to gain the public support among policy-makers, including Ginandjar Kartasasmita, Head of the Economic Planning Board, Bappenas, and former Head of Sekneg's Team 10, and Coordinating Minister for Industry, Hartarto. Ginandjar argued that 'We can no longer rely so heavily on what is usually called comparative advantage in such areas as natural resources, low labour costs and soft foreign loans' (*Economic and Business Review Indonesia* 19 June 1993; *Jakarta Post* 23 March 1993 and 7 June 1993). In a series of public addresses Ginandjar was also to argue the destructive effects of liberalisation on the distribution of wealth and economic power, harnessing populist sentiment to the nationalist policy cause (*Jakarta Post* 29 May 1995: 1 and 10 July 1995: 4). In similar vein, Hartarto was to claim that industry policy had reduced imports in petrochemicals and generated a trade surplus in industrial products (*Tempo* 17 April 1993: 90–1).

In 1995 the 'engineers' were given additional teeth. Under Presidential Decision 6, the day-to-day operation and control of public sector procurements were transferred from the Coordinating Minister for Economics, Finance and Development to the Economic Planning Board, Bappenas, under Ginandjar (Nasution 1995: 5), an ominous move in the light of Ginandjar's previous period in charge of public sector procurements under Sekneg from 1978–88.

In the contest between liberal and economic nationalists, the petrochemical industry, in particular the massive Chandra Asri Olefins project, was to be a focal point for policy conflict. Relying heavily on its expected position as supplier of ethylene to the domestic chemical producers, PENI and Tri Polyta (in which the principles of Chandra Asri were also major shareholders), and on guarantees of Pertamina feedstock, Chandra Asri confronted difficulties when both Trypolita and PENI proved reluctant to conclude an agreement so long as they had access to a free international market in which prices looked set to collapse as a world-wide glut of olefins loomed. This confirmed the views of a range of observers who had long argued that Chandra Asri was not viable without protection. Indeed, the principles of Chandra Asri themselves were to request a 40 per cent protection for olefin producers for eight years, Peter Gontha, a director of Chandra Asri stating that the industry could not survive without protection (Gontha 1992).

Within the government, economic nationalists were to join with upstream producers of petrochemicals and synthetic fibres to argue for protection. Minister for State Finance, Sanyoto Sastrowardojo, argued that infant industries served a strategic purpose in reducing imports and building local capacities in technology and engineering, thereby legitimately requiring protection. In the debate on Chandra Asri, Soeharto himself argued that industries justified protection providing they contributed to the creation of a self-supporting national industrial structure, enjoyed protection for a limited period and the process did not harm downstream producers (*AWSJ* 6–7 January 1995: 4; Hobohn 1995: 29). In effect, Chandra Asri was viewed as virtually a state corporation, part of a broader national industry strategy.

The Finance Ministry and other liberals were reluctant to support protection because they feared that facilitating high production costs would cause potential harm to downstream producers. In particular, a tariff on ethylene, currently without tariffs, would directly affect the synthetic fibre industry. A number of Japanese and other North Asian investors indicated that they would seek other sites for investment if tariffs raised production costs (*Tempo* 30 October 1992; *AWSJ* 6–7 January 1995: 4). As late as September 1994, Industry Minister, Tunky Aribowo, announced there would be no tariff.

However, confronted with the threat of bankruptcy to a US$3.26 billion project belonging to the President's son as well as a major setback to plans for a domestic petrochemical industry, the government was to move in favour of protection. In 1994 President Soeharto abolished the COLT team headed by Finance Minister, Mar'ie Muhammad, and replaced it with a team headed by the Coordinating Minister for Trade and Industry, Hartato and including Investment Minister Sanyoto (*AWSJ* 16–17 December 1994:

1). In February and March 1996 the President imposed a 20 per cent surcharge on imports of propylene and ethylene to make polyethylene. These measures were expected to induce both PENI and Tri Polyta to sign supply agreements with Chandra Asri, thereby ensuring its survival (*Jakarta Post* 16 February 1996: 1, 4; *AWSJ* 4 March 1996: 1).

Liberal reformers, nationalist agendas and the interests of politico-business families were also to collide in the automobile industry. Liberal reformers, including Finance Ministers, Sumarlin and Mar'ie Muhammad, together with the World Bank and private liberal critics, had long pressed for deregulation as the answer to high prices and overcrowding in the industry (*Kompas* 20 November 1990; World Bank 1981; Sjahrir 1988). Reformers faced not only powerful and well-connected corporate groups within the auto sector but also an effective alliance between these interests and economic nationalists. Industry Minister, Hartato, and Research and Technology Minister, Habibie, had long been more concerned with the continuing low levels of domestic components in cars assembled in Indonesia and with the problems of technology transfer rather than with opening markets and lowering prices. They favoured an industry policy that would give market advantages to those companies willing to develop high levels of domestic content (*Suara Pembaruan* 25 May 1991; *Kompas* 8 November 1991; *Jakarta-Jakarta* 16–22 March 1991; *FEER* 10 December 1992: 39).

Industry lobbyists attached themselves to the nationalist argument. Citing prospective floods of foreign vehicles and loss of a potential domestic industry with its flow-on into technology, the auto industry association (Gaikindo), as well as leading assemblers and components manufacturers, including Willem Soeriadjaja, Subronto Laras and Sofjan Wanandi, invoked the Japanese experience which, they argued, had involved a thirty-year incubation period within strategic protection to achieve an internationally competitive local base of industry and technology (*Jakarta-Jakarta* 16–22 March 1991; 15 June and 15 September 1991).

The stand-off meant that the auto industry remained insulated as reforms transformed other sectors of the economy. When an initiative did arrive, it was not a decision for deregulation but to build a national car. Presidential Instruction No. 2, 1996 granted exemptions from import duties and luxury sales taxes for automobile manufacturers which produced cars with Indonesian brand names and which were developed with domestic technology, engineering and design. Two Ministerial Decrees allocated this facility to P. T. Timor Putra, a company 70 per cent owned by Soeharto's youngest son, Hutomo Mandala Putera (Tommy), and 30 per cent by Kia of Korea. Timor was required to guarantee 60 per cent domestic components by the end of three years. In return, no other company was to be granted the same package of exemptions for this period at least (*AWSJ* 20 February and 18 March 1996: 1; *Jakarta Post* 29 February 1996: 1).

The policy was defended by Industry Minister, Tunky Aribowo, and other officials in terms of the need to acquire technology transfer and a domestic capability in engineering production, as well as reducing the US$3.6 billion used each year to import auto components. Kia's involvement with Tommy's business group, Humpuss, was likened to the way Austin and Renault were linked with Nissan and Hino in Japan in the 1950s as the basis for the development of a competitive Japanese industry (*Tiras* 14 March 1996: 28; *Asia Times* 18 March 1996: 8 and 16). More difficult to explain was the appointment of Tommy Soeharto, with little experience in the auto industry, as the sole beneficiary of the policy. Indeed, without any transparent tendering process the appointment could not be defended on the basis of any formal criteria and questioning on this issue was simply ignored.[13]

The new policy produced a new raft of policy conflicts. Firms dominant within the industry, Astra, Indomobil and Krama Yudha, were all confronted with declining sales, a slump in profits and, in the case of Astra, a fall in share prices. Astra announced the shelving of Rp.1.2 trillion in investment plans while Indomobil postponed a Rp.600 million program (*Asiaweek* 22 March 1996: 50; *Jakarta Post* 6 March 1996: 8). The Japanese ambassador argued that the policy transgressed standards of transparency fairness and equity expected by foreign investors, while the Japanese government was reportedly considering taking the case to the World Trade Organisation (*Jakarta Post* 16, 23 March 1996). USA companies, General Motors, Ford and Chrysler, planning a comeback into the Indonesian market, immediately indicated they would also reassess or cancel planned investments (*AWSJ* 4 April 1996: 9, and 2 April 1996: 11).

While other local car-makers were clearly shocked, they immediately announced new strategies to deal with the challenge. Indomobil stated its intention of producing a car priced at around Rp.20 million within four or five years based on the exemptions already available under the May 1993 package. Claiming that only itself and Astra had the capacity to achieve local content requirement, Indomobil clearly indicated its doubts about the seriousness of Timor's intentions to fulfil its obligations (*Jakarta Post* 16 March 1996). One of the ironies of the situation was that it pitted members of the Soeharto family against each other. The President's step-brother, Probosutejo, was involved in a joint venture with General Motors to produce Opel and Chevrolet, while Bambang Trihatmojo's Bimantara group was allied with Hyundai. Like Indomobil, Bimantara announced its intention of producing a cheap car to compete with Timor on the basis of existing exemptions. Unlike other competitors, Bambang criticised the new regulations as 'inadequate and monopolistic', vowing to press for the same privileges as Timor (*Jakarta Post* 7 March 1996: 1; *Asia Times* 3 April 1996: 5).

What is less certain is that the industry policy objectives will be achieved. There has been considerable scepticism within the press, in parliament and within the auto industry that Timor has either the capacity or the intention of meeting the domestic content provisions, and similar scepticism that the government will be able to effectively monitor and enforce the provisions of the regulation. With only seven years until Indonesia is forced to open itself to competition from the Malaysian and Thai auto industries under the AFTA provisions, further crowding of the small Indonesia sedan market of 350 000 vehicles by such a costly program of protection for a small, new and vulnerable producer is also seen as a questionable move. This is particularly so when the previous emphasis has been on multi-purpose vehicles, where Astra's Kijang already achieves over 50 per cent local content (*Kompas* 1 March 1996: 1; *Jakarta Post* 1, 4 and 18 March 1996: 4).

Conclusion

Indonesia has undergone a successful shift from an oil-based economy to one more reliant on non-oil exports and revenues, private sector investment and increased international integration. Nevertheless, the architecture of economic and corporate power and access continues to be forged within institutional structures and relationships dominated by the President and his family and the politico-bureaucratic hierarchies of the state apparatus. It is the contest of power and the formation of coalitions and alliances, rather than systems of rules and procedures about contracts, property, exchange and procurement, that determine the market. In strategic instances, political power is drafted to override decisions based in law and formal procedure by state managers. Agendas promoting economic nationalism, social and political stability, the political ascendancy of the state and its politico-bureaucrat strata, the growth of the politico-business families and their corporate allies continue to prevail. While these are subjected to increasing structural pressures, expectations that economic nationalist policy frameworks and predatory politico-bureaucrat controls over economic activity would be logically replaced with generalised liberal market institutions have not eventuated.

It had been expected that the costs of subsidising systems of gatekeeping and rents through the state banking system and other state corporations in an era of increasing competition for fiscal resources would be an undermining factor. However, although debt servicing of official loans now exceeds receipts for loans and grants within the state budget, there has been no fiscal crisis. The state banks and BPIS, both expensive burdens on state resources, continue to be funded. Expectations that markets and transparent and predictable rules for business would be imposed by the need to attract foreign investors have not been fulfilled. While there are continuing complaints about bureaucratic exactions and lack of transparent and predictable rules for business the rate of foreign investment continues to accelerate. Among the largest investors, Taiwanese and Koreans have proven to be adept at operating within the politico-business networks and nationalist industry policy frameworks. British, USA, Australian and European companies too, have also been successful in adjusting, particularly in the new infrastructure sector.

While conflicts between different elements of domestic capital over general policy agendas and the politics of market access has increased, these have tended to be related to particular cases rather than commitments for or against the principle of market institutions and regulatory frameworks based in law. Textile producers, for example, have protested against import and upstream monopolies in the petrochemicals sector but have been among the biggest beneficiaries of protection and the prevailing systems of allocating state bank credit. Within the auto industry, critics of the Timor project are motivated by the shift in advantage towards Tommy rather than the principles of general rules and a level playing field. It does not appear that the risks of fighting for advantage within a rent-seeking system are yet so high that they are forcing demands for a regulatory system based in law. The rate of growth and the rapid emergence of new markets are producing continued opportunities in the domestic business world.

While structural pressures have been accommodated within the complex amalgam of markets, rents, protection and politics which constitutes Indonesia's market institutions, the real threat to the system lies in possible unravelling of the political and social order itself. The architecture of the New Order will be difficult to recreate in the post-Soeharto era precisely because of the degree to which it has revolved around Soeharto and the accumulated resources and alliances he has commanded. His departure will release a range of powerful social and economic interests seeking to gain influence and authority. Without an institutionalised mechanism for managing the transition from one government to another, such as a Leninist-type party in the PAP style, or a system of contending patronage networks as in Thailand or the Philippines, the process will rely heavily on the military and on the ability of individuals to construct deals. An unravelling or loosening of the highly centralised nature of state power seems more likely than a shift towards liberal democracy or a more highly structured legal-rational state apparatus. Institutions of rent-seeking and protection will, in this event, become more fluid and unpredictable.

---

[1] *Yayasan* were formed as ostensibly charitable organisations and were not subjected to public audit. Military commands and the President (as Soeharto) were to operate numbers of these, receiving funds from private conglomerates which were used to invest in private businesses and political activities (Robison 1986: 345–6; Vatikiotis 1990: 62–4).

[2] Detailed discussion of the issue of cartels and monopolies is to be found in *Kompas* 5 July 1995: 2; *Warta Ekonomi* 3 July 1995: 18–22; *Gatra* 15 July 1995: 21–30.

[3] According to figures released by the Coordinating Minister for Economics, Finance and Industry, Radius Prawiro, in September 1991, US$31.3 billion of the US$80 billion was private investment. Of the total investment, petrochemicals accounted for US$10.4 billion, pulp and paper, US$9.9 billion, and mining, fertilisers, industrial estates, property development and other manufacture, US$11 billion, infrastructure and transport, US$21.4 billion. Pertamina projects, mainly in refining and petrochemicals constituted US$22.9 billion of the state investment (Muir 1991: 19). In a detailed survey of mega-projects, *Warta Ekonomi* estimated the total value of the largest thirty-three at US$35 billion (29 April 1991: 22–39).

[4] The major players are Prajogo Pangestu (Barito), Eka Tjipta Widjaja (Indah Kiat and Tjiwi Kimia), Sukanto Tanoto (Inti Indorayon utama and Riau Pulp and Paper) and Bob Hasan (Kertas Kraft Aceh and Santi Murni Plywood). The Soeharto family was to enter the industry in a substantial way in alliance with these groups and with state firms. Siti Hardijanti Rukmana joined Prajogo in P. T. Tanjung Enim Lestari, a US$1 billion pulp and rayon complex with associated forestry and plantation interests. Sigit Hardjojujanto and Bob Hasan joined the government in P. T. Kertas Kraft Aceh while Tommy was involved in the takeover of the government-owned paper mill, Kertas Gowa, and was associated with Eka Tjipta Widjaja in Indah Kiat. In 1994 Bimantara joined with Nusamba and the military Yayasan, Kartika Eka Paksi in a US$600 million venture in pulp production.

[5] For example, the US$1 billion Tanjung Enim Lestari pulp and rayon venture of Prajogo and Tutut, originally receipient of a US$450 million state bank loan, was reportedly to proceed with a US$700–750 overseas loan and with a 20–30 per cent shareholding by Marubeni. Public floats of forestry industry companies on the Jakarta Stock Exchange proved highly successful. Among the listed companies were the forestry giant, Barito, which raised US$295 in a controversial 1992 float, the paper maker, Pabrik Kertas Tjiwi Kimia, the pulp and paper venture, Indah

Kiat, which raises US$348 and Inti Indorayon Utama, which raised Rp.253 billion in 1991. (*FEER* 11 April 1991: 56–7; *AWSJ* 25, 26 May 1990: 1, 4; *Warta Ekonomi* 29 April 1991: 30–9; *AWSJ* 11 April 1994: 1, 7).

6 In 1991 Sigit Hardjojujanto was to negotiate a 24 per cent share with British Petroleum in P. T. Petrochemia Nusantara Interindo (PENI), a US$400 million polyethelene venture, while Bambang Trihatmojo established a US$260 million polypropolene plant, P. T. Tri Polyta, with Prajogo Pangestu in which he was to hold a 32 per cent share. Hutomo Mandala Putera's attempts to establish a huge aromatics plant at Arun were finally abandoned and the project awarded to Mitsui. Nevertheless, he is reportedly to secure substantial sub-contracts for the project and, in the meantime, established a US$300 million methanol plant, P. T. Kaltim. Bob Hasan's P. T. Parama Witara Widya and a company from Siti Hardiyanti Rukmana's Citra Lamtorogung group are acting as Mitsui contractors in Pertamina's US$1.9 billion export refinery at Balongan. Despite vigorous opposition from several *Pribumi* investors the licence for local contractor in a US$1.8 billion export-oriented refinery at Dumai was awarded to Liem Sioe Liong, while a similar licence was given to Prajogo for a US$1.7 billion hydrocracker at Cilacap (*Warta Ekonomi* 29 April 1991: 30–9; *AWSJ* 17, 18 September 1990: 1, 6, 7; *FEER* 7 January 1993: 52 and 11 August 1994: 70; *Jakarta Post* 23 June 1995: 1; *Tempo* 24 August 1991: 89–92; *FEER* 2 May 1991: 40–1).

7 The lists were simply headed 'Information on state banks', see *Kompas* 24 June 1993: 1). Although denied by several of the named debtors, especially the top-of-the-list Prajogo Pangestu, there was a remarkable consistency in the lists. Judging from the revelations to come out of the Bapindo trial and recent statements of the Head of Bank Indonesia (that state bank bad debt stood at US$9.2 billion in November 1993), the lists would appear to be remarkably accurate.

8 For details of the Bimantara and Humpuss holdings, see Schwarz 1992: 56–9; McBeth 1994: 47–9; *AWSJ* 29 May 1995: 1, 7; *Swa* August 1995: 12–55).

9 More prominent figures include former Finance Minister, academic and businessman, Sumitro Djojohadikusumo, consultants and former student activists, Sjahrir and Rizal Ramli, Lawyer, Mulya Lubis, academics/consultants, Iwan Jaya Aziz and Anwar Nasution, former business figures now involved in reformist politics, Kwik Kian Gie and Laksamana Sukardi, former journalist and business intelligence consultant, Christianto Wibisono, and researchers in the privately funded Centre for Strategic and International Studies, Mari Pangestu and Hadi Soesastro.

10 Since 1987–88 debt servicing has exceeded aid receipts in the state budget. For 1994/95 debt servicing of Rp.17.6 trillion compared with receipts of Rp.10.7 trillion. The main creditor consortia (CGI) pledged US$5.26 billion in 1996 of which Japan provided an aid package of US$2.6 billion, almost all in the form of official development loans (ODA).

11 External debt had reached US$100 billion by mid-1995. A decline in merchandise trade of US$3 billion in 1995/96 was almost entirely in non-oil exports, the biggest declines occurring in plywood and garments (World Bank 1996: 10).

12 Those dismissed included Soeparno of Garuda who had been required to purchase MD11 planes through a Bimantara company, allegedly an expensive and unnecessary process (*Jakarta Post* 17, 18 January 1992; *FEER* 21 September 1993: 71–2). Isshadi of TVRI had criticised the award of an educational channel to Tutut which used TVRI time and facilities as well as generating advertising revenue (*Tempo* 25 August 1990: 75–8; *Matra* 7 August 1992). Others included Cacuk Sudarijanto of Telkom (*Jakarta Post* 10, 12 October 1992; *FEER* 5 April 1990), and Ermansjah of PLN (*Jakarta Post* 22 April, 2 May 1992; *Prospek* 22 June 1992).

13 When asked why it was he who obtained the monopoly, Tommy suggested reporters should ask the person who issued the decree in the first place, his father (*Tiras* 14 March 1996: 23–6).

References

Bhattacharya, Anindya and Pangestu, Mari (1992) 'Indonesia: development and transformation since 1965 and the role of public policy', paper prepared for the World Bank Workshop on the Role of Government and East Asian Success, East West Center, Hawaii, November.

Econit (1996) *Dampak Kelangkaan Semu Pupuk Urea Terhadap Impor Beras dan Kesejahteraan Petani Tahun*, Jakarta.

Gontha, Peter (1992) 'Dampak positip dan negatip: 'Tight money policy' dan Pembatasan offshore loan', unpublished paper, 22 April.

Habibie, Bakharuddin J. (1983) 'Some thoughts concerning a strategy for the industrial transformation of a developing country', address delivered to the Deutsche Gesellschaft fur Luft-und Raumfahrt Bonn, Federal Republic of Germany.

Hobohn, Sarwar (1995) 'Survey of recent developments', *BIES* 31 (1): 29.

Jones, Steven (1986) 'Soeharto's kin linked with plastics monopoly', *AWSJ*, 25 November.

Jones, Steven and Pura, Raphael (1986) 'Suharto-linked monopolies hobble economy', *AWSJ*, 24 November.

Kwik Kian Gie (1993) 'A tail of a conglomerate', *Economic and Business Review Indonesia*, 5 June: 26–7 and 12 June: 26–7.

MacIntyre, Andrew (1993) 'The politics of finance in Indonesia: command, confusion, and competition', in Stephen Haggard, Chung H. Lee and Sylvia Maxfield (eds) *The Politics of Finance in Developing Countries*, Ithaca: Cornell University Press.

McBeth, John (1994) 'Market to market', *FEER*, 25 August: 47–9.

Muir, Ross (1991) 'Survey of recent developments', *Bulletin of Indonesian Economic Studies*, 27 (3) December: 19.

Nasution, Anwar (1995) 'Survey of recent developments', *Bulletin of Indonesian Economic Studies*, 31 (2).

Pangaribuan, Robinson (1995) *The Indonesian State Secretariat 1945–1993*, Perth: Asia Research Centre, Murdoch University.

Pura, Raphael (1986) 'Suharto family tied to Indonesian oil trade', *AWSJ*, 24 November.

Pura, Raphael (1995) 'Bob Hassan builds an empire in the forest', *AWSJ*, 20–21 January: 1, 4.

Pura, Raphael (1995) 'Indonesian plywood cartel under fire as sales shrink', *AWSJ*, 23 January: 1, 4.

Ramli, Rizal (1992) 'Kayu', *Tempo*, 13 June: 77.

Robison, Richard (1986) *Indonesia: The Rise of Capital*, Sydney: Allen & Unwin.

Schwarz, Adam (1992) 'All is relative', *FEER*, 30 April: 54–8.

Schwarz, Adam (1994) *A Nation in Waiting*, St Leonards: Allen & Unwin.

Sjahrir, (1988) 'Ekonomi politik deregulasi', *Prisma*, (9): 29–38.

Soesastro, Hadi (1989) 'The political economy of deregulation', *Asian Survey*, 29 (9).

Supriyanto, Enin (1989) 'Growth-oriented strategy and authoritarianism' in Committee to Revitalise Student Activities, *Bertarung Demi Demokrasi*, Bandung: Bandung Institute of Technology.

Tan, Tjin-Kie (ed.) (1967) *Sukarno's Guided Indonesia*, Brisbane: Jacaranda.

Thomas, K. D. and Panglaykim, J. (1973) *Indonesia—The Effect of Past Policies and President Soeharto's Plans for the Future*, Melbourne: CEDA.

Trobek, David M. et al. (1993) *Global Restructuring and the Law*, Working Paper no. 1, Global Studies Research Program, University of Wisconsin, Madison.

Vatikiotis, Michael (1990) 'Charity begins at home', *FEER*, 4 October: 62–4.

Winters, Jeffrey (1996) *Power in Motion: Capital Mobility and the Indonesian State,* Ithaca: Cornell University Press.

World Bank (1981) *Indonesia: Selected Issues of Industrial Development and Trade Strategy, Direct Private Foreign Investment in Indonesia*, Annex 5, Jakarta: Country Department III, East Asia and Pacific Region.

World Bank (1993) *Indonesia: Sustaining Development*, Jakarta: Country Department III, East Asia and Pacific Regional Office, 25 May.

World Bank (1994) *Indonesia: Stability, Growth and Equity in Repelita VI*, Country Department III, East Asia and Pacific Region, 27 May.

World Bank (1995) *Indonesia: Improving Efficiency and Equity—Changes in the Public Sector's Role*, Jakarta: Country Department III, East Asia and Pacific Region, 4 June.

World Bank (1996) *Indonesia: Dimensions of Growth*, Jakarta: Country Department III, East Asia and Pacific Region, 7 May.

Journals, Newspapers and Weeklies

Asia Times

Asian Wall Street Journal (AWSJ)

Asiaweek

Bisnis Indonesia

Detik

Economic and Business Review Indonesia

Editor

Eksekutif

Far Eastern Economic Review (FEER)

Forum Keadilan

Gatra

Independen

Infobank

Jakarta Post

Jakarta-Jakarta

Kompas

Matra

Media Indonesia

Prospek

Suara Pembaruan

Swa

Tempo

Tiara

Tiras

Warta Ekonomi

3

Pressure on Policy in the Philippines

Jane Hutchison[1]

Introduction

Until very recently, the Philippines has failed to achieve the levels of sustained economic growth which have occurred elsewhere in the region. Indeed, at several points in the mid-1980s and early 1990s, the most common measure of national economic development—annual percentage changes to the GDP—turned negative, with manufacturing output being hit particularly hard (see table 3.1). In the 1950s the Philippines did experience a period of rapid and significant import-substitution industrialisation but, since then, employment in the industrial sector has been virtually unchanged, despite the expansion of labour-intensive export manufacturing after the 1970s. Moreover, as the distribution of national income has become more unequal, the standard of living of the bulk of the population has stagnated or declined over the past thirty years. Thus, the country has one of the highest incidences of poverty in the Asia-Pacific region—a scenario not altered by a turnaround in levels of economic growth and investment since 1993 (de Dios 1995: 286).

Table 3.1 Average annual growth rate in the Philippines (%)

	1960–70	1970–80	1980–93	1994
GDP	5.1	6.0	1.4	4.5
Agriculture	4.3	4.0	1.2	2.9
Industry	6.0	8.2	–0.1	6.0
Manufacturing	6.7	0.1	0.8	5.2
Services	5.2	5.1	2.9	4.0

Sources: World Bank (1979) 128 and (1995): 164; Asia Pacific Economic Group (1995): 201.

Rent-seeking

Neo-classical economists have typically attributed the poor economic record of the Philippines to a policy mix which 'misallocates' resources to areas of inappropriate and inefficient economic activity (Power and Sicat

1971; Ranis 1974; Bautista 1989; World Bank 1989; Shepherd and Alburo 1991). In particular, they argue that for most of the post-war era, protectionist trade and macroeconomic policies have encouraged capital-intensive manufacturing for the domestic market, while disadvantaging areas of 'comparative advantage' in agriculture and labour-intensive export manufacturing. Moreover, concomitant government involvement in the economy is said to have encouraged a form of business behaviour known as 'rent-seeking'. Rents are defined in the economic literature in a number of ways, however, in this context, they are understood as additional returns from government intervention in the market. Rent-*seeking* is therefore the effort applied to competition in the political arena over sources of rent, particularly quotas, tariffs and public monopolies (Krueger 1974; Buchanan 1980). The neo-classical criticism is that 'by substituting political for economic criteria in many allocative decisions', rent-seeking is antithetical to the operation of market forces and, hence, to the development of an internationally competitive economy (Shepherd and Alburo 1991: 155–6).

The role of rent-seeking in the political economy of the Philippines has also been pursued by a group of writers whose intellectual debt to Max Weber on bureaucracy places them within the new institutional political economy (Hutchcroft 1991 and 1994; McCoy 1993; Rivera 1994a and 1994b; Villacorta 1994). These writers concur with the notion that rent-seeking detracts from market performance, but they are more concerned with the consequences such behaviour has for the state's ability to align market forces with national economic goals. Whereas in Indonesia, for example, rent-seeking has generally been the province of state officials, in the Philippines the practice is more often associated with a socio-economic elite whose primary base is '*outside* the state' (Hutchcroft 1994: 220). Thus, the neo-Weberians' core criticism of rent-seeking in the Philippine context is that it has produced a '*weak* state'—a polity which cannot steer the course of national economic development because it is constantly subject to the particularistic policy and resource demands of a politically and economically powerful oligarchy. Drawing on 'statist' perspectives on the East Asian newly industrialised countries, they argue that the country's economy will continue to flounder unless the public sector assumes a more corporate culture. In other words, whereas the neo-classical view is that development is dependent on a 'free' market, the neo-Weberians hold that development is contingent on a 'strong state'.

These two perspectives differ over the appropriate role of the state. Yet, in that each link the Philippine's poor economic performance to persistent rent-seeking, both consider the solution lies in the greater *rationalisation* of economic and political decision-making processes—

whereas the neo-classical perspective looks to rationality in the market, the neo-Weberians see it in an 'ideal type' bureaucracy. This convergence has important implications for how the policy program of economic liberalisation pursued by the Ramos regime since its election in 1992 is interpreted. From the neo-classical standpoint, economic liberalisation is a necessary and sufficient condition for the end of rent-seeking and, hence, for improvements in the country's economic performance. The neo-Weberians, on the other hand, raise questions about the extent to which efforts to strengthen the market are associated with a strengthening of the state. Given the close relationship which has existed between rent-seeking and interventionist trade and financial policies, deregulation of the economy can be viewed as a break with the past in that it is held to depend on the state being more insulated from social forces behind protectionism (Malaluan 1994). The puzzle for the neo-Weberians is that liberalisation appears to be proceeding without the type of institutional changes which would suggest an actual diminution in the political economy of 'patrimonialism'. Indeed, the administration has persisted with a program of political decentralisation which likely conflicts with enhanced state autonomy and capacity (Doronila 1994; Villacorta 1994).

This chapter argues that recent moves to liberalise the Philippine economy are better not understood as a rationalisation process *per se*, but rather as an outcome of various domestic and international pressures on the state. First, structural changes to the Philippine economy have not led to improvements in the distribution of wealth, but they have been associated with new forms of wealth-making and, hence, with new social forces. Because the oligarchy has largely survived these changes, the neo-Weberian perspective (with its focus on state-elite relations) has been inclined to overlook them. In particular, there has emerged a more diversified bourgeoisie, significant sections of which are historically unconnected to the rent-seeking elite and or are engaged in export industries (Tiglao 1991: 16; Pinches 1996: 106). Coupled with this is the experience of fourteen years of authoritarian rule under President Marcos. In a number of important senses, the subsequent political and economic policies of both the Aquino and Ramos regimes have been a reaction to the concentration of state power in that era. Finally, recent fiscal problems associated with foreign debt and the withdrawal of the USA military bases have exposed the Ramos regime to international pressures to deregulate the economy. In other words, consistent with a weak polity, the impetus for change has been largely external to the state.

The first part of this chapter outlines the pattern of economic development in the Philippines this century. The social and political dynamics of this pattern of development are then discussed before returning to the specific issue of policy formation and the range of domestic and international factors which help to explain the shift towards economic liberalisation in the post-Marcos decade. Finally, the chapter comments on prospects for a redistribution of wealth and power which emerge as a consequence of the continuities and changes in the form of government in the Philippines.

Historical Background

Export Agriculture

The Philippines was a colony of mercantilist Spain for three and a half centuries, until 1898. For most of that time, administration of the nation was directed at religious conversion and trade regulation, rather than the exploitation of natural resources. Beginning in the late sixteenth century, merchant ships carried Chinese silks from Manila to Acapulco in Mexico, returning with silver. Although highly profitable, the volume of this trade was restricted by the Crown so as to protect existing monopolies. It was not until the late eighteenth century—more than two hundred years later—that the Spanish turned their attention to the commercial development of agriculture in the Philippines. With the opening of the country's ports to international trade in the first half of the nineteenth century, the growing of crops such as sugar, abaca and tobacco for overseas markets was stimulated (Cushner 1971). At the same time, the Philippines became an outlet for manufactures from Britain and the USA. Large imports of textiles in particular caused the decline of the indigenous hand-loom industry (Owen 1978; McCoy 1982).

The Philippines continued to exhibit this so-called colonial pattern of trade for the first half of the twentieth century. After the defeat of Spain, the Philippines became a colony of the USA. Following the expiry of the Treaty of Paris, a 1909 trade agreement granted Philippine agricultural products preferential entry into the large and lucrative USA market. Thus, the area of land put to export crops further expanded, also the production of rice and corn for domestic consumption. Notably, the sugar industry in the mid-1930s contributed some 30 per cent of national income, 43 per cent of government revenue (directly or indirectly) and 65 per cent of the value of total exports (Brown 1989: 204). Industrial manufacturing was consequently first centred on the processing of export crops—sugar milling, vegetable oil production, rope and cigar and cigarette making. However, the industrial production of a variety of consumer and intermediate goods also commenced (Resnick 1970: 62; Yoshihara 1985: 34–5).[2]

Import-Substitution Industrialisation

During the Second World War, the Philippines was occupied by the Japanese and its economy virtually destroyed. After the war, despite formal independence, the political economy of the nation continued to be influenced by ongoing ties with the USA. In particular, a 1946 bilateral trade agreement to renew free trade reproduced the pre-war pattern of economic integration between the two countries. However, export agriculture was relatively slow to recover at a time when the inflow of war rehabilitation funds helped to fuel a surge in imports of manufactures. In three short years a resultant imbalance in trade threatened to bankrupt the nation by draining its reserves of foreign currency. In response, the government introduced import and currency exchange controls.

The impact of these controls was dramatic. The country entered an era of rapid industrialisation as manufacturing expanded from 8 per cent of GDP in 1950 to 20 per cent in 1960. At the same time, the contribution of agriculture fell from 42 to 26 per cent (Jayasuriya 1987: 85) (see table 3.2). In the region at the time, this level of structural change and economic growth was unprecedented and the Philippines was touted as Asia's second newly industrialising nation after Japan. However, this form of protectionism altered the composition of imports, rather than reduced their actual levels (Ranis 1974: 4). As an instrument to rectify an imbalance in trade, controls were principally directed at the importation of consumer items which were regarded as 'non-essential'. It was these same items which became the target of investment: while imports of finished goods fell, imports of raw materials and capital goods continued to grow (Sicat 1972: 17). At the end of the decade the Philippines thus continued to experience balance of payments difficulties. Meanwhile, growth rates in manufacturing slowed once the limits of the domestic market were reached as income levels for the bulk of the population failed to rise.

Table 3.2 Structure of GDP of the Philippines (%)

	Agriculture	Industry	Manufacture	Services
1950	42	14	8	44
1960	26	28	20	46
1970	28	30	23	42
1980	23	37	25	40
1985	25	32	24	43
1990	22	35	25	43
1994	22	35	na	43

Sources: Jayasuriya (1987): 85; World Bank (1995): 166; Asia Pacific Economic Group (1995): 214.

In the early 1960s controls on foreign exchange were lifted and the peso devalued by almost half against the US dollar. These changes had a 'serious but not devastating effect' on the manufacturing sector (Snow 1983: 26). Profits fell and a number of firms were put out of business, but there seems to have been no major sectoral or intra-firm restructuring as a result. Import-substitution manufacturing thus 'stagnated rather than declined' into the late 1960s, largely because tariffs were introduced which reproduced the structure of protection in favour of consumer items over intermediate and capital goods (Shepherd and Alburo 1991: 206). While the lifting of controls and devaluation favoured exports of primary products, they coincided with a fall in the world price of these commodities. As a result, the value of imports continued to rise against the value of exports (Boyce 1993: 253). At the same time, inflation kept real wages in industry and agriculture down, thereby cheapening the value of labour.

Export Restructuring

During the 1970s various transformations in the world economy converged to influence the course of economic development in the Philippines into the early 1990s. The most important of these were: the changing international prices for major Philippine exports and imports, an increase in the international subcontracting of manufactures and labour from the Philippines and, finally, the growth in overseas borrowings which greatly expanded the country's foreign debt. Domestically, these changes coincided with the authoritarian Marcos regime (1972–86).

Between 1962 and 1985 the value of Philippine exports fell by more than half in relation to the value of imports, largely due to the general, ongoing decline in world prices for primary products (Boyce 1993: 73–4). As well, in 1974 the USA government responded to its own domestic interests by not extending the quota on sugar imports from the Philippines. When the world price for sugar fell from 67 cents per pound in 1974 to a low of 7 cents in 1978, the industry was hit hard. Billig (1994: 666) reports that the people of the southern island of Negros 'recall these years as the time when [unsold] sugar was stored in swimming pools, on basketball courts, and even in schools and churches'. World sugar prices rose again in the early 1980s, but by that time Marcos had installed his 'crony', Roberto Benedicto, as head of the government sugar marketing authority. Price increases were not passed on to the growers, production was halved and sugar workers squeezed to the point of starvation (Billig 1994: 666–7).

To a small extent, the decline in the value of traditional exports such as sugar and coconuts has been compensated by the expansion of exports of plantation crops—particularly pineapples, and fish and livestock. However, of far greater significance has been the transformation in the structure of exports which has resulted from the rapid growth of 'non-traditional', labour-intensive manufacturing, particularly garments and electronics. Between 1970 and 1985 the contribution of 'non-traditional' manufactures to the value of total exports increased from 7 to 60 per cent, at the same time as the value of traditional exports fell from 92 to 28 per cent (Montes 1989: 71). In the early 1990s the contribution of non-traditional manufactures has continued to expand to just over 70 per cent of total exports (Asia Pacific Economics Group 1995) (see table 3.3). Significantly, the expansion of the garments industry under Marcos was largely a result of the 'quota hopping' activities of Taiwanese and Hong Kong firms.[3] Under Aquino, efforts to rationalise the distribution of export quotas also contributed to the growth of the industry as utilisation rates improved and more openings were created for local entrants in the subcontracting chain.

Table 3.3 Structure of exports in the Philippines (%)

	1970	1993
Fuels	23	7
Other primary commodities	70	17
Machinery, transport, equipment	0	19
Other manufacturing	8	58

Source: World Bank (1995): 190.

Finally, in the last two decades, labour itself has become of the Philippines' major exports (see table 3.4). For a number of years professionals (especially in medicine) had been leaving the country for work in the USA. However, during the 1970s the oil boom generated demand for construction and domestic workers in the Middle East. More recently, overseas contract workers from the Philippines have been absorbed more into the economies of the Asian region, and an increasing proportion (currently, about half) are women in domestic and personal services such as entertainment (Eviota 1992: 144; Tan 1993). By 1995 official statistics put the number of Filipino workers employed overseas at some 4.3 million and the value of their remittances at over US$2.6 billion per year in the first four years of the Ramos regime (Tiglao 1996: 26). This inflow of funds has caused national income (GNP) to expand ahead of national production (GDP) and contributed to economic recovery by lifting consumer spending (de Dios 1995: 273).

Table 3.4 Structure of merchandise exports in the Philippines (1993)

	Agriculture-intensive	Labour-intensive	Capital-intensive	Mineral-intensive
Philippines	18	51	24	7
Malaysia	20	11	57	11
Thailand	27	27	40	5
Indonesia	30	26	13	32
Singapore	7	7	71	14
Vietnam	32	3	30	34

Source: Asia Pacific Economics Group (1995).

Foreign Debt

In the 1970s growth in the Philippine economy was largely financed by international loans (Montes 1989: 75–9; Boyce 1993: 245–77). Public overseas borrowing began in the mid-1960s, but was greatly extended in the following decade as commercial sources of funds increased and the state under Marcos sought to provide credit for various economic development programs. However, many of the loans were dispensed for political ends— a large proportion simply disappearing overseas as capital flight. Because such loans were government guaranteed, when private individuals defaulted their debt was transferred to the public sector (Boyce 1993: 279, 319–21). By the end of the 1970s sources of cheap commercial loans were disappearing and the Philippine government found it more difficult to cover existing debt with new borrowings. Thus, it became increasingly dependent on the International Monetary Fund (IMF) and the World Bank and, hence, more vulnerable to pressures for structural reform. A crucial juncture was reached in 1983 when, in the midst of a general economic and political crisis, the government was forced to place a 90-day hold on its debt repayments. In 1984 and 1985 GDP fell by 6 and 4.3 per cent—industry faring worst with falls of 10.2 per cent in both years (World Bank 1989: 3). President Marcos was finally driven from office in early 1986 but, in the subsequent decade, both the Aquino and Ramos regimes have continued to pay off the debt so as to retain access to multilateral and Japanese bilateral funds (Jayasuriya 1992: 57).

Economic recovery in the post-Marcos era has been fitful. Foreign debt has reached more manageable proportions but a return to growth in the late 1980s was cut short by political uncertainties, a string of costly natural disasters and breakdowns to the power supply. Labour-intensive exports have continued to grow, although their heavy reliance on imports weakens the balance of trade. On the other hand, much of other manufacturing and traditional agriculture exhibits productivity levels below that of the region (World Bank 1989: 11, 65). In 1995 shortages in the main food staple, rice, contributed to a rise in the cost of living. Even the once-powerful sugar industry now produces largely for the domestic market, and then it has not always been able to meet demand.

In comparative terms, the Philippines is less integrated into the region through trade than are its neighbours, and conversely, more oriented towards the USA market. However, changes to the geography of production and consumption in the Asian region are seeing this alter. For example, in the last decade, some thirty-six industrial estates have been established in rural and semi-rural hinterlands through domestic, Japanese and other Asian investment (Koike 1993; Marasigan 1995). As well, there have been several urban property and associated construction booms and considerable private investment in infrastructure and utilities, such as telecommunications, transport and energy. These developments have proved profitable for new and established capitalists, but, as Emmanuel de Dios (1995: 286) points out they are areas which do not offer 'massive and sustained employment' opportunities. Moreover, such 'catch-up' economic activity largely reflects 'how far the country has fallen behind'—there is still much need for follow-up investment in the tradeable sector of the economy (de Dios 1995: 282–3).

Social Structure and Development

The pattern of economic development just outlined has been associated with a social structure dominated by wide disparities in the ownership and control over productive assets and personal wealth. This facet of development is central to a proper understanding of the origins of rent-seeking in the Philippines. However, in the literature on the subject, rent-seeking is often couched as a *cultural* phenomenon, as a form of behaviour which is but one manifestation of a more general proclivity towards particularistic modes of social interaction among Filipinos (see Stifel 1963; Hawes 1992; Buendia 1993). However, this explanation tends to reify culture by treating it as an independent determinant. The alternative approach prefers to view rent-seeking as co-determinant with—and, hence, contingent upon—a range of other social, economic and political factors (Billig 1994; Wolters 1984). The conditions of existence of rent-seeking which this chapter is particularly concerned with is 'class capacity', that is, with 'the capacities of a given class to act in relation to others and the forms of organisation and practice they thereby develop' (Therborn 1983: 38). In other words, rent-seeking is here explained as an outcome of the class power of the oligarchy.

Land and Labour

The class power of the oligarchy has its origins in the private ownership of land for commercial agriculture. In the late nineteenth century, British and American interests invested in the commerce and infrastructure associated with export cropping, but not directly in land. Instead, particularly in the important sugar and rice growing areas, land came under the control of a small section of the indigenous population. This control was codified in law when, in the early part of this century, the USA government oversaw the sale of Church estates rather than return them to the direct producers

(Wolters 1984: 12–14). As well, legal limits on the size of corporate land-holdings discouraged large-scale foreign investment in plantation cropping. Thus, in the period up to the Pacific war, there was in the Philippines a level of domestic capital accumulation, centred on the commodification of land and its produce, which was unparalleled in South-East Asia at that time (Rivera 1994a: 26).

As commercial cropping expanded and traditional household manufacturing declined, the proportion of the Philippine labour force in agriculture climbed from half to almost three quarters in the first four decades of this century (Resnick 1970: 64–5). From the start, forms of labour control which the landowners exerted varied according to region, crop type, farm size and so on. In Central Luzon, for example, work in rice growing was originally organised largely on a *kasamá* or share-cropping (share-harvest) tenancy basis. Wolters (1984: 24) argues that landowners in Luzon preferred this system to wage labour as it gave the tenant farmers an interest in maximising the output of their labour, under conditions where it was otherwise difficult to directly supervise production. Over time, as less new land was opened up and the supply of labour increased, the lot of tenants is judged to have worsened with landowners in a more powerful position to appropriate a greater share of the harvest (Kerkvliet 1979: 17–25; Wolters 1984). Consequently, tenant–landlord relations became more overtly conflictual and even violent. In the 1930s a peasant resistance movement emerged in Central Luzon which, just after the Second World War, waged a seven-year armed rebellion until eventually crushed by government forces (Kerkvliet 1979). In an attempt to ameliorate such unrest, some land reforms were subsequently introduced, but these did little to alter the basic balance of class power in the region. From the 1950s numbers of landlords introduced more machinery-intensive forms of rice growing, largely to circumvent their labour control problems. These changes to the agricultural labour process set in place a trend towards greater landlessness and usage of wage labour in the industry which accelerated under Marcos (Kerkvliet 1991: 34–58).

In the 1970s international funds were deployed for a 'green revolution' in agriculture through the introduction of high-yield varieties of rice. As these new rice strains required more inputs of water, chemical fertilisers, herbicides and pesticides they transformed the social relations of production—generating new sources of surplus extraction which have tended to decentre the landlord–tenant relationship (Wolters 1984: 207–10; Ofreneo 1987: 85–93; Lim 1990: 123–6). In particular, while the main lines of credit and indebtedness for the direct producer still involve landowners, they now also include rural banks and commercial traders in agricultural inputs and outputs. Greater economic differentiation has also occurred among the direct producers themselves. Some peasants have prospered from improved yields and been able to take up more conducive lease arrangements or consolidated their own small holdings. Others, on the other hand, have lost their land and/or become low-paid contract workers in order to supplement their income from cropping.

This more complex pattern of class relations in rice agriculture has generated new alliances and conflicts which make it difficult to coalesce direct producers around the issue of land reform. From the late 1960s violent opposition to socio-economic privilege resumed in Central Luzon with the formation of the Maoist Communist Party of the Philippines (CPP) and its New People's Army (NPA). But, in the late 1970s Willem Wolters (1984: 214–15) observed the latter organisation attempted to resolve the disparate interests of the landless and small landholders through 'a broad appeal for struggle against the state apparatus in general and ... President [Marcos] in particular'. In the post-Marcos era, these organisational difficulties continue at the same time as divisions have arisen in this radical movement over political strategy—especially over whether or not to adjust to the new political circumstances of electoral democracy and a decentralising state (Magno 1993; Rocamora 1994).

There is not the space to discuss in detail class relations in other sectors of agriculture. However, it is worth noting that the important sugar industry in the southern Visayas region has also been historically characterised by a highly concentrated pattern of land (and mill) ownership, albeit wage labour was always more wide spread than in rice growing (McCoy 1982). Given the relative abundance of cheap labour, the industry remained labour-intensive—by overseas standards—for a long period. In the 1960s cheap loans from state and international sources were thus made available to lift industry productivity through mechanisation. The employment and income consequences for workers were severe and, subsequently, worsened with falling prices on the world market—so much so that, under Marcos, various international aid organisations intervened with emergency food relief (Boyce 1993: 195–8). In the other important export industry, coconuts, the pattern of land ownership has been less concentrated and so small-scale farming is more common. The use of wage labour is also growing, but the level of political organising in this sector is less than in the sugar industry (Boyce 1993: 189–92). Finally, many of the newest sectors in agriculture—bananas and other fruits, fish farming and poultry and livestock production—have high levels of corporate involvement and contract farming, often through vertically integrated, transnational agribusinesses (Boyce 1993: 199–203).

Industrialisation

Import-substitution industrialisation in the 1950s largely reproduced the oligarchic socio-economic structure of the Philippines (Doronila 1986: 42). Given the importance of access to import licences and foreign exchange through the state, between one-third and a half of the new industrialists were members of the landed oligarchy—the other major group having an existing foot in commerce or manufacturing (Carroll 1965; Rivera 1994b: 159–64). Thus, industrialisation in the 1950s and 1960s did not unseat the oligarchy, rather it brought new opportunities for investment and rent-seeking in the finance, real estate and other services sector and construction, as well as in manufacturing. Yet, the homogeneity and relative cohesion of the dominant economic class was at the same time disrupted by the entry into manufacturing of individuals with different social and economic backgrounds and interests.

These developments intensified tensions over policy in export agriculture between the industrial bourgeoisie and landowners in the years before martial law. Often that conflict was framed in economic nationalist terms: the argument of interests behind protected manufacturing being that state intervention to limit import competition was essential for the development of a domestic industrial base. This position was echoed by the radical Left in Philippine politics. However, the nationalism of the bourgeoisie lacked a concomitant call for fundamental socio-economic reforms and, hence, it did not develop a mass base (Doronila 1986: 44–5). Indeed, consistent with the patrimonial nature of politics, contestation over policy concealed particularistic competition over the spoils of public office (Hawes 1992: 153; Doronila 1992: 125). Moreover, many of the import substitution industrialists had substantial links with foreign capital, hence, one of the main targets of bourgeoisie economic nationalism were, in fact, local Chinese business interests.

In the decades before the war, industry had been a site of social conflict as independent, and often radical, trade unions formed in the sectors where workers were congregated in their hundreds and even thousands—in large-scale agricultural processing (especially the sugar mills), cigar and cigarette factories and in land transport and stevedoring (Kunihara 1945: 72). During the 1920s the number of unions and their reported membership increased rapidly (Wurfel 1959: 584). Cuts to wages and rising unemployment in the 1930s resulted in unprecedented levels of industrial action. Significantly, during this era there was an almost fourfold increase in the ratio of organised to spontaneous strikes (Kunihara 1945: 65). By 1940 an estimated 5 per cent of the workforce was organised, although less than 2 per cent were in registered unions (Kunihara 1945: 72). Just after the war, the labour movement was briefly reconstituted under radical leadership until, in the early

1950s, at the height of state action against the peasant rebellion in Central Luzon, the peak organisation was declared illegal and its affiliated unions deregistered (Infante 1980: 110).

With import-substitution industrialisation, the proportion of the labour force in manufacturing grew, peaking at almost 12 per cent. This stimulated an increase in the number of unions, although many were company controlled, ineffective or simply inactive (Snyder and Nowak 1982: 48–53). The 1950s saw some pick-up in real wages in industry, but in the subsequent decade they fell by more than one half, signalling a redistribution of income away from the working class. By the mid-1980s real wages were one quarter of their value twenty years earlier (Doronila 1986: 42; Shepherd and Alburo 1991: 150; Boyce 1993: 27). These trends demonstrate the relative class capacities of capital and labour which underpin the intensification of income inequalities in the post-war era. They are also a consequence of disparities among industrial workers in terms of wages, working conditions and levels of unionisation. In a post-war legislative environment where industrial disputation has been principally enterprise-based, there is a strong correlation between positive scores for such indicators and enterprise size—as defined by the number of employees in a particular workplace. Given that some 82 per cent of the non-agricultural workforce (90 per cent in manufacturing) are currently outside the formal sector (Balisacan 1994: 127; Buenaventura-Culili 1995: 274), it is clear that the vast majority of workers are in little position to exert pressures for socio-economic reform through industrial conflict, particularly as population growth continues to swell the available labour force.

The rise of export manufacturing the 1970s has not been reflected in an expansion of manufacturing employment (see table 3.5). Nevertheless, production for an overseas market has been associated with a deepening of capitalist relations of production. In the 1970s the export garments industry was found to be more capital-intensive and management was reported to be paying greater attention to the organisation and control of the work process (Abad 1975). Although, in the policy literature on the subject, labour-intensive industries are often associated with small-scale enterprises, Gwendolyn Tecson et al. (1990) argue that the expansion of the garments industry contributed to the general increase in the number of large-scale enterprises in manufacturing in the 1980s. Pressures from overseas buyers to maintain quality have been a factor in this as they have set limits on local subcontracting out of production (Hutchison 1992). Such changes can have an important impact on levels of union organisation in manufacturing, but they are often counterposed by political controls and ideological divisions within the labour movement.

Table 3.5 Structure of employment in the Philippines (%)

	Agriculture	Industry	Manufacture	Services
1970	54	16	11.9	29
1980	51	16	10.9	33
1985	50	14	9.5	36
1990	45	15	9.7	40
1992	45	16	10.6	39

Sources: International Labour Organisation *Yearbook*, various years.

Middle and New Forces

In agriculture and industry in the post-war era the number of medium-sized landholders, salaried workers in the public and private sectors, professionals and so on has grown. This heterogeneous stratum is often termed 'middle-class' according to level of income, occupation type and status, pattern of consumption, etc. In a more structural sense, the economic location of the professional middle class is thought to bring it into a 'contradictory' ideological and political relationship with capital and labour (Wright 1985; Robison and Goodman 1996: 7–11). In the Philippines this stratum has been largely socially and economically independent of the traditional oligarchy, yet not strong enough to fundamentally alter the balance of class power. In the 1960s and 1970s the economic interests of this stratum fell on both sides of the political divide which ushered in the Marcos era: some in the middle class were outspoken participants in the radical nationalist movement, but many also found the mounting social conflict threatening and were therefore inclined to support authoritarianism (Crouch 1984; Doronila 1986: 51; Pinches 1996: 110–14). During the 1980s this stratum was prominent in the organised opposition to Marcos, often providing 'a conduit for the expression and mobilisation of peasant and working-class discontent' (Pinches 1996: 114). Moreover, in the transition to formal democracy, the more moderate and liberal expressions of middle-class activism usurped the influence of the radical Left, contributing to its subsequent marginalisation. Pinches (1996: 115) argues that it was largely 'this grouping that came to be associated with "people power" and with what many journalists in Manila described as a middle-class revolt'.

Significantly, that section of the population which gained the *economic* hand under Marcos—the Presidential 'cronies'—did not generally come from the ranks of the traditional oligarchy, but nevertheless also developed or expanded their business operations through privileged access to state-controlled licences, monopolies and loans (Koike 1989: 127–9). In a relatively short period of time, they were a dominant force in most major sectors of the economy, an important exception being labour-intensive export manufacturing where 'fierce international competition made it difficult to translate local monopoly into international advantage' (Hawes 1992: 157). Yet, as the fall of the dictator has demonstrated, most of the traditional oligarchy was 'harassed but not destroyed' by authoritarian rule (Doherty 1982: 30). Under more favourable political circumstances, many of the 'old rich' have returned to economic prominence, in tandem with a handful of surviving cronies. However, in the 1990s they no longer necessarily monopolise key areas of the economy as there has been a diversification of the bourgeoisie in sectors such as agribusiness, finance, retailing, real estate and property development (Koike 1993: 444). Among the first and most important of these have been investors of Chinese descent who, once the target of the nationalist tendencies of the traditional elite, have benefited from the easing of discriminatory legislation under Marcos, and now have a significant foot in most sectors of the economy (Tiglao 1990). This group is important because it has largely been without the same political connections as the traditional oligarchy or the Marcos 'cronies' and, therefore, has not been as clearly associated with rent-seeking on major scale. As well, it is socially and culturally well-positioned to gain from ethnic Chinese investment networks in the region (Pinches 1996: 120).

Less spectacularly, but in the long term not less significantly, the middle strata of Philippine society has been a vital source of medium-scale investment in expanding export manufactures, particularly garments. Through international and local subcontracting arrangements it has been possible for relatively small investors to set up in manufacturing, sometimes as a means to supplement or substitute professional incomes (Pinches 1996: 121). This development has forged a group within industry whose economic interests are linked to the world market and domestic state 'in new ways' (Snow 1983: 78). The business affairs of this section of the bourgeoisie are less reliant on political patronage than home market manufacturers. As well, there is the observable trend for second-generation participants in family firms to be schooled in management and, therefore, better prepared for business in a competitive international market. In other words, export manufacturing has broadened the social base of investment in industry and engendered pressures behind more productive forms of work organisation.

In sum, industrialisation and the subsequent decline of traditional export crops has induced a more complex and, at times, more overtly conflictual social landscape in the Philippines. Central to this has been the processes of capitalist development in agriculture, industry and services and the concomitant commodification of socio-economic relations. These changes have not significantly enhanced the class capacities of the direct producers: peasants and rural and urban workers. Yet, in the last decade, there has been a diversification of interests in the capital-owning class which has implications for policy directions in the 1990s. As well, the end to authoritarian rule has been an impetus for new modes of populist political engagement, exemplified by the rapid growth in numbers of non-government and people's organisations (NGOs and POs). Before outlining these developments, it is important to say more about the 'weak' nature of the Philippine state.

A 'Weak State'

There are two aspects to the notion of a 'weak state' (see Migdal 1988; Evans 1989). The first is a quantitative one which centres on the relative distribution of power between state and society—or, more particularly, between the state and dominant interests in the private sector of the economy—hence the question of *state autonomy*. As James Caporaso and David Levine (1992: 191) put it, this view assumes that causality can be located either 'inside' or 'outside' the state: a weak state is one which is consistently acted upon by 'external' social forces. The second aspect is a qualitative one and refers to the particular organisational form of the state—often expressed as *state capacity*. As previously discussed, this draws heavily on the distinction Weber makes between patrimonial and rational-bureaucratic polities. On this basis, a weak state is one in which relatively little distinction is made between the personal interests and official duties of decision-makers in the executive, legislature and/or bureaucracy and, therefore, one in which the policy-making process is constantly stymied by arbitrary demands.

'Elite Democracy'

The Philippine state can be characterised as 'weak' in both senses of the term. Democratic political institutions were introduced into the Philippines early in the twentieth century by the USA colonial administration, but they did not deliver representative government. Without concomitant socio-economic reforms the landed oligarchy was able to exercise its wealth and influence to dominate the national legislature. In power, members of this elite protected their own class and individual interests, not least by appropriating the resources of the state in particular ways. As Paul Hutchcroft explains, economic returns from public office in the Philippines thus largely fell to politicians of private means, rather than to an elite within the bureaucracy: 'In contrast to "bureaucratic capitalism", where a powerful bureaucratic elite is the major beneficiary of patrimonial largesse and exercises power over a weak business class, the principal direction of rent extraction is [in this case] reversed: a powerful oligarchic business class extracts privilege from a largely incoherent bureaucracy' (1994: 230).

For this reason, the Philippine state has not itself been an effective appropriator of domestic economic surpluses—through the taxation system, for example—nor has it effectively regulated the relations of production in agriculture and industry which generate such surpluses. Instead, the political economy of rent-seeking in the Philippines has been very much associated with the role of the state in the *international* arena, particularly the role it has in deciding and implementing policies to do with trade and currency exchange and its function as a conduit for foreign development assistance and foreign loans (Hutchcroft 1994: 222). Given the enormous drain on public resources which rent-seeking causes, the state has only been able to survive as a consequence of the access it has had to overseas sources of funding: 'while the state is plundered internally, it is repeatedly rescued externally' (Hutchcroft 1994: 226).

For the duration of the modern state, opposition to the socio-economic base of political power has not been sectionally represented in the formal processes of national-level government. The party system has been notoriously characterised by affiliations centred on the shifting terrain of personality-based loyalties, rather than on 'policies aimed at meeting the interests of particular categories of citizens' (Crouch 1984: 43). Pressures for significant reform have, as a consequence, mainly taken extra-parliamentary—and even extra-legal—forms, as seen in the previous discussion of successive peasant rebellions on the island of Luzon. The political landscape of the Philippines has hence also long included a large number and variety of 'cause-oriented' groups in both the rural and urban areas. Indeed, in the 1990s the country is acknowledged to have the most extensive and active social movements in the region, ranging in their activities from radical protest and grass-roots organising to self-help and the provision of alternative service delivery mechanisms. Significantly, many of these organisations are heavily dependent on international funding sources (Goertzen 1991; Silliman 1994).

Authoritarianism

In the post-war period the state's brokerage role in the distribution of international funds contributed to the rise and fall of political authoritarianism. Initially, in the 1960s and 1970s, greater access to such funds combined with the growing importance of capital in the domestic economy to create certain centralist tendencies in Philippine politics (Wolters 1984: 194–5; Doronila 1992). As summed up by Gary Hawes (1992: 150), the locus of power:

> was shifting in favor of the government; and within the government the balance of power was shifting towards the executive branch. The president had always controlled the release of government funds, but with an increase in

the role of economic planners, the new emphasis on technical expertise in the control of the economy, growing economic and military assistance from abroad, and an increased resort to foreign borrowing ... the power of the executive branch grew, and for the first time began to extend out into the countryside and into local politics.

This structural trend was greatly magnified by the political ambitions of Marcos. After his election to the Presidency in 1965 he used public funds to win Congressional support, at the same time reorganising the bureaucracy to facilitate more direct links with the rural masses, then some 70 per cent of the population (Doronila 1992: 127–31). When, in 1972 he faced constitutional barriers to a third term in office, he entrenched himself in office by declaring martial law—in the process shutting down the legislature, limiting press and civil freedoms, expanding the police and the military and bringing significant sections of the economy under state control.

Importantly, the direct political power of the oligarchy was broken under Marcos although, with the exception of some key families, they remained sufficiently influential to protect their private economic base from land and tariff reforms (Crouch 1984). The key actors in the Marcos era were, instead, his closest political supporters who (as previously pointed out) were not, as a rule, originally from the oligarchy. The interests of this group was not firmly behind a coherent policy agenda, rather their interests lay in the rent-seeking opportunities provided by favoured political treatment. Indeed, as competition for such treatment was limited and magnitude of rents increased by monopolistic controls and foreign borrowings, the scale at which public resources were transferred to private hands reached unprecedented levels. During this time the state itself became a 'direct instrument of surplus extraction' in that the public sector marketing monopolies in agriculture extracted a levy or tax from the direct producers. Yet, the political economy of the Marcos regime was ultimately tied to the availability of overseas funds, hence the factor which contributed to the concentration of political power eventually played a key role in the demise of the dictator. Moreover, it continues to influence the policy pronouncements of the Ramos regime.

Political Decentralisation

In reaction to the centralism of the Marcos era, the Aquino regime reintroduced democratic elections, restored the legislature, disbanded the agricultural monopolies and privatised a significant proportion of public enterprises, thereby reconstituting the institutional conditions for 'the restoration of elite democracy' (Bello and Gershman 1990). In the 1987

Congressional elections, most of those elected were from families with a prior record of political incumbency. However, in keeping with the post-war trend, it is the case that the main economic base of these 'dynasties' is no longer in agriculture but in industry, banking and real estate (Wurfel 1979; Doronila 1992: 83–9). More broadly, the Aquino regime did little to enhance the class capacities of social movements seeking more fundamental change. Her administration did not push through with land reform in agriculture and, in industry, disputation and wage bargaining was returned to the regional and enterprise levels, moves which have contributed to the fragmentation of the labour movement. As well, the regime oversaw an escalation in the use of violence against political and community activists as it failed to control the proliferation of paramilitary and vigilante organisations. Towards the end of her term in office, Aquino attempted to broaden the support base of her regime with overtures towards the rapidly growing NGO–PO sector. However, there was little substance to the claimed motive of 'people empowerment' and such moves did not do much to counter assessments that the Aquino administration had presided over the return of the traditional political and economic elite (Clarke 1994).

Yet, Aquino also introduced the legislative conditions for what some commentators are terming 'a new form of governance' (Magno 1993; Tigno 1993; Törnquist 1993). Under the Local Government Code of 1991, NGO–PO representation in various facets of local government is institutionalised, at the same time as sizeable resources (including up to 40 per cent of internal revenue), decision-making and regulatory powers have been devolved from the national to the local level of the state. In principle, these changes mean that the opportunities now exist for a more participatory mode of democracy and, hence, for greater accountability and responsiveness on the part of the public sector (Brillantes 1994: 44). Perhaps most significantly, this process of political decentralisation 'irreversibly diminishes the role of the national executive' in government (Magno 1993: 15). However, optimistic assessments of the democratic potential of the Local Government Code tend to uncritically rest on the liberal dualism of state and civil society which normatively associates the former with constraint and the latter with freedom (Rodan 1996). Jorge Tigno's (1993: 64) claim, that a strong NGO–PO sector 'means the gradual end of the dominant and domineering state', fails to capture the specifics of the Philippine state in the post-Marcos era. Alex Brillantes' (1994: 42) point that such NGO–PO activity is a response to the state's 'general inability to deliver basic services and to effectively enforce laws', more closely acknowledges the problems associated with a *weak* state and the need to consequently develop alternative modes of resource distribution.

Ramos Regime

In the 1992 election President Ramos drew support from sections of business and the military known to be antagonistic to the oligarchy (Pinches 1996: 116). Subsequently, he and his advisers have pursued a policy agenda of liberalising the economy which, they claim, is an overt challenge to the *modus operandi* of the oligarchy. In conformity with academic assessments of the Philippine's poor economic record, members of the Ramos regime have stated that the rent-seeking behaviour of the oligarchy lies at the heart of the nation's difficulties and that, therefore, actions to reduce government intervention will directly target the political conditions of rent-seeking (Villacorta 1994: 71; Tiglao 1995b). Pronouncements such as these give some support for the assessment of an association between policies designed to strengthened the market and greater state autonomy in the Philippines. However, the personal leadership qualities of the President have not generally been matched by more fundamental indicators of institutional change. Indeed, the evidence points the other way to policy shifts being induced by external pressures on the state: causality still operates to a significant degree *upon* the state.

As already outlined, transformations in the regional and global economy, and the Philippine's place within it, have engendered a greater diversity of interests in the business community and the professional middle class. While it is a mistake to assume that such interests are therefore unambiguously aligned with a 'free market', it is also the case that they can no longer simply be encompassed by the general category of 'crony' or 'patrimonial capitalism' (Pinches 1996: 124–7). Hence, there is a significant constituency behind attempts to dismantle the political economy of rent-seeking in the Philippines. That attempts at reform have centred on the pull-back of the state—the removal of mechanisms of government intervention in the economy—is a legacy of the years of authoritarian rule under Marcos. The phenomenon of persistent cronyism, despite the greater relative autonomy of the state from the traditional elite, has created a domestic environment in which there is wide-spread support for the dismantling of state power (Malaluan 1994: 38–9). This is manifest in the popular movement's preparedness to be a party to the processes of political decentralisation flowing from the Local Government Code and the concomitant decline in support for authoritarian left alternatives.

Apart from the Ramos regime's commitment to a more decentralised polity, no real attempt has, in fact, been made to reform other aspects of the political system which impact on the issue of state capacity. The workings of the bureaucracy have not been overhauled to inculcate a culture of public service over private advantage (Villacorta 1994: 86). The party system continues to be dominated by 'personality-based coalitions of politicians with independent power bases'. Twelve months after the 1992 elections some seventy-one members of the House of Representatives had already switched to the party of the President (Riedinger 1994: 140). As well, in his dealings with individual members of Congress over the passage of controversial legislation—such as that for the introduction of a value-added tax (VAT)—Ramos has regularly resorted to the 'pork barrelling' of state resources to garner support (Tiglao 1995a; Agpalo 1995: 268). In other words, government processes are still widely subject to particularism and patronage; economic liberalisation has not been accompanied by institutional changes to strengthen the state. As such, recent efforts to deregulate the economy are better understood as a response to the fiscal crisis of the state; consequently, they are in conformity with the post-war pattern of policy making.

Crisis Management

Economic policy making in the Philippines has been rendered relatively incoherent by the weak character of the state. This means that change is often piecemeal and driven by political expediency. However, beyond the centralist tendencies previously outlined, there have been episodes when the policy has taken a different direction in response to economic crises. The introduction of import and exchange controls in the early 1950s is a good example of this—they were not initiated as an instrument of long-term economic planning, but as a device to deal with the balance of payments problem. Arguably, each subsequent period of economic *deregulation* have been associated with similar cases of crisis management.

The introduction of martial law in 1972 was widely interpreted as a political response to sectional opposition to policies designed to liberalise the economy. It was argued that Marcos strengthened the state for the specific purpose of attracting foreign investment and for reorienting the economy towards export-oriented manufacturing (Stauffer 1974; Crowther 1986; Hawes 1987; cf. Hawes 1992). The actors considered central to this policy turn-around were the 'technocrats'—neo-classical economists in the bureaucracy whose influence, it was predicted, would increase with the centralisation of state power and stronger links to the World Bank and the IMF. In practice, the regime had little commitment to a more open economy; developments which did occur—notably the expansion of export manufacturing—were often not a direct consequence of government initiatives (Lindsey 1992: 80; Shepherd and Alburo 1991: 215–16). Indeed, against the policy preferences of the 'technocrats' in the planning bureaucracy, the executive favoured a program of second-stage import-substitution industrialisation at the same time as the 'easy' phase of domestic manufacturing was continued: in the 1970s the tariff system was unaltered while controls on imports even increased (Bautista 1989: 9). More particularly, the implementation of policy was regularly undermined by the politics of favourism, as Richard Doner (1992) demonstrates in the case of the car industry.

However, by the early 1980s circumstances had changed, the government was obliged to agree to the conditions of a World Bank 'structural adjustment' package and introduced 'significant' measures to liberalise the economy (Bautista 1989). Crucially, this was 'an unwanted and forced agenda', made necessary by the need for ongoing access to foreign funds (Montes 1989: 80; Jayasuriya 1992: 60; Lindsey 1992: 85). In other words, the fiscal dependency of the state forced the government to push through with an otherwise politically unpalatable policy program. In the post-Marcos era similar pressures are at work as a result of foreign debt and the withdrawal of financial assistance from the USA following the 1992 closure of the military bases. As Paul Hutchcroft (1994: 234) has argued, the political economy of rent-seeking in the Philippines has long been sustained by the presence of the bases, their withdrawal will therefore add to pressures for the economy to become more internationally competitive. Far from the policy program of the Ramos regime being evidence of a stronger state, it demonstrates continuities in the weak nature of the state, in an environment which is increasingly less conducive to 'patrimonial capitalism'.

Significantly, it seems that members of the traditional oligarchy have been able to make their own adjustments. For example, the Ramos regime's actions to deregulate the telecommunications industry required it to overcome the concerted opposition of the powerful Cojuangco family. In the end, the family retained control of the Philippine Long-Distance Telephone company, albeit now in a competitive environment (Tiglao 1995: 38). Major interests behind the new companies in the industry are existing key players in the Philippine private sector and, together, they demonstrate its 'interlocking' character: 'the Ayala Corporation (land development and manufacturing), the Eugenio Lopez clan (the nation's largest broadcast network and Manila Electric Company), and John Gokongwei (shopping malls, hotels and manufacturing) are leading shareholders in three [of the four] new firms ...' (Riedinger 1994: 144).

The significance of this is twofold. One, it helps to underline a weakness in the neo-liberal perspective on rent-seeking which almost exclusively concerned with the role of the state, and not with the obverse concentration in the private ownership and control over productive assets. Two, it reveals an opportunistic aspect to the business activities of the old (and new) oligarchy which is often overlooked by the 'patrimonial' model of the political economy of the Philippines. It demonstrates the importance of understanding the particular conditions of existence of certain forms of economic behaviour and, hence, the circumstances under which they change (Billig 1994).

Conclusion

Protectionist trade and financial policies in the Philippines have had a powerful political constituency in the traditional oligarchy. It is, therefore, reasonable to assume that government efforts to liberalise the economy require the enhanced autonomy of the executive arm of the state. However, as this chapter has sought to argue, the policy agenda of the Ramos regime does not signal a strengthening of the Philippine state. Instead, this agenda is better understood as an outcome of a series of domestic and international pressures on a *weak* state. While these pressures may indeed see a turnaround in the political economy of development in the Philippines—in the sense that key sectors of the economy will be required to operate in a more competitive environment—there is no guarantee that 'the market' will redistribute wealth and power to the majority of the population. There is some hope that NGO–PO participation in local government will forge a new state–society nexus in which popular demands are better represented. At this stage, however, it remains an open question as to whether or not this 'alternative route' (Tigno 1993) will deliver the same relative socio-economic equalities as stronger states in the region have done in the past.

---

1 I wish to thank Michael Pinches and Herb Thompson for their assistance in the preparation of this chapter.

2 Conversely, traditional household industries—notably hand-loom weaving—declined from 60 to 13 per cent of total value-added between 1902 and 1938 (Resnick 1970: 63-4). Yet, not all household industries declined with trade. Hat-making and embroidery, for example, developed as export industries, the latter receiving government assistance in the areas of training and marketing (Rutten 1990: 16–17).

3 Up until the formation of the World Trade Organisation, most trade in garments (and textiles) from developing to developed countries was regulated by various Multi-Fibre Agreements (MFAs). These MFAs consisted of a series of bilaterally negotiated, product-specific restrictions on trade. In the 1970s the Philippines was one of a number of developing countries which had a low quota utilisation rate.

As a consequence, it attracted companies from countries (such as Hong Kong and Taiwan) which consistently filled their quotas (Trela and Whalley 1990: 17–21).

References

Abad, Ramon (1975) The Garment Industry: A Study Prepared for the NEDA–PDCP–UP Special Course in Corporate Management and Industry Evaluation, Manila.

Agpalo, Remigio E. (1995) 'The Philippines: remarkable economic turnaround and qualified political success', *Southeast Asian Affairs*, Singapore: Institute of Southeast Asian Studies: 259–72.

Asia Pacific Economic Group (1995) *Asia Pacific Profiles 1995*, Canberra: Research School of Pacific and Asian Studies, Australian National University.

Balisacan, Arsenio M. (1994) 'Urban poverty in the Philippines: nature, causes and policy measures', *Asian Development Review* 12 (1): 117–52.

Bautista, Romeo M. (1989) *Impediments to Trade Liberalization in the Philippines*, London: Trade Policy Research Centre and Aldershot: Gower.

Bello, Walden and Gershman, John (1990) 'Democratization and stabilization in the Philippines', *Critical Sociology*, 17 (1): 35–56.

Billig, Michael S. (1994) 'The death and rebirth of entrepreneurism on Negros Island, Philippines: a critique of cultural theories of enterprise', *Journal of Economic Issues*, XXVIII (3): 659–78.

Boyce, James (1993) *The Philippines: The Political Economy of Growth and Impoverishment in the Marcos Era*, London: Macmillan.

Brillantes, Alex B. (1994) 'Decentralization: governance from below', *Kasarinlan*, 10 (1): 41–7.

Brown, Ian (1989) 'Some comments on industrialisation in the Philippines during the 1930s', in Ian Brown (ed.) *The Economies of Africa and Asia in the Inter-War Depression* pp. 203–20, London and New York: Routledge.

Buchanan, James M. (1980) 'Rent seeking and profit seeking', in James M. Buchanan, Robert D. Tollison, and Gordon Tullock (eds) *Towards a Theory of the Rent-Seeking Society* pp. 3–15, Texas: Texas A&M University Press.

Buenaventura-Culili, Venus (1995) 'The impact of industrial restructuring on Filipino Women Workers' in Helene O'Sullivan (ed.), *Silk and Steel: Asian Women Workers Confront Challenges of Industrial Restructuring*, pp. 268–97, Hong Kong: Committee for Asian Women.

Buendia, Rizal G. (1993) 'Colonialism and elitism in Philippine political development: assessing the roots of underdevelopment', *Philippine Journal of Public Administration*, XXXVII (2) April: 141–74.

Caporaso, James A. and Levine, David P. (1992) *Theories of Political Economy*, New York: Cambridge University Press.

Carroll, John (1965) *The Filipino Manufacturing Entrepreneur: Agent and Product of Change*, Ithaca: Cornell University Press.

Clarke, Gerard (1994) 'Kabisig: Aquino's lasting legacy?', *Philippine Quarterly of Culture and Society*, 22: 37–45.

Crouch, Harold (1984) *Domestic Political Structures and Regional Co-operation*, Singapore: ASEAN Economic Research Unit, Institute of Southeast Asian Studies.

Crowther, William (1986) 'Philippine authoritarianism and the international economy', *Comparative Politics*, 18 (3): 339–55.

Cushner, Nicholas (1971) *Spain in the Philippines*, Quezon City: Ateneo de Manila University Press and Tokyo: Charles E. Tuttle.

de Dios, Emmanual (1995) 'The Philippine economy: what's right, what's wrong', *Southeast Asian Affairs 1995*, pp. 273–88, Singapore: Institute of Southeast Asian Studies.

de Dios, Emmanuel (1993) 'Poverty, growth, and the fiscal crisis', in Emmanuel de Dios (ed.) *Poverty, Growth, and the Fiscal Crisis*, pp. 3–75, Manila: Philippine Institute for Development Studies.

Doherty, John F. (1982) 'Who controls the Philippine economy: some need not try as hard as others', in Belinda Aquino (ed.) *Cronies and Enemies: The Current Philippine Scene*, pp. 7–35, Philippine Studies Occasional Paper No. 5, Honolulu: University of Hawaii.

Doner, Richard F. (1992) *Driving a Bargain: Automobile Industrialization and Japanese Firms in Southeast Asia*, Berkeley and Oxford: University of California Press.

Doronila, Amando (1986) 'Class formation and Filipino nationalism: 1950–70', *Kasarinlan*, 2 (2): 39–52.

Doronila, Amando (1992) *The State, Economic Transformation, and Political Change in the Philippines, 1946–72*, Singapore: Oxford University Press.

Doronila, Amando (1994) 'Reflections on a weak state and the dilemma of decentralization', *Kasarinlan*, 10 (1): 48–54.

Evans, Peter (1989) 'Predatory, developmental, and other apparatuses: a comparative political economy perspective on the Third World state', *Sociological Forum*, 4 (4): 561–87.

Eviota, Elizabeth (1992) *The Political Economy of Gender: Women and the Sexual Division of Labour in the Philippines*, London: Zed Books.

Goertzen, Donald (1991) 'Agents for change', *Far Eastern Economic Review*, 8 August: 20–1.

Hawes, Gary (1987) *The Philippine State and the Marcos Regime: The Politics of Export*, Ithaca: Cornell University Press.

Hawes, Gary (1992) 'Marcos, his cronies, and the Philippines' failure to develop', in Ruth McVey (ed.) *Southeast Asian Capitalists*, pp. 145–60, New York: Cornell University Press, Southeast Asian Program.

Hutchcroft, Paul D. (1991) 'Oligarchs and cronies in the Philippine state: the politics of patrimonial plunder', *World Politics*, 43: 414–50.

Hutchcroft, Paul D. (1994) 'Booty capitalism: business–government relations in the Philippines', in Andrew MacIntyre (ed.) *Business and Government in Industrialising Asia*, pp. 216–43, Sydney: Allen & Unwin.

Hutchison, Jane (1992) 'Women in the Philippines' garments export industry', *Journal of Contemporary Asia*, 22 (4): 471–89.

Infante, Jaime (1980) *The Political, Economic, and Labour Climate in the Philippines*, Industrial Research Unit, University of Pennsylvania.

International Labour Organisation *Yearbook*, various years, Geneva: International Labour Organisation.

Jayasuriya, Sisira (1987) 'The politics of economic policy in the Philippines during the Marcos era', in Richard Robison, Kevin Hewison, and Garry Rodan (eds) *Southeast Asia in the 1980s: The Politics of Economic Crisis*, Sydney: Allen & Unwin.

Jayasuriya, Sisira (1992) 'Structural adjustment and economic performance in the Philippines', in Andrew J. MacIntyre and Kanishka Jayasuriya (eds) *The Dynamics of Economic Policy Reform in South-east Asia and the South-west Pacific*, pp. 50–73, Singapore: Oxford University Press.

Kerkvliet, Benedict J. (1979) *The Huk Rebellion: A Study of Peasant Revolt in the Philippines*, Quezon City: New Day Publishers.

Kerkvliet, Benedict J. (1991) *Everyday Politics in the Philippines: Class and Status Relations in a Central Luzon Village*, Quezon City: New Day Publishers.

Koike, Kenji (1989) 'The reorganization of Zaibatsu groups under the Marcos and Aquino regimes', *East Asian Cultural Studies*, XXVIII, March: 127–43.

Koike, Kenji (1993) 'The Ayala group during the Aquino period: diversification along with a changing ownership and management structure', *The Developing Economies*, XXXI (4): 442–63.

Krueger, Anne O. (1974) 'The political economy of the rent–seeking society', *American Economic Review*, LXIV (3): 291–303.

Kunihara, Kenneth K. (1945) *Labour in the Philippine Economy*, Stanford: Stanford University Press.

Lim, Joseph Y. (1990) 'The agricultural sector: stagnation and change', in Emmanuel S. de Dios and Lorna Villamil (eds) *Plans, Markets and Relations: Studies for a Mixed Economy*, pp. 118–32, Manila: Kalikasan Press.

Lindsey, Charles (1992) 'The political economy of international economic policy reform in the Philippines: continuity and restoration', in Andrew J. MacIntyre and Kanishka Jayasuriya (eds) *The Dynamics of Economic Policy Reform in South-east Asia and the South-west Pacific*, pp. 74–93, Singapore: Oxford University Press.

Magno, Alexander R. (1993) 'A changed terrain for popular struggles', *Kasarinlan* 8 (3): 7–21.

Malaluan, Nepo A. (1994) 'Philippines 2000 and the politics of reform', *Kasarinlan* 9 (2)(3): 37–53.

Marasigan, Cynthia G. (1995) 'Tycoons: young and restless', *Far Eastern Economic Review*, 6 July.

McCoy, Alfred (1982) 'Introduction: the social history of an archipelago', in Alfred McCoy and Ed de Jesus (eds) *Philippine Social History: Global Trade and Local Transformations*, pp. 1–18, Quezon City: Ateneo de Manila University Press.

McCoy, Alfred (1993) 'Rent-seeking families and the Philippine state: a history of the Lopez family', in Alfred McCoy (ed.) *An Anarchy of Families: State and Family in the Philippines*, pp. 429–536, Madison: University of Wisconsin.

Migdal, Joel S. (1988) *Strong Societies and Weak States: State-Society Relations and State Capabilities in the Third World*, New Jersey: Princeton University Press.

Montes, Manuel F. (1989) 'Philippine structural adjustments, 1970–1987', in Manuel Montes and Hideyoshi Sakai (eds) *Philippine Macroeconomic Perspective: Developments and Policies*, pp. 45–90, Tokyo: Institute of Developing Economies.

Ofreneo, Rene (1987) *Capitalism in Philippine Agriculture*, 2nd edn, Quezon City: Foundation for Nationalist Studies.

Owen, Norman (1978) 'Textile displacement and the status of women in Southeast Asia', in G. P. Means (ed.) *The Past in Southeast Asia's Present*, Ottawa: Canadian Society for Asian Studies.

Pinches, Michael (1995) 'Entrepreneurship, consumption and ethnicity in the making of the Philippines new rich', paper presented to the Cultural Constructions of Asia's New Rich Workshop, Asia Research Centre, Murdoch University.

Pinches, Michael (1996) 'The Philipppines' new rich: capitalist transformations amidst economic gloom', in Richard Robison and David S. G. Goodman (eds) *The New Rich in Asia: Mobile Phones, McDonalds and Middle-Class Revolution*, pp. 105–33, London and New York: Routledge.

Power, John and Sicat, Gerardo (1971) *The Philippine Industrialization and Trade Policies*, Oxford: Oxford University Press.

Ranis, Gustav (1974) *Sharing in Development: A Programme of Employment, Equity and Growth for the Philippines*, Geneva: International Labour Organisation.

Resnick, Stephen (1970) 'The decline of rural industry under export expansion: a comparison among Burma, Philippines, and Thailand, 1870–1938', *Journal of Economic History*, 30 (1): 51–73.

Riedinger, Jeffrey (1994) 'The Philippines in 1993: halting steps toward liberalization', *Asian Survey*, 34 (2): 139–46.

Rivera, Temario C. (1994a) *Landlords and Capitalists: Class, Family, and the State in Philippine Manufacturing*, Quezon City: University of the Philippines Press.

Rivera, Temario C. (1994b) 'The state, civil society, and foreign actors: the politics of Philippine industrialization', *Contemporary Southeast Asia*, 16 (2): 157–77.

Robison, Richard and Goodman, David S. G. (1996) 'The new rich in Asia: economic development, social status and political consciousness', in Richard Robison and David S. G. Goodman (eds) *The New Rich in Asia: Mobile Phones, McDonalds and Middle-Class Revolution*, pp. 1–16, London and New York: Routledge.

Rocamora, Joel (1994) *Breaking Through: The Struggle Within the Communist Party of the Philippines*, Manila: Anvil Publishing.

Rodan, Garry (1996) 'Theorising political oppositions in East and Southeast Asia', in Garry Rodan (ed.) *Political Oppositions in Industrialising Asia*, London and New York: Routledge.

Rutten, Rosanne (1990) *Artisans and Entrepreneurs in the Rural Philippines: Making a Living and Gaining Wealth in Two Commercialised Crafts*, Amsterdam: VU University Press.

Shepherd, Geoffrey and Alburo, Florian (1991) 'The Philippines', in Demetris Papageorgiou, Michael Michaely and Armeane Choski (eds) *Liberalizing Foreign Trade*, vol. 2, pp. 133–308, Cambridge, Mass. and Oxford: Basil Blackwell.

Sicat, Gerardo P. (1972) *Economic Policy and Philippine Development*, Quezon City: University of the Philippines Press.

Silliman, G. Sidney (1994) *Bilateral Programs of Official Development Assistance to Non-Government Organizations*, Pomona: International Center, California State Polytechnic University.

Snow, Robert T. (1983) 'Export-oriented industrialization, the international division of labour, and the rise of the subcontract bourgeoisie in the Philippines', in Norman Owen (ed.) *The Philippine Economy and the United States: Studies in Past and Present Interactions*, Michigan Papers on South and Southeast Asia no. 22.

Snyder, Kay A. and Nowak, Thomas C. (1982) 'Philippine labour before martial law: threat or nonthreat?', *Studies in Comparative International Development*, 17 (3–4): 44–72.

Stauffer, Robert B. (1974) 'Political economy of a coup: transnational linkages and Philippine political response', *Journal of Peace Research*, 3: 161–77.

Stifel, Laurence (1963) *The Textile Industry: A Case Study of Industrial Development in the Philippines*, Southeast Asian Program Data Paper, no 49, Ithaca: Cornell University.

Tan, Edita A. (1993) 'Labour emigration and the accumulation and transfer of human capital', *Asian and Pacific Migration Journal*, 2 (3): 303–28.

Tecson, Gwendolyn, Valcarcel, Lina and Nunez, Carol (1990) 'The role of small and medium-scale industries in the industrial development of the Philippines', in Richard Hooley and Muzzafer Ahmad (eds) *The Role of of Small and Medium-Scale Industries in Industrial Development: The Experience of Selected ASEAN Countries*, pp. 313–422, Manila: Asian Development Bank.

Therborn, Göran (1983) 'Why some classes are more successful than others', *New*

Left Review, 138.

Tiglao, Rigoberto (1990) 'Gung-ho in Manila', *Far Eastern Economic Review*, 15 February: 68–72.

Tiglao, Rigoberto (1991) 'March of pluralism', *Far Eastern Economic Review*, 5 September: 16–17.

Tiglao, Rigoberto (1994) 'Strength in numbers', *Far Eastern Economic Review*, 21 July: 60–1.

Tiglao, Rigoberto (1995a) 'Progress on parade', *Far Eastern Economic Review*, 6 July: 36–8.

Tiglao, Rigoberto (1995b) 'Right-hand man', *Far Eastern Economic Review*, 2 November: 25–8.

Tiglao, Rigoberto (1996) 'Welcome exchange', *Far Eastern Economic Review*, 29 February: 24–6.

Tigno, Jorge V. (1993) 'Democratization through non–governmental and people's organizations', *Kasarinlan*, 8 (3): 59–73.

Tîrnquist, Olle (1993) 'Democratic empowerment and democratisation of politics: radical popular movements and the May 1992 Philippine elections', *Third World Quarterly*, 14 (3): 485–515.

Trela, Irene and Whalley, John (1990) 'Unraveling the threads of the MFA', in Clive Hamilton (ed.) *Textile Trade and Developing Countries: Eliminating the MFA in the 1990s*, pp. 11–45, Washington DC: World Bank.

Villacorta, Wilfrido V. (1994) 'The curse of the weak state: leadership imperatives for the Ramos government', *Contemporary Southeast Asia*, 16 (1): 67–92.

Wolters, Willem (1984) *Politics, Patronage and Class Conflict in Central Luzon*, Quezon City: New Day Publishers.

World Bank (1989) *Philippines: Towards Sustaining the Economic Recovery*, Country Economic Memorandum, Washington: World Bank.

Wright, Erik Olin (1985) *Classes*, London: Verso.

Wurfel, David (1959) 'Trade union development and labour relations policy in the Philippines', *Industrial and Labour Relations Review*, 12 (4): 582–608.

Wurfel, David (1979) 'Elites of wealth and elites of power: the changing dynamic', *Southeast Asian Affairs 1979*, pp. 233–45, Singapore: Institute of Southeast Asian Studies.

Yoshihara, Kunio (1985) *Philippine Industrialization: Foreign and Domestic Capital*, Manila: Ateneo de Manila University Press and Singapore: Oxford University Press.

4

Thailand: Capitalist Development and the State

Kevin Hewison[1]

To suggest that Thailand is undergoing a capitalist revolution is not the debatable proposition it once was. There has been a reluctance to use the term 'capitalism' to describe the forces which have plucked Thai society from its agricultural past and plunged it into the industrial present. However, this has disappeared as rapidly as state socialism evaporated in the creation of the post-Cold War world. Capitalism's ascendancy as a worldwide system is seen by some as a victory, heralding an 'end to history' or a sign of entry to the post-modern world. While not having to go this far, it is clear that Thailand's capitalism is a part of a world capitalist system that is closer to a global system than it has ever been.

While still subject to significant contention, most writers on Thailand tend to use 'capitalism' as a shorthand designation for the process of industrialisation which has taken place within the free enterprise system. To consider industrialisation as a process is important. It indicates the history of the transformation of an essentially agricultural economy, where production was mainly for subsistence, to an economy where production involves the application of investment capital in enterprises which produce primarily for profit. Trade and commerce take on an increasingly significant role, while the industrial system sees producers separated from their products as they become waged workers, employed by the owners and managers of enterprises who control those products. That capitalist industrialisation takes place within a free enterprise environment means that business is owned by individuals or companies. Competition between businesses is usually considered the norm. The notion that these elements constitute a system is also significant. Capitalism, as well as being a system of economic production, also structures social organisation. The organisation of free enterprise demands a particular social system, where a class of workers have their labour for sale, purchased by a small class of business owners and managers. While many political economists will see important shortcomings is this characterisation, for the purposes of this chapter it is not necessary to extend it further.[2]

While some might want to argue about the character of Thailand's capitalist transition, and suggest that it has not always matched the history of capitalist development in Europe, the basic forces at work and the nature of the transformation include much that is recognisable in the development of capitalism in many countries which have experienced a similar transition.[3] The last century or so of Thailand's history has certainly seen these processes at work. While this chapter will discuss the development of Thailand's capitalism, we may leave aside these theoretical issues in order to concentrate on debates regarding the role that the state has played in Thailand's capitalist development.

A Sketch of Theoretical Approaches

Some time ago I began a discussion of the role of the state with the observation that the most influential texts on Thailand's politics seldom had the state at the centre of their analysis (Hewison 1985: 266). This observation is no longer accurate. The recent literature on Thailand's politics and economics has seen a concerted effort to understand the role that the state has played in one of the world's most consistent economic growth performances.

Earlier writings, heavily influenced by structural-functional and modernisation approaches to society and politics, gave little attention to the economic role of the state.[4] Writing in the 1960s, Fred Riggs characterised Thailand as a 'bureaucratic polity', which included a conception of the relations between business and government. In summary, Riggs (1966: 252) saw business as having little political influence, arguing that government had no reason to protect business interests.

This perspective consolidated a position which saw society as loosely structured, where the mass of the population lived an apolitical existence, and politics was seen to revolve around a politicised elite presiding over the masses. Hierarchy and status were important factors in binding the masses to the elite through a vast network of patron–client relations. The political elite was dominated by civil and military bureaucrats who took a predatory attitude towards business, making business people their clients.

Orthodox dependency theorists considered this approach inadequate as it placed too much emphasis on elite behaviour, obscuring the nature of international capitalism, where transnational companies and aggressive rich countries were seen to exploit Thailand and its people. Local business and state elites become the tools or compradors of internationalised corporations, assisting in the exploitation of their own country (for example, see Suthy 1980).[5] While their critique of modernisation approaches was well placed, and their attention to the international context of development appropriate, this radical perspective has also been criticised (Hewison 1985; 1989). Basically, these criticisms argue that dependency approaches have underestimated the significance of local capitalists and the potential for the local state to develop policies supportive of local business. Another problem, for both the modernisation and dependency perspectives, is that they fail to reflect the realities of Thailand's rapid economic and political development, especially since the mid-1980s.

Whereas these earlier assessments effectively argued that the state—or at least rapacious officials—was an impediment to capitalist industrialisation, recent attempts to explain economic growth have been more interested in assessing the role of the state in less negative terms.

Non-Marxist political economists and political scientists have tended to regard the 'bureaucratic polity' as an appropriate characterisation of the 1932–73 period, but of less utility for the period since then (see Anek 1992: ch. 1; McVey 1992: 20–2). In short, they argue that the bureaucratic polity has been replaced with a system of corporatism marked by a high degree of mediation between government and business interests, where the latter have far more autonomy than under the bureaucratic polity. The importance of organised business, the rise of which is seen in the creation and operation of representative business associations and their influence on government policy, can no longer be denied. This suggests that business–government relations are now more a partnership, resembling that of the East Asian NICs, although it is usually argued that Thailand's system is 'less statist' than those of East Asia (Anek 1992: 13–15).

Economic growth theory, always present as a pillar of public policy (Higgott 1983), has seldom been theoretically concerned with explaining the role of government in development (for a recent comment see the *Economist* 25 May 1996: 23–5). However, the influential World Bank (1993) *East Asian Miracle* study is notable for taking the state's role seriously, attempting to delineate the manner in which state policies and actions have had an impact on economic development.

That the Bank considered this issue was notable in itself, but it also reflected the impact of work by a developing 'school' of economists and political scientists who had powerfully demonstrated that the state had played an important role in East Asian development (see Wade 1990; Amsden 1989).

In summary, as noted in the first chapter of this volume, the *Miracles* team concluded, in line with growth theory, that the East Asian economies were successful because they got the basics right; and intervention, systemic and through multiple channels, was important in fostering development, including that of specific industries. The report emphasised that South-East Asian governments had played a much less prominent and less constructive role. The Thailand background paper for the *Miracles* study confirmed this.

The Thailand report maintains that the country's development has been characterised by 'a strange mix of strong macroeconomic management and poorly coordinated sectoral interventions' (Christensen et al. 1993: 19). In this context, it is argued that the Thai case adds little to the debate on the role of the state because 'neither the minimalist nor the activist analysis accurately describes Thailand's economic policy and state interventions'. Indeed, it is suggested that rather 'than being an example of the state guiding the market, ... [Thailand is] a case where the market has compensated for government failures' (Christensen et al. 1993: x, 1).

The report indicates, as does the final *Miracles* report, that Thailand's economic policy, characterised as conservative, has maintained a macroeconomic environment conducive to investment, trade, and the development of private enterprise (Christensen et al. 1993: ix–x, 1). But it also argues that the state, while having been 'fairly activist' at the industry and firm level, has been less effective in implementing sectoral objectives, especially as patronage and rent-seeking have dominated. It is suggested that the 'institutional skills' required by an interventionist state have not been available to Thai policy-makers, and along with economic stability it has been the dynamism of the private sector which has overcome the weaknesses on the sectoral side. Further, it is argued that development would have been *more rapid* if industrial policy had not focused on particular industries but had promoted increased labour intensity and agriculture-related processing, which was the basis of comparative advantage (Christensen et al. 1993: 1–8).

As noted in the Introduction to this book, it is significant that the World Bank's theoretical inclinations have been scrutinised. However, this has been an exercise in revising growth theory rather than challenging it. The focus on government has, as noted above, revolved around notions of getting the policy basics right. The problem with this is the tendency to conceptualise government in neutral terms: 'good' policy is developed by governments relatively insulated from political influences, suggesting that democratic or representative politics is an impediment to good policy. The Thailand report concurs, describing the legislature as having been unproductive 'in making laws, especially when members of parliament are elected ...' (Christensen et al. 1993: 19). The coup is seen to 'perform an important function', as it 'break[s] the legislative logjam developed in previous elected parliaments' (Christensen et al. 1993: 20). While it is acknowledged that bureaucrats are not immune to extra-bureaucratic demands, this is an afterthought.

It is these 'extra-bureaucratic demands' which are of significance in the approach of Marxist political economists. While there has been a long retreat from a reliance on instrumentalist approaches,[6] their argument has been that the state is an 'expression, or condensation, of social-class relations, and [that] these relations imply domination of one group by another' (Carnoy 1984: 250). In referring to the 'capitalist state' these political economists identify the nature of domination in the contemporary period. Having said this, it is important to conceive of a basic distinction between state and capital—business people do not *need* to inhabit positions in government for a state to be identified as capitalist. But, as the state operates within a system of ideological and economic domination, there are limitations placed on the theoretical independence of the state (for details see Miliband 1983; Hewison 1989: 26–31).

Again over-simplifying, the state—or more correctly, state managers—operates within an international context which, to a greater or lesser extent, limits the autonomy of policy-making. Their independence is constrained by the fact that as state managers they must secure the conditions required for private investment. This is necessary as the fiscal position of the state is tied to the health of the economy and, hence, growth in the private sector. Thus, private property is protected and the relation of domination of labour by capital is maintained. These minimum conditions are required, and more than this is a bonus for capitalists. Capitalists will often negotiate their bonuses using, for example, their increased mobility to gain benefits, while the imperatives associated with the financial base and capacity of the state will encourage state managers to provide such bonuses. In other words, it is not essential to seek personal, family or class links between state officials and capitalists to explain why policy will support the development and expansion of private investment.[7]

The *Miracles* approach seeks to explain what factors encourage some governments to offer conditions and incentives which promote growth. The problem is that this analysis tends to treat politics in essentially formal terms, examining an 'independent' state, formal political structures and legal institutions. This largely ignores what Robert Scalapino (1996: 227), in classic pluralist terms, identifies as 'informal politics', meaning the 'patterns of political behaviour between and among individuals and groups ...'.[8] While the pluralist approach, especially as it has been applied to Thailand, has been criticised (see Hewison 1992), this identification of the need to go beyond the formal structures of politics is important.

In this chapter an interpretation of the development of Thailand's capitalism and the role of the state will be outlined. It relies on the characterisation of an accumulation regime in each of the epochs that are identified and examined. Marxist political economists have been particularly interested to explain 'informal politics', although it must be emphasised that such a distinction between formal and informal structures is not seen to be particularly illuminating (see Hewison and Rodan 1994). For Marxists, the significant issue is not so much to identify 'good' and 'bad' policy choices, but which policies emerge under particular accumulation regimes. They seek to understand the class relations involved in the development of capitalism, the changing international context of development, the role of the state in this, and the way in which policy reflects class domination (or by particular elements or fractions of a class).

The Political Economy of Early Industrialisation

Thailand's path to capitalist development is conveniently signposted by the Bowring Treaty of 1855, which opened the country to Western trade (see Ingram 1971; Suthy 1980). While emphasising this treaty underestimates important social and economic changes already taking place within the country (see Hong 1984; Terwiel 1989; Nidhi 1982), the period of the mid-1800s is a convenient point to begin the discussion of the emergence of modern Thailand (or Siam, as the kingdom was then known).

Prior to the 1850s much of the relatively small population was engaged in essentially subsistence economic activity. Rice cultivation, collecting forest products, fishing and the like, took place within family and village units. The links between these units and the royal state became increasingly remote as distance from Bangkok and other administrative centres increased. The control the state did have tended to be translated as obligations, to be met either as corvée labour, military service or in the delivery of important and valuable commodities to the state. Many of these commodities were for external trade, controlled by aristocrats who were also state officials. The class structure of this society was dominated by the group of royals and nobles who drew their wealth from their control of labour and land, while the majority of the population could be described as a peasantry (for details on pre-modern class structures, see Akin 1969).

Political organisation reflected the structure of class domination, with the state controlled by the monarchy, its nobles and aristocracy. The monarchy, while sometimes in competition with the leading noble families, attempted to keep close control of government, and this saw the development of a highly personalised state, focused on the monarch.

This picture began to change under a range of pressures. First, international trade patterns began to alter, being increasingly driven by the industrial production of the West and the sustenance of its colonies. Second, colonial expansion in Indochina, Burma and on the Malayan Peninsula demanded that Thailand reform its administration. Third, these changes created opportunities for a rapid expansion of exports, beginning with rice, and later in timber, rubber and tin; these commodities remained Thailand's main exports until the 1960s. In response, and to facilitate trade, the state was reformed to better fit the colonial model. Territorial boundaries were more carefully defined and national and provincial administration was reformed (see Tej 1977; Thongchai 1994).

These changes also saw significant transformations to class relations. Slavery and the corvée were eventually discarded, resulting in the emergence of a free peasantry which, while still ruled by the nobility, increasingly came to rely on land, taxation and the control of business for their wealth and that of the state (Hong 1984: ch. 2). The development of business, focused on the export trade, saw the nobility and state—there was still little distinction between the assets of the state and those of the monarchy and nobility—entering into alliances with foreign business and the Chinese. The latter became an important part of the reforms, becoming tax farmers, business leaders and labourers in the new economic system.[9] The tax farming system allowed the state to convert its revenue system to one based on money rather than products and labour services. In business, Chinese merchants, royals and aristocrats, and Western businesses developed a supportive but still competitive business structure (Suehiro 1989). Additionally, a myriad of Chinese merchants and traders spread across Thailand, establishing businesses and shops and enhancing commodity trade.

Wage-labour also became an area of Chinese activity. Labour was in short supply in developing Bangkok, in the mines of the south and in a range of small industries, especially as Thais tended to work in agricultural production, where there were also enhanced opportunities from trade. The result was that from the late 1880s to the 1930s there was an addition of one million Chinese to the population.

Administration saw the emergence of a centralising and modernising state, where the royal family established personal control or supervision of the bureaucracy, with the monarchy developing a state which was similar to that in neighbouring colonies. By 1900 the monarchy was particularly strong, supported by a modernising military and the bureaucracy. Opposition to the regime was vigorously suppressed, and political decision-making was tightly controlled within the court (see Chatthip 1984; Kobkua 1988).

Further change, a result of domestic and international events, came in the early 1930s. Within Thailand, rising expectations of political reform were thwarted by King Prajadhipok and his advisers, and the desire for economic reform was unfulfilled, especially as the state faced serious fiscal problems following a profligate period under the previous monarch (see Batson 1984; Copeland 1994). When combined with the impact of the world economic depression and international political instability, the stage was set for change. In June 1932 a small group of commoners, organised as the People's Party, in a swift and well-planned coup, seized the state from the monarchy and established constitutional rule.[10] Among other things, the overthrow of the absolute monarchy represented a triumph of new economic ways.

While not a fully-fledged capitalist system, the commercialisation, monetisation and commodification of the economy was well established, and the overthrow of the monarchy's accumulation regime, dominated by a personalised economic system, was an important step towards a modern economic system. It also represented a rejection of the prevailing political model, and while being unable to fully establish the alternative constitutional form, was the first successful move against unrepresentative political forms.

While initially brought together by their opposition to the absolute monarchy and its state and by economic nationalism, it is significant that the new government soon became split over economic policy and political representation, setting the more radical elements against the conservatives. No agreement could be reached on fundamental economic and political tasks, but the regime maintained itself through opposition to various royalist-inspired plots and rebellions. While it was significant that the constitutional government received public support, equally important was the political stature gained by the military during this period.

The period from 1932–57 saw considerable conflict over politics and economic policy. Political conflict continued between Royalists and the People's Party throughout this period, while debate on economic policy, still dominated by nationalist concerns, revolved around the need for state intervention to stimulate industry and to improve the lot of the farming majority. Initially bitter, with accusations of 'Bolshevism', these debates came to focus on not the need for state intervention in order to achieve economic progress, but over the degree of intervention. The result was a poorly developed populist set of policies promoting the non-agricultural employment and investment of ethnic Thais. On many occasions this meant investments by the state and its enterprises.

This economic approach reached its policy crescendo in the decade after the military seized power in 1947 and placed Field Marshal Plaek Phibunsongkhram (Phibun) back into the prime minister's position he had lost towards the end of the Second World War. However, nationalisation— seen mainly as Thai-ification—and opposition to private enterprise did not become policy in any form other than a rather haphazard approach to promoting industrialisation. There does not appear to have been any pattern to state investments other than a desire to reduce foreign imports and to 'set an example for the private sector', with the vague hope that the state's lead would be followed by the private sector (Amnuay 1967: 272).

If there was a common theme that emerged in these investments, it was the involvement of military leaders in many of the enterprises (see Sungsidh 1983). It is this period that Fred Riggs (1966) identified as being the clearest expression of the bureaucratic polity, with powerful military and bureaucratic figures tapping into the resources of Chinese business for both personal gain and also to finance political activity (Girling 1981: 75-6). At the same time, there were benefits for those businesses which were linked to powerful political leaders, which gave them competitive advantages.[11]

For the capitalist class as a whole—as opposed to the individuals and firms linked to powerful political figures, thereby gaining great advantage—this accumulation regime meant an uncertain investment climate was created by such haphazard and increasingly personalised arrangements (Hewison 1985: 276). When the Korean War boom began to weaken in the mid-1950s, there were increasing demands for the investment role of the state to be limited to infrastructure. Such calls received the support of foreign business. US companies, the largest foreign investors, had taken the lead in attempting to force the government to be more receptive to foreign capital, using aid to encourage positive attitudes to investment (Hewison 1985: 277).

The twin coups of 1957–58, led by General Sarit Thanarat, marked the end of a political era, pushing aside the last vestiges of the People's Party. Motivated by the political competition between Sarit and his rivals in the Phibun government, the coup established a system of 'despotic paternalism'. The government was highly authoritarian and determined to establish order and stability and to make Thailand progressive and 'civilised' (Thak 1976; 1979). In addition to its authoritarianism, the government was distinguished by its determination to promote private investment. This enthusiasm was not evident immediately, but developed out of a realisation that shaky state enterprises could not be saved, and coincided with a range of reports by international organisations recommending increased support for the private sector, import substituting industrialisation (ISI), and a role for the state in infrastructure development (see, for example, World Bank 1959; van Rijnberk 1961).[12] These reports echoed a common sentiment regarding development strategy which had emerged among a rising group of young civilian officials who took important positions in newly created economic agencies (see Amnuay 1967).

The Era of Import Substitution Industrialisation

Sarit and his advisers were receptive to the World Bank's advice that the role of manufacturing needed to be expanded through increasing incentives to foreign and domestic investment. The report argued that an ISI strategy be adopted with generous promotional privileges (World Bank 1959: 94–106). The regime's first national development plan took up the Bank's proposals and accepted USA and World Bank assistance in implementing the plan and establishing a revised investment promotion Act (Hewison 1985: 278–9).[13] The plan presented a philosophical position promoting private initiative, while making it clear that state investments would be limited to infrastructure development (Abonyi and Bunyaraks n.d.: 21–4).

The first plan placed much of its emphasis on agricultural development, while including a section on industry which was influenced by the ISI strategy (National Economic Development Board 1964). However, when this was combined with the impetus of the investment promotion Act, which targeted the promotion of industry and foreign investment, private investment was directed to the industrial sector. For local business this approach meant that there was more room to invest, free of state competition, while industrialists gained the tariff protection they needed for domestic manufacturing. Foreign investors were also keen to establish manufacturing behind protective barriers, while the government sought foreign investment to promote access to capital, technology and entrepreneurial skills (Hewison 1985: 280–1).

ISI was supported by the establishment of a number of government agencies charged with implementing the strategy, including the Board of Investment (BoI), the National Economic and Social Development Board (NESDB) and a revamped Ministry of Industry. Their position, and that of the Bank of Thailand and local manufacturers, was reflected in the first (1961–66) and second (1967–71) development plans, where the majority of capital invested with government promotional privileges went into import substituting industries. The taxation and export of agricultural production and the extraction of savings from households into the commercial banking sector provided a pool of funds for industrial development (see Silcock 1967b; Jansen 1990). ISI brought important changes to the manufacturing sector, with increased production of consumer goods, and manufacturing's contribution to GDP rising significantly between 1960 and 1971, with real annual growth rate averaging 11 per cent (see table 4.1). High rates of protection encouraged domestic investment which, in turn, further strengthened local finance and banking.

Table 4.1 Thailand's GDP by industrial origin, 1960, 1971, 1980 (%)

Sector	1960	1971	1980
Agriculture	39.8	29.8	24.9
Mining and Quarrying	1.1	1.5	1.6
Manufacturing	12.5	17.5	20.7
Construction	4.6	5.5	5.7
Electricity and water supply	0.4	1.8	1.9
Transport and communications	7.5	6.7	6.4
Wholesale and retail trade	15.2	17.1	16.5
Banking, insurance, real estate	1.9	4.2	6.0
Ownership of dwellings	2.8	1.9	1.5
Public administration and defence	4.6	4.3	4.2
Services	9.6	9.7	10.6

Source: Bank of Thailand, *National Income of Thailand*, various years.

The big ISI winners were the developing industrial conglomerates and their financial partners, and most notably the fifteen to twenty families who dominated the highly protected domestic commercial banks (Hewison 1989: ch. 8). These families, most of them Sino-Thai, had established remarkably profitable operations in the key areas of agriculture, and the finance and export of primary commodities and import substituting industry. Their control of finance—the stock market was in its infancy and raising capital overseas was tightly controlled—allowed them to build oligopolies in a range of economic sectors (Krirkkiat 1981; Pasuk and Baker 1995: 125–45). They had also established excellent links with powerful political figures.

There was considerable inertia in pressing policy reform from such a comfortable and protected position, especially while profits were maintained. Funds deposited in the banks grew rapidly, and these were used to finance the export of commodities (a traditional bank activity and highly profitable, especially after government controls were lifted in the 1950s) and to establish protected industries. The protection of manufacturing, and the banks themselves, ensured profitability. Support for ISI also came from some influential foreign investors, especially the Japanese, with investments in textiles and auto assembly and parts manufacture (see Pasuk and Baker 1995: 130–8). The accumulation regime that consolidated after 1958 was tied to ISI, and it required considerable external 'shock', threatening profits and expansion, for the move to EOI to be seriously contemplated.

For all of its apparent success and support, ISI came under attack from critics who argued that the relatively small domestic market was saturated and that ISI created disincentives to export (see Hewison 1989: 118–21). Increasingly, calls were made to promote and expand manufacturing for export. However, there was considerable resistance to the dismantling of protection and ISI. While there were state agencies and technocrats arguing for a change to EOI from the late 1960s, and especially from the mid-1970s (see, for example, NESDB 1977), there was no great pressure for change while economic growth continued. In fact, under pressure from domestic capitalist groups, protection for import substituting manufacturing actually increased through the 1970s and into the 1980s (Pasuk and Baker 1995: 144–5). It was not until the mid-1980s that EOI policy—meant to be based on a nation's advantage in producing commodities for a world market, utilising cheap labour—was established. The required shock came with the economic downturn of the mid-1980s.

This challenge to the cosy ISI club came from a confluence of events beginning in the late 1970s. First, from the late 1970s the *baht*, being tied to the dollar, began a steady climb, making primary commodity exports less attractive on the world market. Second, the nature of international investment was changing, with the beginning of the relocation of East Asian firms. Third, international prices for agricultural commodities began a steady descent from the late 1970s. Fourth, the second oil crisis saw the government seeking loan funds, significantly raising public sector debt. Fifth, military assistance had declined from the mid-1970s. Finally, the military embarked on a spending spree as counter-insurgency became less of a concern and regional conflict justified new kinds of arms purchases (see Hewison 1987: 61–76; Muscat 1994: ch. 6; Pasuk and Baker 1996: ch. 4).

For business, the downturn had a substantial impact. Growth predictions were the lowest for years, bankruptcies mushroomed, investment dropped precipitously, unemployment increased, and even the biggest and strongest companies reported flat profits or losses. The sharp decline in agricultural commodity prices was a major contributing factor to the downturn. But there was a silver lining in these dark economic clouds, as Thailand's manufactures became more attractive on the world market. The downturn also indicated significant problems for the state policy and its fiscal and monetary position. Budget deficits ballooned, official debt reached unprecedented levels, and trade and current account balances were increasingly negative (Hewison 1987: 61–9). Among the poorer classes, farmers faced very low prices and wages were eroded by inflation and increased government charges.

As Pasuk and Baker (1996: 65–6) point out, the technocrats were split on the appropriate response, and even entreaties from the powerful banking and textiles sectors, as well as the World Bank, brought few decisions. It was the belated recognition that agricultural commodity prices were not about to save economic growth that brought a savage devaluation and a concerted move to EOI.

The Rise of Export Oriented Industrialisation

In policy and production terms, EOI has grown and strengthened since the mid-1980s. Indeed, the economic results have been spectacular. This success can be seen in the rapid expansion of exports, from average annual growth rates of 6 per cent in the 1960s and 11 per cent in the 1970s, to over 16 per cent in the 1980s. Annual rates in excess of 10 per cent have been maintained in the first half of the 1990s (*Thailand Statistical Yearbook*, various years). The wider economic significance of the move to EOI can be seen in the data presented in table 4.2. Figure 4.1 (next page) provides an indication of the growth that has resulted.

The change in the relative shares of the manufacturing and agricultural sectors is most significant (see figure 4.2, next page). The rapidly declining significance of agriculture is remarkable: in 1960 it was the most important economic sector. It accounted for almost 40 per cent of GDP, most exports, and employed the bulk of the population. At the same time, manufacturing made a relatively small contribution to the country's production and employment. The transition took place in the early 1980s, in part related to the downturn discussed above. By 1993, while agriculture was probably still the largest employer of labour, it produced just 10 per cent of GDP, ranking lower than manufacturing, trade and services. Rapid industrial growth has seen manufactured exports expanding from just 1 per cent of total exports in 1960 to a huge 75 per cent by the early 1990s (Thailand Development Research Institute 1992: 6).

Table 4.2 Thailand's GDP by industrial origin, 1985–93 (%)

Sector	1985	1990	1991	1992	1993
Agriculture	15.8	12.7	12.7	12.0	10.0
Mining and quarrying	2.5	1.6	1.5	1.5	1.5
Manufacturing	21.9	27.2	28.4	28.0	28.5
Construction	5.1	6.2	6.7	6.7	6.9
Electricity and water supply	2.4	2.2	2.1	2.3	2.4
Transport and communications	7.4	7.1	7.0	7.2	7.5
Wholesale and retail trade	18.3	17.6	16.9	16.6	16.6
Banking, insurance, real estate	3.3	5.5	5.3	6.5	7.3
Ownership of dwellings	4.2	3.0	2.8	2.7	2.6
Public administration and defence	4.6	3.5	3.4	3.8	3.8
Services	14.5	13.3	12.9	12.8	13.0

Source: Thailand Development Research Institute (1995).

It is clear that Thailand has emerged as an industrially-oriented economy. While a slim majority of people continue to engage in agricultural pursuits, there has been a decline from 82 per cent of the economically active population in 1960 to an official figure of 60 per cent in the early 1990s.[14] These figures understate the magnitude of change as many agricultural families

Figure 4.1 Growth of trade in Thailand, 1985–94 (in billions of *baht*)

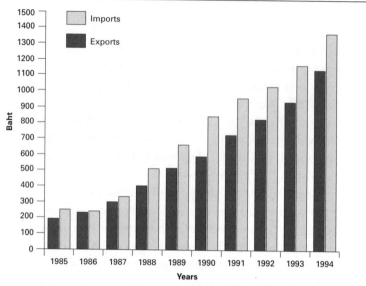

Source: Thailand Development Research Institute (1995).

Figure 4.2 Changes in percentage shares of Thailand's GDP, 1960–93

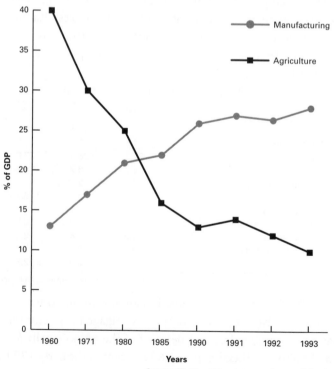

Source: Thailand Development Research Institute (1995).

now rely on income from non-farm sources to maintain their farms. While employment in manufacturing industry remains relatively small (Christensen et al. 1993: 6), employment in non-agricultural activities has grown significantly.

This growth has been driven by the private sector. As noted above, in the 1950s public investment was important to industrial investment. However, following the Sarit coup, private investment has been predominant. As can be seen in table 4.3, private investment has become increasingly dominant in recent years. The table also indicates the boom in private sector investment which occurred following the policy changes of the mid-1980s. Real gross fixed capital formation grew by an average 20 per cent annually between 1986 and 1991, and the value of new business registrations and capital expansions grew by an average annual rate of more than 50 per cent over the same period. These levels have reduced since 1991, but remain high by international standards (Board of Investment, various issues).

Table 4.3 Investment levels in Thailand, 1975–93

Year	Private investment % of GDP	Public investment % of GDP
1975	17.7	5.2
1980	16.3	8.9
1985	14.6	9.1
1988	21.9	5.2
1989	25.3	5.2
1990	29.1	6.4
1991	34.7	7.2
1992	31.2	8.1
1993	32.4	7.6

Source: Board of Investment, various issues.

Behind these figures is the reality of a significant and powerful domestic capitalist class. In the 1980s the development of this class was especially rapid, producing a class that is increasingly diverse. While the domestic capitalist class took the dominant role in the post-1985 investment boom, there was also considerable stimulus from foreign investment.

The promotion of foreign investment has been an important task for the state since 1958, and has been concentrated in the leading sectors of the economy. Domestic investors have been favourable to foreign investment as they have tended to prefer joint ventures when moving into new business sectors. The level of foreign investment has been seen as a barometer of business confidence, meaning that foreign investment has been politically significant. As seen in table 4.4 (next page), the inflows of foreign investment increased substantially through the late 1980s, but political instability, recession in the developed countries, and heavy profit repatriation saw a decline in growth, although the stock of investment has remained high, supported by increased portfolio investment.

Table 4.4 Net flow to Thailand of foreign direct investment, 1986–94

Year	Value ($US million)	% Change on previous year
1986	276.3	55.5
1987	361.7	30.9
1988	1118.5	209.2
1989	1827.9	63.4
1990	2587.8	41.5
1991	2055.6	−20.6
1992	2150.6	4.6
1993	1674.9	−22.1
1994	598.1	−64.3

Source: Thailand Development Research Institute (1992: 18; 1995: 16).

The sources of foreign direct investment (FDI) have diversified with the expansion of investment. In 1984 the USA was the largest investor, followed by Japan, Singapore and Hong Kong. By 1986 Japan had become the leading investor, and the levels of net inflows from Japan increased nine-fold between 1987 and 1990 (Thailand Development Research Institute 1995: 17).

While strategically important, investment by local business has always overshadowed that by foreign capitalists. For most of the period between 1960 and 1994, while the government has had relatively generous foreign investment laws, and though foreign investment has increased steadily, its contribution to gross capital formation has usually remained less than 10 per cent (Hewison 1989: 112; Pasuk and Baker 1996: 35). While not an accurate measure of total FDI, BoI data show that between 1960 and 1993 approximately one-third of the registered capital in promoted firms was identified as foreign (Board of Investment 1995a: 11). While these levels are low when compared to other ASEAN states, foreign investment has still had a pivotal economic and political position. Even so, as Pasuk and Baker (1996: 35) explain: 'Foreign investment may have sparked the [post-1985] boom ... [but] Thai investments made it a big boom'.

In promoting investment, the government, acting through its BoI, makes no distinction between foreign and domestic investors. This means that all of the promotional privileges provided to foreign firms are also available to domestic investments, although a few areas of employment remain reserved for Thais. These privileges are generous. In addition to guarantees against nationalisation, state monopolies or government competition, the BoI offers substantial incentives, such as tax holidays and other taxation relief, import bans on competing products, tax deductions, repatriation of profits and substantial and additional benefits for exporters (see, for example, Board of Investment, 1995a).[15]

The boom, pushed by ballooning exports, pulled the domestic market along. This was especially noticeable in Bangkok, but urban centres throughout the country experienced an investment boom, and saw markets expand and diversify in real estate and wholesale and retail trade. This consolidated the emergence of a more diverse capitalist class. Pasuk and Baker (1995: ch. 5; 1996: chs 3–4) argue that there were a number of changes which were significant.

The major change was the challenge that emerged to the financial dominance of the big banks.[16] As noted above, the commercial banks virtually controlled the supply of funds to the domestic investment market. This was challenged by a range of factors. First, the post-1985 increase in the FDI boom saw increased numbers of foreign investors seeking local partners, and this demand was such that it went well beyond the boundaries of the bank-dominated clique.

Second, policy changes to the financial sector saw state controls on the flow of funds eased, meaning that domestic borrowers were able to go beyond the domestic banks. An important innovation was the ability to borrow overseas. At the same time, increased numbers of foreign banks were established in Thailand and became more aggressive in their corporate and business lending. In addition, merchant banking developed remarkably, and a number of small finance companies were in a position to expand their activities, especially as they were also freed from reliance on the commercial banks.

Third, an important basis for the challenge to the banks dominating Thailand's capitalism was found in the Securities Exchange of Thailand (SET). The SET had existed for many years, but had been hampered by a lack of investor confidence following the collapse of a finance company in 1979 (see Hewison 1981). Following the Wall Street crash of October 1987, however, the SET took off, and capitalisation and turnover expanded markedly (see table 4.5). While still volatile and seen by many as a surrogate casino, the SET has become increasingly attractive to both local and international investors. The SET has mobilised large amounts of capital for the private sector, loosening the grip of the banks on finance and industry. An important factor in this was the establishment of international securities and brokerage companies in Bangkok (Pasuk and Baker 1996: 39).

Table 4.5 Selected statistics on the Securities Exchange of Thailand, 1985–93

Indicator	1985	1990	1991	1992	1993
Annual turnover ($US billion)	0.6	25.1	31.7	74.4	88.0
Average daily turnover ($US million)	2.5	101.6	129.5	301.2	359.4
SET index (end of period)	134.9	612.9	711.4	893.4	1682.8
Market dividend yield (%)	8.2	3.6	3.6	2.9	2.1
Number of quoted companies	97.0	214.0	276.0	320.0	347.0
Market value capitalisation ($US billion)	2.0	24.5	35.9	59.4	133.0

Source: Board of Investment (1990, 1992 and 1995 issues).

For many capitalists the expansion of the SET was liberating, allowing a range of new companies and groups to emerge to challenge those who had brought business through the ISI period. Many of the business people involved saw the SET as an unlimited source of funds and an expression of the free-wheeling spirit of capitalism. Manipulation was not unusual, especially as regulation was loose (Pasuk and Baker 1996: 40–1).

The result was a more diverse capitalist class. No longer were banking and industry the areas where the dominant capitalist groups concentrated. These areas remained central, but the widened financial sector, telecommunications, real estate, tourism and a range of services produced remarkably wealthy capitalist groups (see Pasuk and Baker 1996: ch. 3). Huge profits have been made, and while much has been reinvested, consumption spending has also increased markedly, further expanding the domestic market.

The accumulation regime which emerged during the ISI era saw a small capitalist group with close relationships between politicians and technocrats being maintained. This relationship was disrupted by the growth of the EOI period, where, while such links remained, the nature of Thailand's capitalism was such that technocrats were more concerned to manage the economy which established and enhanced the necessities for expanded accumulation. This was a more relevant strategy for dealing with a much expanded and diversified business community. However, the development of electoral politics, where success depended on access to very large amounts of money, saw business and elected politicians establishing relationships as significant as those between officials and business under ISI (for details, see Pasuk and Baker, 1995 and 1996).

There is no doubt that Thailand's remarkably consistent and rapid economic growth has been a boon to capitalists and to the state. However, the economic performance has not been unreservedly positive. A range of negative consequences has been identified (see several chapters in Medhi 1995), with one of the most challenging being increased income and wealth disparities.

Darker Clouds: Distributional Issues

While there have been significant increases in per capita incomes and reductions in absolute poverty, those in urban areas have done far better than those in the countryside (Medhi 1993: 403-16). Income and wealth distribution have become increasingly skewed and, since 1976, while poverty has been marginally reduced, wealth disparities have widened (see figure 4.3). In addition, it seems that the rate of poverty reduction is now negligible. Even in urban areas, where industrial workers have made significant contributions to economic growth, they have yet to reap adequate rewards for their labours. Indeed, worker's lives have been characterised by low wages and poor conditions. Related to this, the widespread exploitation of women, particularly as prostitutes, and children in small factories and sweatshops is also well-known (see Mathana 1995).

Figure 4.3 Income distribution in Thailand, 1975–91

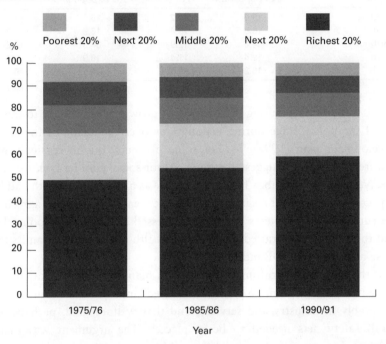

Source: Pranee (1995: 222) and Thailand Development Newsletter, 27–8, 1995.

Many analysts and policy-makers had expected and repeatedly restated their belief that the benefits of growth would trickle down to all levels of society; few had predicted increased inequality. In the words of one influential economist: 'Measured against the past, much has been accomplished … Measured against what is possible to achieve with wealth generated by five decades of growth, much is wanting' (Ammar 1996: 5).

Distribution is most inequitable when rural and urban incomes are compared. The principal reason for this is that peasant-based agriculture is increasingly an unattractive and marginal way of making a living. This is illustrated by comparing the population and productivity of the various regions, as shown in table 4.6 (next page), where it is seen that Bangkok dominates the economic scene. Most industry is clustered in and around the capital, making the area highly productive. Generally, as distance from the centre increases and agricultural activities become more significant, productivity decreases. This is especially noticeable for the north-east, the most populous region, where incomes are lowest. Interestingly, despite governmental efforts to decentralise development, the gaps have widened in recent years.

Table 4.6 Regional productivity in Thailand, 1989, 1993

Region	1989		1993	
	% of population	% of GDP	% of population	% of GDP
Bangkok	16.0	48.2	16.2	55.4
Central	16.8	18.4	16.9	16.7
North-east	34.6	12.9	34.3	10.6
North	19.4	11.4	18.9	9.1
South	13.2	9.1	13.6	8.2

Source: Thailand Development Research Institute (1992; 1995).

This issue of distributing the benefits of growth was brought to the centre of the policy stage during negotiations over the shape of the seventh development plan (1992–96). For some planners, this increasingly bleak picture of inequality suggested a potential for social conflict (see, for example, *Nation*, 14 December 1992). This came as a shock, as growth had been expected to solve most social problems, not create them. It was agreed that the rural sector's problems had to be addressed at the national policy level, and the seventh plan raised income and wealth distribution as an issue for close attention (NESDB n.d.).

At the time of the drafting of the seventh plan, the mood was that something needed to be done to provide increased employment for rural people, preferably in industry and services, and that welfare (and, perhaps, electoral) safety nets needed to be in place.[17] The argument was that an adequate income must be provided to all in order to avoid social and political conflict and thus maintain the stability required for sustained economic growth (*Nation* 14 December 1992).

The integration of these issues in the seventh plan gave them a national profile. The plan's objectives were threefold: first, to sustain economic growth while maintaining stability; second, to distribute income and the benefits of growth to provincial areas; and third, to enhance human resources, improve the quality of life, environment and natural resources. In essence, government planners felt that unless the benefits of development could be more equitably distributed, growth itself may be threatened (NESDB n.d.: 1–24). But this approach also created something of a policy dilemma as poverty reduction programs were seen to require some reduction in growth rates.

In order to achieve broad distributional objectives and solve the policy dilemma, no major change to development policy was envisaged. Indeed, the plan continued to promote rapid growth based on expanded industrialisation. The change was to recognise that agriculture would continue to stagnate and that incomes in rural areas would almost inevitably continue to fall behind. The answer to this dilemma was to encourage industrial activity in the provinces (NESDB n.d.: 99–104). Particular emphasis was given to improving opportunities to poor farmers and agricultural labourers, small business, the self-employed, the low-waged and 'those who cannot help themselves'. In other words, these people could be assisted by providing industrial employment opportunities. It was felt that this would lead to an increased trickle-down effect in provincial areas. Thus, growth was to be achieved with better distribution, through further industrialisation.

Central to this approach was the further expansion of export oriented industry. The logic was that export industry required higher skill levels, therefore demanding better educated labour. Workers with higher education and better skills would be more productive, thus bringing higher wages to them and their families (see Narongchai et al. 1991; NESDB n.d.: 143–64). The plan urged that Thailand move out of low-wage/low-skill production into higher value-added industry utilising higher skilled human resources. As the government was to promote the development of human resources through the expansion of education in rural areas, better educated poor families would have access to higher incomes.

In short, the seventh plan continued the tried and trusted logic of development: the state has a directive function and a legitimate economic policy and planning role, but also creates space for the private sector to drive growth. It also began the process of establishing a role for itself in solving the social and welfare problems emerging from rapid capitalist development. The basic welfare issue was that the agricultural sector was not growing fast enough. The reason for this was that commercialisation had not gone far enough. Hence, government policy aimed at the promotion of agro-industry, seeing this as having significant advantages over the small peasant farmer. Agro-industry was seen as innovative, as having capital to invest and as being research and technology-oriented. This is in line with a general belief that the small-scale farmer is an anachronism and that these people will be better off in industry.

That the approach has not achieved any major short-term impact on these problems has been demonstrated in the development of the eighth plan (1997–2001). The continued failure has seen a move to include a wider range of technical and social expertise in drafting the plan, as well as public participation in the process (*Thai Development Newsletter*, 27–8, 1995: 22–5). While the eighth plan emphasises increased redistribution, the nature of the current accumulation regime is unlikely to allow movement beyond the emphasis on increasing the productive capacity of the population (see Thailand 1995). That this may include expanded welfare measures will be based on decisions regarding the political acceptability of such an approach (see the comments by Anand Panyarachun in Thailand Development Research Institute n.d.: 3).

Conclusion

Thailand's economic development can be judged successful in growth terms. It has recently been the subject of considerable speculation regarding the reasons for this achievement, and especially the role of the state in development success. The World Bank has concluded that the Thailand case offers little to the theoretical search for a strong interventionist or minimalist state. Indeed, Christensen and his colleagues (1993) argue that Thailand's growth might have been even higher had it not been for some poor state interventions.

This chapter has attempted to demonstrate that this search is rather narrow, based on sterile historical analysis and a poor theorisation of the state. In general terms, it is not surprising that Thailand's development strategies and plans, seldom subject to public scrutiny, have supported the expansion of capitalist accumulation. These policies were adopted not necessarily because politicians, bureaucrats and technocrats were listening to capitalists or including them in policy-making—sometimes they did—but because the structural imperatives and fiscal constraints on the state essentially demanded that capitalist accumulation be supported.

Within this general approach, the chapter has shown that the history of Thailand's development has seen the establishment of particular accumulation regimes. These regimes are an amalgam of economic, social and political power, and structure the outcome of much policy development. In other words, policy outcomes are not the measure of the abilities or intellectual capacities of a state or its officials. Rather, policy results from an interplay of a range of political and economic forces which, in Thailand, have resulted in the development and diversification of a powerful class of domestic capitalists.

---

[1] Thanks to Andrew Brown for comments on an earlier draft.

[2] For more details see Bottomore (1985). A discussion of capitalism related to Thailand is in Hewison (1989: 16–32, and the references cited there), and Pasuk and Baker (1995).

3 For an innovative discussion of the issues involved in the comparative study of capitalist development in Asia and Europe, see Rutten (1994).

4 For a discussion and critique of these approaches see Hewison (1989: 5–13) and Tamada (1991).

5 While writing from a quite different theoretical perspective, it is interesting to note that Yoshihara (1988), in arguing that South-East Asian capitalism is *ersatz*, develops an analysis which produces conclusions not far removed from that of dependency theorists.

6 'Instrumentalist' is used here to refer to perspectives influenced by the discussion in *The Communist Manifesto*, but more closely associated with Lenin's approach, which saw the state as a tool of class rule (see Miliband, 1978). For an important critique of this perspective see Nicos Poulantzas (1969).

7 This is not to suggest that state managers cannot operate against the interests of some capitalists and in support of sectional interests or even their own interests; clearly they often do.

8 See the special issue of *Asian Survey*, 36 (3) (1996) for more information on informal politics in East Asia.

9 The history of the Chinese is best analysed, in what have become truly classic studies, by Skinner (1957; 1958). While these have been updated in Suehiro (1989) and Hewison (1989), they remain unsurpassed.

10 The 1932 change is often portrayed as simply a replacement of one elite by another (see Thawatt 1972; Riggs 1966). It was far more than this, however, as the People's Party, while not a cohesive political organisation with a clear ideology for political or economic development, did establish the fundamentals of the political and economic landscape for the next twenty-five years and, arguably, longer. At the same time, by definition, it established a readily identifiable political opposition—royalists—and brought the military to political prominence for the first time. It can also be argued that these elements of economic management, modernisation and progress, constitutionalism, representation, opposition and the role of the military, remain fundamental.

11 Some Chinese and foreign businesses were greatly disadvantaged by the Phibun government's combination of anti-Chinese legislation and support for state enterprises. At the same time, many private enterprises benefited enormously, including commercial banks linked to powerful officials, including the Bangkok Bank (see Hewison 1989: 192–5).

12 At the same time, Sarit had fewer links to state enterprises than leaders of the previous government, and was personally linked to powerful domestic business (see Silcock 1967a: 20).

13 The USA had provided substantial aid to Thailand prior to Sarit's coups, but this had been concentrated on military assistance, prompted by anti-communism (see Darling 1965: 95–113, 131–8). US military aid from 1950 to 1957 was estimated

at US$262.2 million, while other aid for the period 1946 to 1957 was $152.9 million. Following Sarit's coups, from 1958 to 1963, this assistance was $274.9 million and $180.5 million respectively (Muscat 1990: 295, 328). For a discussion of planning in the 1950s and 1960s, see Muscat (1994).

[14] This is a somewhat misleading figure. Most of Thailand's population statistics draw on household registration documents, and because people do not regularly update these, rural populations and the numbers involved in agricultural pursuits are over-estimated.

[15] It is worth noting that the BoI has recently embarked on a program which promotes foreign investment by Thai companies. Recent data has not been made available, but in 1991, US$172 million was officially reported as having been invested overseas (Pasuk and Baker, 1996: 49–51; *International Business Asia*, 21 June 1996).

[16] This is not to imply that the commercial banks opposed the direction of the boom. The big banks have done well from the boom, and have been aggressive in financing exports (Pasuk and Baker, 1996: 38).

[17] Interestingly, this focus on distribution became politically significant when electoral politics was dominant. One of the main reasons for this is because rural areas, which lag behind urban areas, provide the majority of national members of parliament.

References

Abonyi, George and Bunyaraks Ninsananda (n.d.) *Thailand: Development Planning in Turbulent Times*, Toronto: University of Toronto–York University Joint Centre for Asia Pacific Studies.

Akin Rabibadhana (1969) *The Organisation of Thai Society in the Early Bangkok Period, 1782–1873*, Ithaca: Cornell University Southeast Asia Program, Data Paper no. 74.

Ammar Siamwalla (1996) 'Two and a half cheers for economic growth: an assessment of long–term changes in the Thai economy', *TDRI Quarterly Review*, 11 (1): 3–5.

Amnuay Viravan (1967) 'Economic policy and political structure in Thailand', *Warasan phanitayasat lae kanbanchi*, 5 (3): 270–5.

Amsden, Alice H. (1989) *Asia's Next Giant: South Korea and Late Industrialization*, New York: Oxford University Press.

Anek Laothamatas (1992) *Business Associations and the New Political Economy of Thailand: From Bureaucratic Polity to Liberal Corporatism*, Singapore: Institute of Southeast Asian Studies.

Awanohara, Susumu (1993) 'The magnificent eight', *Far Eastern Economic Review*, 22 July: 79–80.

Bank of Thailand *National Income of Thailand* (various years).

Batson, Benjamin A. (1984) *The End of the Absolute Monarchy in Siam*, Singapore: Oxford University Press.

Board of Investment, *Key Investment Indicators Thailand* (various years, with the year of publication indicated in each reference). Bangkok: Office of the Prime Minister.

Board of Investment (1995a) *A Guide to the Board of Investment*, Bangkok: Office of the Board of Investment.

Bottomore, Tom (1985) *Theories of Modern Capitalism*, London: George Allen & Unwin.

Carnoy, Martin (1984) *The State and Political Theory*, Princeton: Princeton University Press.

Chatthip Nartsupha (1984) 'The ideology of the "holy men" revolts in North East Thailand', in Andrew Turton and Shigeharu Tanabe (eds) *History and Peasant Consciousness in Southeast Asia*, Osaka: Senri Ethnological Studies (13): 111–34.

Christensen, Scott, Dollar, David, Ammar Siamwalla and Pakorn Vichyanond (1993) *The Lessons of East Asia. Thailand: The Institutional and Political Underpinnings of Growth*, Washington DC: World Bank.

Copeland, Matthew Phillip (1994) 'Contested nationalism and the 1932 overthrow of the absolute monarchy in Siam', PhD thesis, Australian National University, Canberra.

Darling, Frank C. (1965) *Thailand and the United States*, Washington DC: Public Affairs Press.

Girling, John (1981) *Thailand: Society and Politics*, Ithaca: Cornell University Press.

Hewison, Kevin (1981) 'The financial bourgeoisie in Thailand', *Journal of Contemporary Asia*, 11 (4): 395–412.

Hewison, Kevin (1985) The State and Capitalist Development in Thailand', in Richard Higgott and Richard Robison (eds) *Southeast Asia: Essays in the Political Economy of Structural Change*, pp. 266–94, London: Routledge and Kegan Paul.

Hewison, Kevin (1987) 'National interests and economic downturn: Thailand', in Richard Robison, Kevin Hewison and Richard Higgott (eds) *Southeast Asia in the 1980s: The Politics of Economic Crisis*, pp. 52–79, Sydney: Allen & Unwin.

Hewison, Kevin (1989) *Bankers and Bureaucrats: Capital and the Role of the State in Thailand*, New Haven: Yale Center for International and Area Studies, Yale University Southeast Asian Monographs, no. 34.

Hewison, Kevin (1992) 'Liberal corporatism and the return of pluralism in Thai political studies', *Asian Studies Review*, 16 (2): 261–5.

Hewison, Kevin and Rodan, Garry (1994) in Ralph Miliband and Leo Panitch (eds) 'The decline of the Left in Southeast Asia', *Between Globalism and Nationalism: Socialist Register 1994*, pp. 235–62, London: Merlin Press.

Higgott, Richard A. (1983) *Political Development Theory: The Contemporary Debate*, London: Croom Helm.

Hong Lysa (1984) *Thailand in the Nineteenth Century. Evolution of the Economy and Society*, Singapore: Institute of Southeast Asian Studies.

Ingram, James C. (1971) *Economic Change in Thailand, 1850–1970*, Stanford: Stanford University Press.

Jansen, Karel (1990) *Finance, Growth and Stability: Financing Economic Development in Thailand, 1960–86*, Avebury: Gower.

Kobkua Suwannathat-Pian (1988) *Thai–Malay Relations*, Kuala Lumpur: Oxford University Press.

Krirkkiat Phipatseritham (1981) *Wikhro laksana kan pen chaokhong thurakit khanat yai nai prathet thai*, Bangkok: Thai Khadi Research Institute.

Mathana Phananiramai (1995) 'Thailand national report: employment situation, problems and policy', *TDRI Quarterly Review*, 10 (3): 11–15.

McVey, Ruth (1992) 'The materialization of the Southeast Asian entrepreneur', in McVey (ed.) *Southeast Asian Capitalists*, pp. 7–33, Ithaca: Cornell University, Southeast Asia Program.

Medhi Krongkaew (1993) 'Poverty and income distribution', in Peter G. Warr (ed.) *The Thai Economy in Transition*, pp. 401–37, Cambridge: Cambridge University Press.

Medhi Krongkaew (ed.) (1995) *Thailand's Industrialization and Its Consequences*, Houndmills: Macmillan.

Miliband, Ralph (1978) 'Marx and the state', in Donald McQuarie (ed.) *Marx: Sociology/Social Change/Capitalism*, pp. 253–73, London: Quartet Books.

Milibrand, Ralph (1983) 'State power and class interests', *New Left Review*, 138: 57–68.

Muscat, Robert J. (1966) *Development Strategy in Thailand: A Study of Economic Growth*, New York: Frederick A. Praeger.

Muscat, Robert J. (1990) *Thailand and the United States*, New York: Columbia University Press.

Muscat, Robert J. (1994) *The Fifth Tiger: A Study of Thai Development Policy*, Armonk: M. E. Sharpe.

Narongchai Akrasanee, Dapice, David and Flatters, Frank (1991) 'Thailand's export-led growth: retrospect and prospects', *TDRI Quarterly Review*, 6 (2): 24–6.

National Economic Development Board (1964) *The National Economic Development Board. Second Phase: 1964–1966*, Bangkok: Office of the Prime Minister.

NESDB (1977) *The Fourth Five-Year National Economic and Social Development Plan (1977–1981)*, Bangkok: Office of the Prime Minister.

NESDB (n.d.) *The Seventh Five-Year National Economic and Social Development Plan (1992–1996)*, Bangkok: Office of the Prime Minister.

Nidhi Aeusrivongse (1982) *Wathanatham kradumphi kab wannakam ton rattanakosin*, Bangkok: Thai Khadi Research Institute.

Pasuk Phongpaichit and Baker, Chris (1995) *Thailand. Economy and Politics*, Kuala Lumpur: Oxford University Press.

Pasuk Phongpaichit and Baker, Chris (1996) *Thailand's Boom!* Chiangmai: Silkworm Books.

Poulantzas, Nicos (1969) 'The problem of the capitalist state', *New Left Review*, 58: 67–78.

Pranee Tinakorn (1995) 'Industrialization and welfare: how poverty and income distribution are affected', in Medhi Krongkaew (ed.) *Thailand's Industrialization and its Consequences*, pp. 218–31, New York: St Martin's Press.

Riggs, Fred W. (1966) *Thailand: The Modernization of a Bureaucratic Polity*, Honolulu: East-West Center Press.

Rutten, Mario (1994) *Asian Capitalists in the European Mirror*, Amsterdam: VU University Press.

Scalapino, Robert A. (1996), 'Informal politics in East Asia. Introduction', *Asian Survey*, 36 (3): 227–9.

Silcock, T. H. (1967a) 'Outline of economic development 1945–65', in T. H. Silcock (ed.) *Thailand: Social and Economic Studies in Development*, pp. 1–26, Canberra: Australian National University Press.

Silcock, T. H. (1967b) 'The rice premium and agricultural diversification', in T. H. Silcock (ed.) *Thailand: Social and Economic Studies in Development*, pp. 231–57, Canberra: Australian National University Press.

Skinner, G. William (1957) *Chinese Society in Thailand: An Analytical History*, Ithaca: Cornell University Press.

Skinner, G. William (1958) *Leadership and Power in the Chinese Community of Thailand*, Ithaca: Cornell University Press.

Suehiro Akira (1989) *Capital Accumulation in Thailand 1855–1985*, Tokyo: Centre for East Asian Cultural Studies.

Sungsidh Piriyarangsan (1983) *Thai Bureaucratic Capitalism, 1932–1960*, Bangkok: Chulalongkorn University Social Research Institute.

Suthy Prasartset (1980) *Thai Business Leaders: Men and Careers in a Developing Economy*, Tokyo: Institute of Developing Economies.

Tamada Yoshifumi (1991), '*Itthiphon* and *Amnat*: an informal aspect of Thai politics', *Southeast Asian Studies*, Kyoto, 28 (4): 455–66.

Tej Bunnag (1977) *The Provincial Administration of Siam, 1892–1915*, Kuala Lumpur: Oxford University Press.

Terwiel, B. J. (1989) *Through Travellers' Eyes: An Approach to Nineteenth Century Thai History*, Bangkok: Editions Duang Kamol.

Thailand (1995) 'Thailand's socio–economic development: achievements and challenges', *Thai Development Newsletter*, (27–8): 32–8.

Thailand Development Research Institute (1992) *Thailand Economic Information Kit*, Bangkok: Thailand Development Research Institute.

Thailand Development Research Institute (1995) *Thailand Economic Information Kit*, Bangkok: Thailand Development Research Institute.

Thailand Development Research Institute (n.d.) *Annual Report 1995*, Bangkok: Thailand Development Research Institute.

Thailand Statistical Yearbook (various years), Thailand: National Statistical Office.

Thak Chaloematiarana (1976), 'Khwamkhit thang kanmuang khong chomphon Sarit Thanarat lae rapob kanmuang baep phokhun uppatham', in Sombat Chantornwong and Rangsan Thanaphonphan (eds) *Rak muang thai*, Bangkok: Social Science Association of Thailand, 1: 35–81.

Thak Chaloematiarana (1979) *Thailand: The Politics of Despotic Paternalism*, Bangkok: Thai Khadi Research Institute.

Thawatt Mokarapong (1972) *History of the Thai Revolution*, Bangkok: Chalermnit.

Thongchai Winichakul (1994) *Siam Mapped: A History of the Geo-Body of a Nation*, Honolulu: University of Hawaii Press.

van Rijnberk, W. L. (1961) *Industrial Development Policy and Planning in Thailand*, New York: UN Department of Economic and Social Affairs, Report No. TAO/THA/14.

Wade, Robert (1990) *Governing the Market: Economic Theory and the Role of the Government in East Asian Industrialization*, Princeton: Princeton University Press.

World Bank (1959) *A Public Development Program for Thailand*, Baltimore: Johns Hopkins Press, for the International Bank for Reconstruction and Development.

World Bank (1993) *The East Asian Miracle: Economic Growth and Public Policy*, New York: Oxford University Press.

Yoshihara Kunio (1988) *The Rise of Ersatz Capitalism in Southeast Asia*, Singapore: Oxford University Press.

Newspapers and Journals

International Business Asia

Nation

Thai Development Newsletter

5

Class, Ethnicity and Economic Development in Malaysia

Rajah Rasiah[1]

Introduction

Malaysia is a relatively recent political formation. Malaya's political independence from the UK was achieved in 1957, and shortly after in 1963 Malaysia was formed with the addition of the two states of Borneo—Sarawak and Sabah—and the neighbouring city-state of Singapore. In 1965 Singapore separated from Malaysia. If the political development of Malaysia has been dynamic, then so too has its economic development. From its colonial structure as an exporter of primary commodities, the country has undergone rapid industrialisation since the 1970s, and most particularly since 1987. Throughout this period, Malaysia has had annual gross domestic product (GDP) growth rates of nearly 8 per cent, and around 12 per cent for manufacturing. Indeed, by 1994 manufacturing contributed just under 32 per cent of total GDP and Malaysia now appears poised to join the elite group of newly industrialising countries (NICs).

In conjunction with this impressive industrial development, Malaysia has experienced a spectacular decline in poverty and material inequality. The incidence of poverty fell from 39.6 per cent of households in 1976 to 13.4 per cent in 1993. Meanwhile, the recognised economists' measure of income inequality—the gini coefficient—showed a significant drop from 0.519 to 0.459 for the same period.

Beneath these remarkable figures, however, lies an intricate and interesting debate regarding causality. As the editors suggest in chapter 1, the debate on the nature of this transition has challenged a range of observers. Some, especially those adopting a new institutional political economy perspective, suggest that the socio-economic forces that have shaped the achievements of the past decades are the result of a determination on the part of policy-makers. Important as this may have been, it has not been so simple. Rather, following the more radical political economy perspective, the explanation for the transformation must trace a complex set of historical factors and tensions derived from Malaysia's colonial legacy that have been addressed through economic policy since independence. As will be demonstrated, the struggle between competing class interests has been enmeshed with an ethnic politics that has deep roots in the colonial experience. The colonial division of labour sharply separated Chinese, Malays and Indians, and this was reinforced in the immediate period of self-government, with the passive official acceptance of this division of labour. However, resultant ethnic tensions culminated in riots and bloodshed on 13 May 1969. This precipitated a reconstitution of the state, which saw a Malay political elite attempt a systematic assault on the prevailing ethnic division of labour. The idea was, ostensibly, to redress the situation of the indigenous people—the *Bumiputeras*, or 'sons of the soil'.

This reconstitution of the state involved the institutionalisation of a range of interests, but most notable has been the incorporation within the bureaucracy of a powerful ethnic Malay elite and the promotion of Malays in private business. Moreover, greater inter-ethnic business collaboration has ensued. Since the late 1980s this has created the preconditions for a dedicated industrial upgrading strategy under the leadership of Prime Minister Mahathir, and a strategy which has involved an active attempt to raise levels of foreign direct investment. Mahathir's capacity to effect this policy refinement attests not just to his authority and supremacy within the ruling political coalition, but also to the record of social and economic reform which now affords his government the latitude to concentrate on structural economic objectives. The case for this strategy has been helped by the effects of a recession in the mid-1980s which exposed the limitations of the domestic capital base and structure. Importantly, the state retains a critical role in the unfolding drive towards industrial maturity, even if technocrats exert a more powerful influence over state policy.

Unfolding Ethnic Politics

The Malay Peninsular had been characterised by shifting cultivation and sedentary farming prior to Western intervention. Trade, which revolved largely around Malacca, mainly involved Indian, Indonesian and Chinese goods. Early Western contact with the Peninsular also came through trade. The mode of production was organised around a tribute paying system, which was stagnationist and wasteful as the surplus generated by peasants ran the risk of confiscation by the Malay ruling class who reinvested little in production. The peasantry offered tribute commonly in the form of a farm surplus to the Sultan and Malay chiefs in return for using land controlled by the latter. Ownership rights, however, did not exist. The Malay ruling class also assumed the role of figure heads in religious and cultural functions, and protected their peasantry from external military threats.[2]

Apart from two short cart tracks there were no roads, with rivers being the prime mode of transport prior to colonialism (Everitt 1952; Lim 1967). The Polynesians were the first to inhabit the area, while the Chinese and Indians arrived several centuries later. With the growth of tin mining to meet growing Western demand, the Chinese were attracted as mine labourers and small operators. Prior to British intervention there was a sizeable Chinese community in the Malay Peninsular engaged in tin mining, and pepper, gambier and tapioca farming (Ooi 1961: 345–403; Jackson 1964: 34–83; Wong 1965: 60–4). Although still utilising traditional technology, the Chinese transformed tin mining from part-time to a full commercial venture using chain pumps. As with sedentary agriculture, the ruling class who exacted taxes from the tin trade did not reinvest in generating technical change. Except for very small traditional producers using simple human skills, manufacturing was virtually non-existent in pre-colonial Malaya (Everitt 1952: 2–3; Ooi 1961: 350; Jomo 1986: 37–8).

Direct British intervention in the Malay states began in 1874 in Perak with tin mining being the prime economic motivation. Until the early twentieth century, British involvement in tin mining was largely in facilitating tin exports and collecting revenue from taxation in the process. Subsequent interest expanded to plantation agriculture involving rubber, oil palm, cocoa, coffee, tea and pineapples. Rubber was by far the most important as it overtook tin as the major revenue earner, making Malaya the chief dollar earner within the sterling area (Nanjundan 1953). Indians became the prime source of labour for plantation agriculture (see Sandhu 1969; Jain 1970). As commodity production expanded, there was a concomitant growth in infrastructural, health, education, security, administrative and other services to meet the growing demands generated by the primary sectors. Spin-offs generated from these sectors also helped generate the evolution of the manufacturing sector (see Rasiah 1995: ch. 3). Primary commodity processing, engineering and consumer goods grew quite strongly towards the end of the colonial period.

A specific ethnic division of labour evolved during colonialism. The Chinese and some Indians spread their activities into wholesale and retail trade, and into basic construction, engineering works and food production (Song 1923; Lim 1967; Thoburn 1977). Given the repressive labour control methods practised in estates, Indian mobility into business was limited (Chandra 1993). The few Indian business enterprises that emerged were begun by non-estate merchants. Despite strong inter-kinship rivalries, there was far stronger mobility among the Chinese who grew to control the commercial and business sector opportunities that emerged from mining and agriculture (see Rasiah 1995). The Malays generally did not figure in the mainstream primary production economy. Apart from some civil service jobs, Malay activities remained agrarian in nature. This meant that the Malays were the main source of food production for the economy. It was only in the last few decades prior to independence that the British encouraged Malay participation in smallholding and rural industry—partly to prevent potential support for the communists (see Harper 1992). Under such circumstances, the nature of colonial governance largely accounts for Chinese predominance in commerce and business. Hence, when colonialism ended the three ethnic groups were entrenched in virtually isolated occupational and spatial development, sharing little inter-ethnic cultural attributes.

The Alliance coalition which obtained independence from the British was comprised of essentially three ethnic-based component parties—United Malays National Organisation (UMNO), Malaysian Chinese Association (MCA) and Malaysian Indian Congress (MIC)—and did little to transform this inter-ethnic polarisation. Thus, the colonial division of labour along ethnic lines continued into the 1960s. The formation of Majlis Amanah Rakyat (MARA) and Bank Bumiputera in the 1960s to uplift the economic position of Malays failed to resolve the growing disenchantment mounting among them, especially the youth wing of UMNO, which became increasingly critical of existing government policies, including its own leadership. During colonialism, divide and rule was deliberately aimed at preventing inter-ethnic unity, and in some cases even intra-ethnic unity. While the independent state did not actively pursue this policy, its approach left the colonial division of labour intact.

Driven by an ill-conceived import substitution policy, the Malaysian economy was also beset with sluggish growth in the 1960s.[3] Governmental industrial intervention was limited to tariffs on final goods, with tax incentives offered to all pioneering firms, irrespective of ownership. There was no local content legislation for these firms which largely operated as screwdriver industries behind high tariff walls (see Edwards 1975). There was no strategy of supporting local capital through financial, research, training and market-promotion support. The non-discretionary offer of protection ran against the notion that protection be premised on sheltering infant firms until they reached maturity (Lewis 1955; Myrdal 1957). Hence, foreign multinationals producing only for the domestic market slowed production once the small internal market became saturated (Rasiah 1995). Little wonder that manufacturing's contribution to gross domestic product stagnated at 9 per cent between 1960 and 1965 (World Bank 1980).

The impact of sluggish growth on unemployment and poverty was serious. On average, employment grew at only 3.6 per cent annually in the period 1961–72 (Rasiah 1995: table 4.9). The unemployment rate rose from 6 per cent in 1962 to 6.5 per cent in 1965 and 8 per cent in 1970 (Wong 1979: table 5.6; Malaysia 1976). Ethnic inequality too had widened in this period. The ratio of Chinese to Malay median incomes rose from 2:1 in 1957/58 to 2.2:1 in 1967/68. In Malay:non-Malay terms, it rose from 1.83:1 in 1957/58 to 1.88:1 in 1967/68 (Masruri in Edwards: 11, 16; Salih and Yusof 1989: 40–2; Edwards 1990: table 10). This worsening economic situation generated intense antagonism. With frustrations surfacing especially among the Malay elites and the young turks, and social structures divided sharply along ethnic lines, communal friction became the ultimate reservoir of class antagonisms. Thus, by the end of the 1960s inter-ethnic rivalry had become the prime venom of social tension.

In addition, draconian labour policies that had been continued from the colonial period intensified class conflict. State policy was liberal to capital, including ownership, but harsh on labour. Despite legislative acceptance of unionisation following the Trade Union Act of 1959, unions were barred from pioneering industries. Where unions were permitted, conservative and pliable union officials were preferred to the original militant labour campaigners and their organisations (Jomo and Todd 1994). The Industrial Relations Act of 1967 specifically prevented unions from bargaining for better terms of employment, promotion, transfer, recruitment, retrenchment, dismissal, reinstatement, allocation of duties or for having strikes concerning these matters (Rasiah 1995: 77). The bottling-up of worker protests in the face of low wages and unemployment offered opposition political parties mass support in several constituencies where they made substantial gains in the 1969 polls. In fact, the electoral mood changed, and the Alliance almost lost the elections (Ratnam and Milne 1970).

Politics in Sarawak and Sabah were different, although their political parties, like their counterparts in the Peninsular, were generally subservient to UMNO.[4] Sarawak's main indigenous groups were the Iban, Malays, Bidayu, Kelabit, Melanau, Kenyah and Penans, but the Chinese also had a large presence. Sabah was dominated by the Kadazan, the Chinese, Murut, Bajau, Malays, Kwijau and Illanun (Wee 1995: 2–3). Sabah, also had a big group of Indonesians and South Filipinos. The Indians as a group were insignificant. Given that the colonial power devoted less interest to these states, and their location was out of the main trade routes, they joined Malaysia in a more backward state. The relative situation worsened as the post-colonial state focused development efforts on the Peninsular. Despite generating a major share of the country's commodity rents, the two East Malaysian states remained peripheral to the thrust of economy policy, relying primarily on resource-based exports, including petroleum, gas and timber, as their main revenue generators.

By the end of the 1960s a distinct ethnic polarisation, occupationally and regionally (including rural–urban), was apparent. Thus, the 13 May 1969 crisis was both a consequence of class confrontation as well as inter-ethnic rivalry. If the three ethnic groups met little and developed their spheres of influence along separate lines during the colonial period, internal political rule from the incipient state led to their convergence after independence. Far from sowing the seeds of integration, the confluence brought to light the stark socio-cultural and economic differences between them. Class relations took on ethnic dimensions as a consequence. Given their late entry, more underdeveloped position and their weak influence politically, government policy gave less emphasis to Sabah and Sarawak's concerns. This was the political reality.

Economic Restructuring

It was under circumstances of the Malay ruling elites lacking economic power, ethnic conflict, economic slowdown, poverty and inequality that the New Economic Policy (NEP) was launched. The relative dominance of UMNO within the ruling coalition offered the Organisation the political clout to impose restructuring policies largely favouring its members. Despite losing its two-thirds majority in the 1969 elections, the inclusion of opposition parties such as Gerakan meant that the National Front government always enjoyed a two-thirds majority in parliament. That, along with UMNO's relative hegemony within the National Front, offered it considerable capacity to implement its restructuring policies (Cham 1979; Hua 1983; Toh 1982). However not all sectors were subjected to strong intervention. The primary and service sectors faced the strongest redistribution of ownership, with *Bumiputera* equity entering largely through trusts.[5] The export oriented manufacturing sector was largely spared of such redistributive policies.

The NEP, launched with the Second Malaysia Plan (1971–75), aimed at fostering national unity and nation-building through poverty eradication and economic restructuring so as to eliminate the identification of ethnicity with economic function (Malaysia, Ministry of Science, Technology and Environment 1991: 31). A 30 per cent corporate equity target by 1990 was set for the *Bumiputeras*, to be achieved through new growth, with manufacturing identified as the engine to propel rapid growth. The nature of the introduction of the NEP made intervention inevitable. Regulation became particularly strong in the primary and service sectors, and especially in plantations and banking (Liow 1986; Hing 1984). *Bumiputeras* and *Bumiputera* controlled companies began buying into plantations and penetrating the media. Foreign capital was gradually displaced by *Bumiputera* agencies in several plantations.

The First Outline Perspective Plan (OPP1) set the broad socio-economic framework for the achievement of NEP targets. Restructuring was of prime importance, with the first aim being to reduce poverty for all, irrespective of ethnicity. The target set was a reduction of poverty from 49.3 per cent of the households in 1970 to 16.7 per cent of households in 1990 for Peninsular Malaysia (see table 5.1).[6] The target for the rural–urban breakdown was a reduction from 58.7 per cent and 21.3 per cent respectively in 1970 to 23 per cent and 9.1 per cent respectively by 1990. Ethnically, the incidence of poverty in 1970 was 74 per cent for Malays, 26 per cent for the Chinese, 39 per cent for the Indians and 44.8 per cent for other ethnic groups.

Table 5.1 Poverty eradication targets and achievements, Malaysia, 1970, 1976 and 1990 (%)

	1970	1976	Target 1990	Achieved 1990
Peninsular Malaysia				
Poverty incidence	49.3		16.7	15.0
By location				
Rural	58.7	23.0	19.3	
Urban	21.3	9.1	7.3	
By ethnicity				
Bumiputera	65.0			20.8
Chinese	26.0			5.7
Indian	39.0			8.0
Others	44.8			18.0
Malaysia				
Poverty incidence		42.4		17.1
By location				
Rural		50.9		21.8
Urban		18.7		7.5
By ethnicity				
Bumiputera		56.4		23.8
Chinese		19.2		5.5
Indian		28.5		8.0
Others		44.6		12.9

Source: Adapted from Malaysia, Ministry of Science, Technology and Environment (1991: tables 2–6).

The second aim was to restructure employment, the ownership of share capital in the corporate sector, and the creation of a *Bumiputera* Commercial and Industrial Community (BCIC). This required the expansion of *Bumiputera* participation in the formal sectors. *Bumiputera* employment in agriculture, secondary[7] and tertiary sectors were 66.2 per cent, 12.1 per cent and 21.7 per cent respectively in 1970 (see table 5.2, next page). The NEP aimed at restructuring these figures to 37.4 per cent in agriculture, 26.8 per cent in secondary and 35.8 per cent in tertiary sectors by 1990. When examined across ethnic groups, *Bumiputeras* contributed 67.6 per cent, 30.8 per cent and 37.9 per cent respectively to overall employment in the agricultural, secondary and tertiary sectors in 1970. The NEP restructuring efforts when pursued to plan would have changed *Bumiputera* participation to 61.4 per cent in agriculture, 51.9 per cent in secondary and 51 per cent in tertiary sectors respectively in 1990. *Bumiputera*, non-*Bumiputera* and foreign corporate equity participation was set at 30 per cent, 40 per cent and 30 per cent respectively for 1990 as against 2.4 per cent, 32.3 per cent and 63.3 per cent respectively in 1970. The remaining 2 per cent was held by nominee companies. The specific targets were defined only in 1976.

Table 5.2 Restructuring targets and achievements, Malaysia, 1970 and 1990

Sectoral employment	1970[a] Total	%	Target 1990 Total	%	Achieved 1990 Total	%	Malaysia 1990 Total	%
Bumiputeras								
Agriculture	951.1	66.2	1091.4	37.4	875.2	29.0	1404.6	36.7
(%)	67.6		61.4		71.2		76.4	
Secondary[b]	173.1	12.1	782.7	26.8	918.5	30.5	1038.9	27.2
(%)	30.8		51.9		48.0		49.8	
Services	312.4	21.7	1046.8	35.8	1219.8	40.5	1381.9	36.1
%	37.9		48.9		51.0		50.9	
Non-*bumiputeras*								
Agriculture	454.9	33.5	686.2	27.1	354.0	14.0	433.0	15.5
(%)	32.4		38.6		28.8		23.6	
Secondary[b]	389.7	28.7	725.4	28.7	996.1	39.5	1048.6	37.5
(%)	69.2		48.1		52.0		50.2	
Services	512.5	37.8	1116.6	44.2	1170.5	46.5	1314.0	47.0
(%)	62.1		51.6		49.0		49.1	
Corporate equity ownership (%)								
Bumiputeras[c]		2.4		30.0				20.3
Other Malaysians		32.3		40.0				46.2
Foreigners		63.3		30.0				25.1
Nominee companies		2.0						8.4

Notes:
a   Peninsular Malaysia.
b   Includes mining, manufacturing, construction, utilities and transport.
c   Includes trust agencies and other related institutions.

Source: Malaysia, Ministry of Science, Technology and Environment (1991: tables 2–7).

While the quantitative targets were clear, the NEP also emphasised the creation of a *Bumiputera* business class so that the NEP's equity participation could be backed by managerial control. This qualitative target has been interpreted in various ways by different parties. Some look at it as merely the creation of a group of *Bumiputera* millionaires while others see it as the creation of a dynamic class of efficient *Bumiputera* entrepreneurs.

State trusts, sheltering the interests of *Bumiputeras*, soon acquired a substantial grip on the plantation, mining and service sectors. Public agencies swelled from 109 in 1970 to 656 in 1980[8] (Jomo 1994: 8). Some of the best known of these were Perbadanan Nasional (PERNAS) and Permodalan Nasional Berhad (PNB). Institutions such as the Majlis Amanah Rakyat (MARA) were given more extensive participation in the economy. Public sector employment, a major source of *Bumiputera* employment, grew rapidly from 398 000 in 1970 to 520 000 in 1975 and 757 000 in 1981— 11.9 per cent, 12.9 per cent and 15 per cent of total employment in those years (Ismail and Osman 1991).

Extensive investment flowed into agriculture, rural development and poverty eradication, including the development of irrigation and drainage, especially in the National Front's strongholds. Poverty eradication programs absorbed 32.4 per cent and 30.1 per cent of allocated development expenditure in 1971–75 and 1976–80 respectively. Agriculture and rural development absorbed 29.3 per cent and 21 per cent of allocated development expenditure in 1971–75 and 1976–80 respectively (Rasiah and Ishak 1994: table 8). Parastatals such as the Federal Land Development Authority (FELDA) and Rubber Industry Smallholders Development Authority (RISDA) were expanded to help alleviate rural poverty among *Bumiputeras* (Halim 1990).[9] MARA and related agencies gave direct assistance to Malays by offering generous loans and other assistance in business and education. Through privileged access to education and scholarships, both domestically and overseas, the government raised the educational qualifications of the *Bumiputeras*. Although the Malay peasantry were gradually moving away from the rural hinterland, the NEP accelerated the movement of Malays to civil service, educational institutions and other urban dwellings. Also, the ethnic identification of occupations gradually fell, although for many years estates still held a large Indian presence. Where the Malays traditionally dominated, in paddy cultivation and *quasi* government land development dwellings, however, there was no inflow of non-Malays. The migration of Malays from paddy areas led to a fall in its cultivation and the utilisation of foreign labour in areas still under cultivation. As S. Ikmal's (1992) study shows, the Malay peasantry was almost a thing of the past. The growing manufacturing sector and urban activities also attracted the Indians and Chinese, where the latter had earlier established a strong hold over wholesale and retail trade. Falling real wages in estates were also a push factor that partly accounted for the migration of Indians to towns and urban employment (Jayakumar 1993).

Restructuring efforts were helped by the rapid growth in manufacturing, which became a major source of new employment. State intervention initially took on the form of subsidies for export oriented firms. The Investment Incentives Act (IIA) of 1968 had begun export orientation, but it was only after the enactment of the Free Trade Zone (FTZ) Act of 1971, which offered tariff-less operating platforms, in addition to tax exemptions, that the inflow of foreign investment became massive (Chee 1986; Rasiah 1995). To make manufacturing in Malaysia more attractive, the 1955 Employment Act was amended in 1969 to allow for night shifts by women (Rasiah 1995). Export processing expanded from 1972, propelled by the sustained promotional activities of the United Nations Industrial Development Organization, the Asian Development Bank, Asian Productivity Organization, the World Bank, and transnationals actively seeking low cost developing country sites with appropriate infrastructure, loose environmental controls, labour that was easily controlled and trained, and generous tax breaks (Salih and Young 1985; Rasiah 1987). Textiles and garment firms were also influenced by quota considerations from the Multi-Fibre Agreements. The Generalised System of Preferences was also important. Foreign-dominated export processing sub-sectors, notably electric/electronics and textiles/garments grew rapidly as a consequence (see table 5.3).

Table 5.3 Foreign ownership share in manufacturing, Malaysia, 1968–93[a]

	1968	1970	1975	1980	1985	1990	1993
Food	74	71	55	32	25	30	33
Beverages and tobacco	93	89	79	76	67	62	58
Textile and garment	52	39	63	54	48	61	64
Leather	17	56	48	48	54	59	57
Wood	15	11	8	13	9	19	36
Furniture and fixture	50	71	61	31	19	45	45
Paper, printing and publishing	na	na	16	10	20	14	13
Chemical	53	61	63	53	16	24	25
Rubber	14	14	42	46	42	55	51
Petroleum and coal	78	77	79	78	37	44	50
Plastic	na	na	na	12	13	27	46
Non-metal mineral	57	60	52	19	32	33	39
Basic metal	49	45	42	35	32	17	33
Fabricated metal	66	69	59	26	23	30	56
Machinery	74	58	51	42	35	53	65
Electric/electronics	70	67	84	80	73	89	91
Transport equipment	na	58	51	32	15	25	35
Others	60	67	69	57	53	69	81
Manufacturing	61	59	52	39	33	42	50

Note: a   Based on fixed assets ownership.

Source: Rasiah (1995); Malaysian Industrial Development Authority (unpublished data).

Ethnic restructuring in manufacturing was formalised following the promulgation of the Industrial Coordination Act (ICA) in 1975, giving the Minister of Trade and Industry powers to screen firms for NEP characteristics before registering them. Firms with an employment record of twenty-five workers or more and a paid up capital of at least RM250 000 had to be registered. The share of *Bumiputera* employment and equity were the main characteristics used. Interviews with state officials show that the non-*Bumiputeras*, especially Chinese, most resented the '30 per cent give-aways' they often had to generate to facilitate registration (see Rasiah and Ishak 1994). Some, however, overcame the problem by declaring lower equity levels and holding the rest in loans. Not all non-*Bumiputera* enterprises, however, were disadvantaged by the ICA. A number of Chinese enterprises that enjoyed *Bumiputera* equity, especially those connected to influential politicians, generated substantial favours from the government, thereby amassing wealth at a pace they could not have managed under normal circumstances (Yoshihara 1988; Hara 1991; Rasiah 1995: chs 6 and 7; Rasiah and Anuwar 1995). Nevertheless, as K. Yoshihara (1988) argued, the uncertainty associated with their business operations led the Chinese to participate more in short-term ventures.[10]

Export oriented firms were generally exempted from the ICA conditions, although the government often slowed incentives renewal and other benefits involving firms not meeting employment restructuring goals (Rasiah 1987). Export oriented transnational corporations seeking low cost and abundant labour manufacturing sites used Malaysia as one of their havens (Lim 1978). The only constraint imposed was occasional informal pressure to increase *Bumiputera* employment. The state was, however, also interventionist in the export oriented sector as it distorted relative prices by exempting land and utilities in designated manufacturing zones from tariffs and taxes.

A combination of both ICA limits and a slowdown in the global economy following the 1973 oil crisis led to a dramatic decline in manufacturing investment in the period 1975–79. This, along with persistent complaints by the Malaysian Chinese Association over the plight of Chinese businessmen having to conform to ethnic restructuring, led to the raising of the registration ceiling to RM500 000 and an employment size of fifty or more. A similar situation arose in the mid-1980s when manufacturing and GDP recorded an absolute decline. The registration ceiling was subsequently raised to shareholder's equity of RM2.5 million and employment size of seventy-five.

The relative freedom offered to non-*Bumiputera* capital from 1979 and 1986 may have been a consequence of reconciliatory efforts by the government following the emergence of new rent sources (Snodgrass 1995). High commodity prices towards the end of the 1970s and discovery of new oil deposits enabled the government to launch more aggressive policies to achieve the NEP targets with less constraints on non-*Bumiputera* capital. Dissatisfaction with the slow development of *Bumiputera* business and the additional revenue generated by government, led to a more intensive effort to create the BCIC. The *Bumiputera* Investment Foundation, which was formed in 1978, launched the Permodalan Nasional Berhad (PNB) in the same year to pursue this goal. In addition to opening extensive channels for business ventures, the PNB also operated alongside schemes that scattered share ownership across the *Bumiputera* population.

Commodity rents in the early 1980s presented the government with the opportunity to intervene directly in manufacturing to expand *Bumiputera* participation in business. The government began heavy industry projects to create intersectoral linkages for a more broadly based industrial growth, and to create a platform for developing the *Bumiputera* business class. Attracted by the success of Japanese and South Korean models, Prime Minister Mahathir introduced the Look East Policy as a means of learning from them (Jesudasan 1989; Bowie 1991). The formation of the Heavy Industry Corporation of Malaysia (HICOM) in 1981 resulted in state-led investment in motorcars, cement, steel and motorcycles, with highway projects emerging as complementary programs. The increased emphasis on the creation of a *Bumiputera* business class in manufacturing, while including joint ventures with foreign capital, was marked by a reluctance by the government to renew incentives for export oriented transnationals.[11] Tengku Razaleigh Hamzah, then Minister of Trade and Industry, expressed dissatisfaction with the development of linkages and technological depth involving the foreign-dominated electronics industry which was the single biggest manufacturing industry in the country in terms of exports, employment and output (*New Straits Times*, 18 August 1984).

Through HICOM the government moved into the market to spawn a strong *Bumiputera*-driven capital goods industry. The government's interventions included subsidies, protection, curbing competitors in the domestic market, controlling bids and direct ownership. Kedah Cement, Perak Hanjoong, Perwaja and Proton were some of the firms created this way. These industries were also designed to be major platforms for absorbing and training *Bumiputera* managers, engineers, other professionals and technicians, and umbrella organisations for the development of *Bumiputera* and *Bumiputera*-related small and medium-sized firms.

HICOM's ventures, as is to be expected when minimum scale efficiencies are involved, faced heavy losses in the initial period (Khor 1987) and given that they were heavily funded by government loans, the external debt soared. In addition, except for oil, primary commodity prices fell sharply in 1980–84. The combined effect of these, and the recession in the major export oriented manufacturing sub-sectors, forced the introduction of counter-budgetary strategies. In spite of this, the expensive mega projects continued, with loans from abroad and an internal 'austerity drive'.

The sharp rise in the external debt, and the bureaucratic inefficiencies associated with public institutions servicing the private sector directly, led to the introduction of a privatisation strategy. While government sponsorship continued in the industrial sector, it was decreased in order to reduce public expenditures (Ng and Toh 1992). The relative decline in public institutions, however, did not seriously affect the absolute number of public institutions which actually increased in the 1980s. For example, the number of public institutions grew from 656 in 1980 to 1014 in 1985 and 1149 in 1990 (Ismail and Osman 1991; Rugayah 1995: table 3.2). Public sector employment grew from 757 000 in 1981 to 820 000 in 1985 and 854 000 in 1991 (Rugayah 1995: table 3.1). The proportion of public sector employment in total employment, however, fell from 15 per cent in 1981 to 14.6 per cent in 1985 and 12.5 per cent in 1991.

Privatisation also coincided with the birth of the Malaysia Incorporated program in 1983 which was aimed at injecting the competitive elements of private ownership (see Rugayah 1995: 70; Jomo, Adam and Cavendish 1995: 81). This was launched at a time when public enterprises such as Keretapi Tanah Melayu and Telekom Malaysia were facing losses, and were often accused of restricting the private institutions utilising them. It also coincided with government efforts to eliminate inefficiencies, especially in public enterprises. A. Supian (1988) noted that of the 900 public enterprises which submitted their annual returns, 269 reported an accumulated loss of RM269 million, and in the period 1980–85, public enterprises accounted for 64.3 per cent of the public sector deficit (Nair and Fillipides 1988: table 2.2). The issue of shares to the private sector also eased the financial pressures the state had encountered in capitalising state firms. Hence, from two in 1983, the total number of public institutions privatised by 1993 was seventy-two (Rugayah 1995: table 3.4).

Privatisation also offered the opportunity to expand *Bumiputera* participation in the business sector. The predominance of *Bumiputeras* in public institutions continued as most employees remained after privatisation. Privatisation thus became a channel for transforming the ethnic occupational structure of the private sector which remained dominated by non-*Bumiputeras* at the upper levels. The essential executive, financial and technical jobs in the private sector remained in the hands of non-*Bumiputeras* (see Rasiah 1987; 1995). Eight of the fourteen projects that were privatised in the 1986–89 period went to the *Bumiputeras* (Kassim 1991).

However, these reforms need to be kept in perspective. Several of the privatised institutions enjoyed a write-off of their debts and little was done to assess the possibility of enhancing their performance by reorganisation from within the state domain. Also, most of the privatised organisations are still controlled by the government through majority share ownership (for example, 75 per cent of shares in Syarikat Telekom Malaysia are still held by the government). Others make the argument that privatisation has been an important route for rent-seeking for the politically connected (Gomez 1991; Jomo 1995). However, in view of the fact that most of the enterprises are public utilities, enjoying extensive scale economies and fairly inelastic demand, the economic assessment of privatisation may require more robust tools than those currently available. Given the authoritarian powers of the state and the secrecy with which statistics of privatised institutions are held, it may take a long time before reliable performance assessments can be made.

Since restructuring policies had little direct effect on export oriented manufacturing, there was little integration with import substituting manufacturing. Export oriented activity was characterised by foreign transnationals specialising in rudimentary export-processing using imported inputs. For state policy, its purpose was to generate both investment and employment (Malaysia 1976). The lack of integration meant that the industrial sector developed a dualistic structure (Edwards 1990; Rasiah 1992). While the nature of transnational production in the 1970s and the early 1980s offered little opportunity for the development of domestic linkages, state policy also did not encourage it. Indeed, by tying incentives to both export and import oriented production, the government had effectively imposed statutory limits on the integration of the export oriented and import substituting sectors.

The import substituting sector faced the brunt of ethnic restructuring. Hence, the tariff-based rents were rendered unattractive because of ethnic equity and employment restructuring regulations. With the exception of the high rent-dominated beverages and tobacco industry, foreign capital fell sharply in import substituting industries (see table 5.3). Tax and other fiscal exemptions were also gradually removed. Only government-sponsored heavy industries enjoyed such privileges in the 1980s.

The eclectic nature of industrial policy also exposed manufacturing to the vicissitudes of external shocks. Thus, when recession hit in the mid-1980s the state had few effective counter-cyclical measures to prevent mass retrenchments. Severe unemployment followed when export oriented firms also began retrenching. Several groups of disgruntled workers took to the streets with banners; one read: 'hungry workers will become angry voters'. The electronics industry faced the biggest crisis, made worse by a lack of union representation (Rasiah 1987).

The deliberate restructuring policies which were in place and enhanced during the downturn eventually resulted in substantial socio-economic redistribution. By 1990 substantial strides had been recorded towards the NEP targets. Poverty fell to 15.1 per cent of households, better than the

16.7 per cent target set in the NEP. The fall was also reflected in the urban–rural breakdown, to 7.5 per cent and 21.8 per cent respectively in 1989 (Ishak and Ragayah 1995: table 7). Income inequality also declined. *Bumiputera* participation in the secondary and tertiary sectors reached 30.5 per cent and 40.5 per cent respectively in 1990 (see table 5.2).

Bumiputera corporate equity ownership, however, did not meet the NEP target despite a rise to 20.3 per cent. This was, nevertheless, a great achievement given that it was only 2.4 per cent in 1970. A new restructuring channel quickened the pace in the 1980s through share purchases. The PNB invested RM6200 million in 1985 alone to purchase 158 firms, with individual *Bumiputeras* gaining access to PNB shares through the Amanah Saham Bumiputera (ASB) and Amanah Saham Nasional (ASN). By 1988 it had successfully transferred RM632 million worth of shares to the ASN (Rasiah and Ishak 1994). In October 1988, 2.35 million *Bumiputeras* who accounted for 44.7 per cent of eligible investors, owned ASN shares (Malaysia, Ministry of Science, Technology and Environment 1991). The distribution of shares also helped improve income distribution among the *Bumiputeras* as the percentage share going to households, farmers and labourers were 17.1 per cent, 16.5 per cent and 16.2 per cent respectively.

Towards the end of the 1980s, with the economy growing rapidly—GDP growth exceeding 8 per cent per annum from 1988—and the *Bumiputera* organisations (including PNB) and individuals enjoying substantial capital reserves, it was much easier for non-*Bumiputera* entrepreneurs to seek *Bumiputera* partners to meet the 30 per cent equity than the situation in the 1970s. Moreover, politically-connected officials made such inter-ethnic alliances economically advantageous.

As with the 1960s, the Sarawak and Sabah sub-economies remained peripheral to the major political and economic forces shaping transformation in Malaysia in the 1970s and much of the 1980s. Primary commodity processing continued as the prime revenue earners of the two states. Simple processing remained the main manufacturing activity, with boat-making, cement, iron and steel and automobile assembly having developed on a small scale.

It can thus be said that the Malaysian economy in the 1970s and much of the early 1980s experienced aggressive, ethnically motivated intervention. Class antagonism had been well contained through ethnic politics, apart from occasional non-militant spillovers. Specially designed socio-economic programs particularly beneficial to the *Bumiputera* community were enacted by political elites who enjoyed the autonomy to dictate economic policies with few constraints. Indeed, the trend was towards increased authoritarianism. Resource rents and rapid growth towards the end of the 1980s enabled greater inter-ethnic collaboration in business as the NEP neared maturity. While socio-economic restructuring did take place, the industrial sector remained unintegrated, and the local business class lacked a competitive dynamism. A *Bumiputera* business class did emerge, albeit significantly smaller than that envisaged by the NEP. But the lack of competitive discipline and the rentier nature of business operations controlled by the class limited its potency. Foreign controlled transnationals were the prime generators of wage employment in manufacturing. The export oriented sub-sector grew while the import substituting sector shrank in importance. Sabah and Sarawak experienced relatively little industrial growth, continuing their naturally endowed resource rent appropriation.

Drive Towards Industrial Maturity

A few major developments helped transform the thrust of state policy from the late 1980s and especially into the 1990s. The mid-1980s recession and the initial failure of state-sponsored heavy industries forced a renewed reliance on foreign controlled enterprises, enhanced by an avalanche of capital from East Asia—driven by the yen appreciation, withdrawal of the Generalised System of Preferences from the Asian NICs, and trade measures in major markets against Japan and the Asian NICs (see Rasiah 1987). Tengku Razaleigh's departure following his defeat in the 1987 UMNO elections, and the consequent revision of the power structure in the country which included the judiciary, gave Mahathir the opportunity to direct economic policy almost single-handedly. The creation of a broad-based, competitive and high technology driven industrial sector became the prime target of state planning.

While the creation of a *Bumiputera* business class was still actively pursued, by 1990 the creation of an advanced industrial infrastructure had become more important. Although not explicit, some amount of ethnic deregulation had begun from the late 1980s (see Lubeck 1992; Snodgrass 1995). Incentives were more easily accessible for non-*Bumiputera* businesses. They were also given longer adjustment periods to conform with NEP requirements. Foreign capital enjoyed a new round of generous incentives after the enactment of the Promotion of Investments Act (PIA) in 1986 which followed the launching of the Industrial Master Plan the previous year. The extent of state involvement in the economy was, however, still profound, albeit overall protection levels gradually declined. Subsidisation through the PIA of export oriented and technology deepening firms became stronger (Rasiah 1995; Rasiah and Anuwar Ali 1995).

The thrust of industrial policy, which was foreign-driven in the 1970s and local-driven in the early 1980s, also changed from the late 1980s to incorporate elements of both. Industrial policy, *à la* South Korea and Japan, was mixed with the Singaporean strategy of broadening and upgrading transnational operations. The state began encouraging consultative committees to govern the technological transformation of firms. Unlike earlier periods when there was little coordination between the two, the state was willing to work with the private sector from the late 1980s. To facilitate industry–government coordination mechanisms in governing the private sector, consultative committees were formed.

The East Asian factor, driven by the factors noted above and rising production costs in the NICs, influenced a large inflow of investment to South-East Asia, including Malaysia (Rasiah 1987). The electronics and textile industries in particular received substantial investment so that their share of fixed·assets, employment and output in manufacturing rose substantially between 1985 and 1990 (Rasiah 1995: table 5.3). The rapid growth of the Asia Pacific market also induced other firms to relocate to Malaysia which, apart from Singapore, enjoyed better infrastructure than the other South-East Asian economies. Thus, German scientific instruments and USA disk drive assembly firms expanded from the late 1980s. This rapid growth helped expand manufacturing's contribution to GDP to 31.7 per cent in 1994.

With the exception of Perwaja Steel, most government-sponsored heavy industries experienced a turnaround. With revamped managements, rationalisations, injections of better technologies, and a rapid growth in domestic demand, transport equipment (including Proton and Perodua) and non-metal minerals (including Kedah Cement and Perak Hanjoong) recorded appreciable labour productivity growth. From annual average growth rates of –2.5 per cent and –2 per cent respectively in 1979–85, these sectors recorded annual average growth rates of 14.8 per cent and 3.5 per cent respectively in 1985–90 (Rasiah 1995: table 5.6). Although these measures are inadequate to make real efficiency comparisons, they do suggest an improved performance.

By 1989 the rapid growth of the country had transformed the country's main industrial belts into a labour-scarce economy. The Western states began facing serious labour shortages in agriculture, largely replaced through migrant labour from Indonesia (Azizah 1995). The shortage extended into services and manufacturing, and labour imports from Indonesia, Bangladesh and the Philippines commenced to meet this demand. Skilled labour shortages and widespread poaching along the Western industrial corridor led to the government adopting various measures, including incentives and later penalties, to promote training (the Double Deduction Training Incentive and the Human Resource Development Act). Consistent with government initiatives to involve the private sector, the Human Resource Development Fund

(HRDF) is administered by the Human Resource Development Council (HRDC) which is dominated by private sector officials. The Penang Skills Development Centre (PSDC), a joint private–public sector initiative, was formed in 1989, involving the Penang state government. Similar initiatives involving other West Malaysian states took off in the 1990s. Indeed, there was a major appraisal of human resource institutions by the Ministry of Science, Technology and Environment (MOSTE) in 1995 to help generate the requisite skilled labour to sustain its movement towards an industrialised economy.

Unlike the 1970s and early 1980s, the government also began to place emphasis on technology deepening, albeit without a concrete strategy. The PIA contained specific incentives for firms participating in training and research and development. Sharp labour shortages in the face of continued expansion in new investment presented the state with improved bargaining powers *vis-à-vis* foreign firms. The state used this to step up its industrialisation drive when the Action Plan for Industrial Technology Development was launched in 1990. The plan offered comprehensive recommendations to transform the economy into an industrialised nation by 2020. The development of strategic high technology industries, supporting firms and the requisite human resources became central to its achievement.

The government's industrial widening and deepening efforts included the expansion of locally owned firms. New institutions were created to support their expansion. The state, which became directly involved in promoting the growth of locally owned firms such as Proton and Sapura, including their external market growth, introduced the Malaysian Technology Development Corporation (MTDC) and the Malaysia Industry-Government Group of High Technology (MIGHT) in 1993. MTDC had invested in eight firms by the end of 1993 to assist the commercialisation and advancement of their technologies. MIGHT has been examining mechanisms to identify and promote technology, markets, businesses and investment opportunities for innovative activities. A high technology park was developed in Kulim, and by 1995 several firms, including Intel, had acquired land there to undertake designing activities. The Malaysian Institute of Microelectronics Systems also launched plans to build wafer fabrication plants there.

To integrate the export oriented sector with the domestic economy, the government tied several incentives to localisation initiatives, and launched subcontract programs. From 1991 the government imposed a 30 per cent domestic sourcing and 20 per cent domestic value-added condition for approving certain financial incentives. The small and medium-scale industries received a special boost as rationalisations led to an agglomeration of thirteen ministries and thirty different departments monitoring their role to five lead agencies (Vijaya 1993: 9–12). The umbrella concept of marketing which was introduced in 1984 was revitalised with greater promotion. The subcontract exchange scheme was introduced from the late 1980s. The vendor development program (VDP) began with two anchor companies, Sapura and Sharp, in 1992. The number has since risen to include Tenaga Nasional Berhad, Syarikat Telekom Malaysia, Petronas, Land and General (Malaysia, Ministry of International Trade and Industry 1994: 72). The latter was enhanced through subsidised loan schemes for local firms under the Industrial Technical Assistance Fund series that was introduced from 1989.[12] Localisation has been stronger in government-sponsored industries, with, for example, Proton sourcing over 70 per cent of its inputs from more than 100 local suppliers in 1994. The VDP also helped increase the participation of small and medium-sized *Bumiputera* enterprises. Though less successful, subcontract relationships between foreign transnationals and local firms have also risen from the late 1980s. The latter has also been influenced by changes in production technologies in high technology industries (see Rasiah 1994).

Such was the importance placed on industrialisation that Prime Minister Mahathir used his high international profile to complain about 'creeping protectionism' against Malaysia's manufactured exports. The formation of the North American Free Trade Area and greater integration of Europe raised fears of the emergence of trade blocs. Efforts to incorporate social and environmental clauses into the World Trade Organization's jurisdiction have been viewed as another tool to restrict developing economy exports. Similar concerns by regional partners led to the formation of the ASEAN Free Trade Area (AFTA) in 1993, and the vocal assessment of Malaysia's role in the Asia Pacific Economic Cooperation fora. These international developments have helped strengthen collaboration among ASEAN members. To further strengthen the regional bargaining platform, Mahathir conceived the East Asian Economic Caucus (EAEC), which includes all East Asian economies.

Unlike the 1970s and much of the 1980s, industrial widening became part of the country's industrialisation efforts from the late 1980s. Decentralisation had been emphasised earlier, but generally only through the provision of locational incentives. Only states in western Peninsular Malaysia enjoyed some benefits from such fiscal instruments in the 1970s and early 1980s. The eastern Peninsular Malaysian states had focused more on agriculture and mining. As with development emphasis in general, Sabah and Sarawak strategies did not feature in such efforts. The joint ASEAN urea project utilising direct conversion of natural gas at Bintulu has been the main industrial undertaking in East Malaysia in the 1980s. There was also involvement on a smaller scale in furniture, boat-making, pottery, food processing, fertilisers, automobiles, cement, steel and, from the 1990s, downstream timber activities. The lack of infrastructure and proactive industrial promotion meant that resource rents from oil, timber and plantation agriculture were the prime revenue generators.

Rapid growth led to bottlenecks that offered the federal government the opportunity to extend industrial emphasis to the eastern states. Also, regional concerns, especially about the relative backwardness of the East Malaysian states *vis-à-vis* the Western industrial corridor, had emerged as a major problem for the ruling National Front. Sabah, in particular, had enjoyed strong opposition sentiments, including opposition rule in the early 1990s.[13] When the National Front wrested power from the opposition Parti Bersatu Sabah in 1994, following several crossovers from the latter, there were substantial efforts to develop the state. Hence, from the late 1980s resource-based manufacturing industries began to receive strong emphasis, including in Sarawak and Sabah. Infrastructure development, including the East–West Highway cutting through Selangor and Pahang, has also been planned (Malaysia, Ministry of International Trade and Industry 1994: 128–9). Finally, a second round of the pioneer status and investment tax allowance scheme, which covers labour-intensive operations, has been launched for the Eastern corridor of Peninsular Malaysia and Sarawak and Sabah.

With industrial deepening and widening assuming prime importance, the emphasis on redistribution declined somewhat in the 1990s. Ownership liberalisation in state-supported trusts to foreigners and non-*Bumiputeras*, depressed agricultural prices and relative stagnation in the plantation workers' wages in the face of rapid income growth in other sectors led to a rise in inequalities. Although the incidence of poverty continued to decline, income inequality worsened with the gini coefficient hitting 0.459 in 1993. *Bumiputera* corporate equity fell to 18.2 per cent in 1992. Hence, while state policy has successfully created a group of *Bumiputera* millionaires, the ethnic Malay share in the overall economy had fallen in the 1990s. There was also a fall in Chinese ownership of corporate equity which declined to 37.8 per cent in 1992 (see Malaysia, Ministry of International Trade and Industry 1994: table 3.5). While Mahathir Mohamad had expressed sadness in the failure of *Bumiputeras* to meet the 30 per cent corporate equity ownership in the country, he was, however, not prepared to achieve it at the expense of checking the pace of industrialisation. It can be argued that rapid economic expansion has helped raise real wages and reduced poverty in the country. The overwhelming emphasis on industrialisation, however, has been accompanied by a reversal in equity trends after 1990 which could pose a complex problem for the political management of class relations.

Conclusion

Explaining the dynamics of Malaysia's transition from colonial commodity producer to its current position as an industrialised economy poised to enter the ranks of the NICs requires a sophisticated methodology. This methodology must account for factors of class, ethnicity and the role of the state. Left with a colonial legacy of ethnic divisions that extended into economic and spatial demarcations, Malaysia has journeyed through a difficult phase of socio-economic transformation for much of its independent existence. Given the ethnic divisions created in the social, economic and political structures, class antagonisms were manifested along communal lines. Communalism boiled over towards the end of the liberal phase of governance in the 1960s as a consequence. The NEP, however distressing it was to certain groups, was a conjunctural result of the socio-political and economic events that were clearly evident in the 1960s. Its implementation meant the state had to become actively engaged in the economy. Economic restructuring through direct and complementary strategies and external events helped reduce unemployment, poverty and inequality, and raised *Bumiputera* participation in the economy.

With ethnic redistribution occupying centre stage, and despite its paramount significance in generating employment, industrial policy was made up of a bundle of eclectic strategies. The result was a dualistic manufacturing sector with little integration between the import substituting and export oriented sub-sectors. Industrial policy in the 1970s and much of the 1980s also lacked technological vision. However, in the 1990s the emphasis on the promotion of a *Bumiputera* business class was overtaken by a new industrialisation drive. Technological deepening, industrial widening, greater intersectoral integration and the creation of locally controlled ventures capable of competing in international markets became important objectives. Rapid growth and commodity rents also helped some ethnic deregulation in the economy. Political developments, especially in UMNO, enabled Mahathir to single-handedly orchestrate growth strategies. The direction taken has bolstered the positions of technocratic elites in the public bureaucracies and been favourably received by those internationally competitive sections of the domestic bourgeoisie. Essentially, though, foreign capital has been the big winner in the manufacturing sector.

The shift in state priority coupled with ownership deregulation strategies (through privatisation and other divestment channels) has, however, increased income inequality and reduced the share of *Bumiputera* corporate equity ownership. It may, however, be premature to argue that ethnic restructuring and the expansion of the *Bumiputera* business class will diminish in importance. Sectional interests, especially evident in party politics, could force a return to the old NEP doctrine. Equity reversals since 1990 already point to potential conflict between different fractions of capital in the future. Much, however, will depend on the success of the industrialisation drive and its concomitant impact on servicing the more powerful *Bumiputera* interests.

While Malaysia has drawn on the East Asian 'model', this has been to draw policy lessons which state officials have attempted to graft to a Malaysian base. These efforts have shown that the Malaysian base—its ethnic structure, class structure and its historical legacy—is the soil in which any grafted lessons must grow and develop. Only through an understanding of its complex political economy can Malaysia's experience be understood.

---

[1] Helpful comments from Richard Robison, Garry Rodan, Kevin Hewison and Chris Edwards are gratefully acknowledged. The usual disclaimer applies.

[2] Slaves, generally war prisoners, and debt bondsmen were lower in the social hierarchy: see Jomo (1986) for a more detailed account.

[3] Neo-liberal explanations on the decline in manufacturing in the 1980s as being a result of import substitution distortions do not address the dynamic elements that could have emerged if its implementation had been tied to generating scale-intensive infant firms. Unlike South Korea, there were no disciplining targets to achieve efficiency improvements for eventual external competition (see Amsden 1989).

[4] There have, however, been two exceptions. First, when the Parti Islam SeMalaysia (PAS) and Semangat 46 (splinter group of the previous UMNO led by Tengku Razaleigh Hamzah who narrowly lost in UMNO's general assembly elections to Dr Mahathir Mohamad in 1987) together won the Kelantan state elections in 1990. Second, when the Parti Bersatu Sabah (PBS) led by Dr Joseph Pairin Kitingan left the National Front in 1990. The situation has since reverted to the national tone following several crossovers from PBS to the National Front after the 1994 state elections bringing the state government back to them.

[5] These trusts are state supported corporate agencies formed to assist *Bumiputeras* expand their participation in the economy.

[6] When launched in 1971 the NEP set targets only for the Malaysian Peninsular.

[7] Includes mining, manufacturing, construction and utilities.

[8] There were twenty-two public agencies in 1960 (Jomo 1994: 8).

[9] The introduction of these programs during Tun Razak's reign as Prime Minister in the early 1970s earned him the epithet 'father of rural development'.

[10] Yoshihara (1988) refers to such short-term businessmen as *ersatz* capitalists.

[11] For example, Advanced Micro Devices and Monolithic Memories Incorporated, two American electronics transnationals, threatened to relocate abroad in the mid-1980s as a result (see Rasiah 1987).

[12] A total of four different types of programs have since been launched (Malaysia 1994: 265-66).

[13] Traditionally, federal support for opposition-ruled areas have been less forthcoming.

References

Amsden, A. (1989) *Asia's Next Giant: South Korea and Late Industrialisation*, New York: Oxford University Press.

Azizah, K. (1995) 'From neglect to legalization: the changing state response to illegal inflow of foreign labour to Malaysia', paper presented at the International Conference on Globalization: Local Challenges and Responses, Penang, 19–21 January.

Bowie, A. (1991) *Crossing the Industrial Divide*, New York: Columbia University Press.

Cham, B. N. (1979) 'Toward a Malaysian Malaysia: a study of political integration', unpublished doctoral thesis, Calgary, University of Alberta.

Chandra, M. P. (1993) 'Political marginalization in Malaysia', in K. S. Sandhu and A. Mani (eds) *Indian Communities in Southeast Asia*, pp. 211–36, Singapore: Times Academic Press and Institute of Southeast Asian Studies.

Chee, P. L. (1986) *Small Scale Industries in Malaysia*, Kuala Lumpur: Pelanduk Press.

Edwards, C. (1975) 'Protection, profits and policy: an analysis of industrialisation in Malaysia', unpublished doctoral thesis, Norwich: University of East Anglia.

Edwards, C. (1990) 'State interventionism and industrialisation in South Korea and Malaysia', *mimeo*, Norwich: University of East Anglia.

Everitt, W. E. (1952) 'A history of mining in Perak', *mimeo*, Ipoh.

Gomez, E. T. (1991) *Money Politics in the Barisan National*, Petaling Jaya: Forum.

Halim, S. (1990) 'Exploitation and control of labour within the Malayan rubber industry till 1941', *Kajian Malaysia*, 6 (1): 1–43.

Hara, F. (1991) 'Malaysia's new economic policy and the Chinese business community', *Developing Economies*, 29 (4): 350–70.

Harper, T. (1992) 'Social history of Malaysia', unpublished doctoral dissertation, Cambridge: University of Cambridge.

Hasan, P. (1980) 'Growth, structural change and social progress', in K. Young, W. C. F. Bussink and P. Hasan, op. cit., pp. 23–89.

Hing, A. Y. (1984) 'Capitalist development, class and race', in S. H. Ali (ed.) *Ethnicity, Class and Development, Malaysia*, Kuala Lumpur: Malaysian Social Science Association.

Hua, W. Y. (1983) *Class and Communalism in Malaysia*, London: Zed Press.

Ikmal, S. (1992) 'Development cycle and capitalist farm expansion', *Ilmu Masyarakat*, (22): 81–114.

Ishak, S. and Jomo, K. S. (1986) *Development Policies and Income Inequality in Peninsula Malaysia*, Institute of Advanced Studies Monograph 1, Kuala Lumpur: Universiti Malaya.

Ishak, S. and Ragayah, M. Z. (1995) 'Economic growth and equity in Malaysia: performance and prospect', paper presented at the Fifth Tun Abdul Razak Conference, 21–23 April, Ohio.

Ismail, S. and Osman, R. H. (1991) *The Growth of the Public Sector in Malaysia*, Kuala Lumpur: Institute of Strategic and International Studies.

Jackson, J. C. (1964) 'Smallholding cultivation of cash crops', in Wang Gungwu (ed.) *Malaysia: A Survey*, pp. 246–73, London: Pall Mall Press.

Jain, R. K. (1970) *South Indians in the Plantation Frontier in Malaya*, New Haven: Yale University Press.

Jayakumar, D. (1993) 'Plight of the plantation workers', *mimeo.*

Jesudason, J. (1989) *Ethnicity and the Economy*, Singapore: Oxford University Press.

Jomo, K. S. (1986) *A Question of Class*, Singapore: Oxford University Press.

Jomo, K. S. (ed.) (1993) *Industrializing Malaysia: Policy, Performance and Prospects*, London: Routledge.

Jomo, K. S. and Todd, P. (1994) *Trade Unions and the State in Peninsular Malaysia*, Kuala Lumpur: Oxford University Press.

Jomo, K. S. (ed.) (1995) *Privatizing Malaysia: Rents, Rhetoric and Realities*, Boulder: Westview Press.

Jomo, K. S., Adam, C., and Cavendish, W. (1995) 'Policy', in K. S. Jomo (ed.) op. cit.

Kaldor, N. (1979) 'Equilibrium theory and growth theory', in M. J. Boskin (ed.) *Economics in Human Welfare: Essays in Honour of Tibor Scitovsky*, pp. 273–91, New York: Academic Press.

Kalecki, M. (1976) *Essays on Developing Economies*, Hassocks: Harvester.

Kassim, H. (1991) 'An assessment of the industrial technical assistance fund (ITAF) in upgrading SMIs', in I. M. Salleh and L. Rahim (eds) *Enhancing Intra-Industry Linkages*, Kuala Lumpur: ISIS and FES.

Khong, S. M. (1991) 'Service sector in Malaysia: structure and change', unpublished doctoral thesis, Cambridge: University of Cambridge.

Khor, K. P. (1987) *Malaysia's Economy in Decline*, Penang: Consumers Association of Penang.

Kornai, J. (1979) 'Appraisal of project appraisal', in M. J. Boskin (ed.) *Economics and Human Welfare: Essays in Honour of Tibor Scitovsky*, pp. 75–99, New York: Academic Press.

Lewis, A. (1955) *The Theory of Economic Growth*, London: Allen & Unwin.

Lim, C. Y. (1967) *Economic Development of Modern Malaya*, Kuala Lumpur: Oxford University Press.

Lim, L. Y. C. (1978) 'Multinational firms and manufacturing for export in less developed economies', unpublished doctoral thesis, Michigan: University of Michigan.

Lim, M. H. (1975) 'Multinational corporations and development in Malaysia', *Southeast Asian Journal of Social Sciences*, 4: (1): 53–76.

Liow, B. W. K. (1986) 'Malaysia's new economic policy and restructuring of commercial banks', *Kajian Malaysia*, 4 (1): 1–32.

Lubeck, P. (1992) 'Malaysian industrialisation, ethnic divisions, and the NIC model: the limits to replication', in R. Appelbaum and J. Henderson (eds) *States and Development in the Asian Pacific Rim*, London: Sage.

Malaysia (1971) *The Second Malaysia Plan 1971–1975*, Kuala Lumpur: Government Printers.

Malaysia (1976) *The Third Malaysia Plan 1976–1980*, Kuala Lumpur: Government Printers.

Malaysia (1994) *Mid-Term Review of the Sixth Malaysia Plan*, Kuala Lumpur: Government Printers.

Malaysia, Ministry of International Trade and Industry (1994) *Report on International Trade and Industry*, Kuala Lumpur: MITI.

Malaysia, Ministry of Science, Technology and Environment (1991) *Technology Action Plan*, Kuala Lumpur: Ministry of Science, Technology and Environment.

Mehmet, O. (1986) *Development Policies in Malaysia*, Forum: Petaling Jaya.

Miller, J. (1990) 'Ali Bab and the 40%', *Banker*, 140 (776).

Myrdal, G. (1957) *Economic Theory and Underdeveloped Regions*, New York: Methuen.

Nair, G. and Fillipides A. (1988) 'How much do state-owned enterprises contribute to public sector deficits in developing countries—and why?', *World Development Report*, Washington DC: World Bank.

Nanjundan, S. (1953) 'Economic development in Malaya', in B. K. Madan (ed.) *Economic Problems of Underdeveloped Countries in Asia*, New Delhi: Oxford University Press.

New Straits Times (1984) 'Jolt for electronic firms from Razaleigh', 18 August: 3.

Ng, C. Y. and Toh, K. W. (1992) 'Privatisation in the Asian-Pacific region', *Asian Pacific Economic Literature*, 6 (2): 42–68.

Ooi, J. B. (1961) 'Mining landscapes of Kinta', in T. H. Silcock (ed.) *Readings in Malayan Economics*, Singapore: Eastern University Press.

Rasiah, R. (1987) *Pembahagian Kerja Antarabangsa: Industri Semikonduktor di Pulau Pinang*, Kuala Lumpur: Malaysian Social Science Association.

Rasiah, R. (1992) 'Foreign manufacturing investment in Malaysia', *Economic Bulletin for Asia and Pacific*, 63 (1): 63–77.

Rasiah, R. (1993) 'Competition and work in the textile and garment industries in Malaysia', *Journal of Contemporary Asia*, 23 (1): 3–23.

Rasiah, R. (1993a) *Transnational Corporations and Backward Sourcing in the Electronics Industry: A Study of Subcontracting Links with Local Suppliers in Malaysia*, Bangkok: United Nations Economic and Social Commission for Asia and Pacific.

Rasiah, R. (1993b) 'Free trade zones and industrial development in Malaysia', in K. S. Jomo (ed.).

Rasiah, R. (1994) 'Flexible production systems and local machine tool subcontracting: the case of electronics components transnationals in Malaysia', *Cambridge Journal of Economics*, 18 (3): 279–98.

Rasiah, R. and Ishak, S. (1994) 'Malaysia's new economic policy in retrospect', paper presented at the Social Science Seminar, Kota Kinabalu.

Rasiah, R. and Anuwar, Ali (1995) *Industri dan Pembangunan Ekonomi di Malaysia*, Kuala Lumpur: Dewan Bahasa dan Pustaka.

Rasiah, R. and Anuwar, Ali (1995) 'Governing industrial technology transfer in Malaysia', *mimeo*.

Rasiah, R. (1995) *Foreign Capital and Industrialisation in Malaysia*, London: Macmillan.

Ratnam, K. J. and Milne (1970) 'The parliamentary election in West Malaysia', *Pacific Affairs*, 43 (2): 203–26.

Rugayah, M. (1995) 'Public enterprises', in K. S. Jomo (ed.) op. cit.

Saham, J. (1980) *British Industrial Investment in Malaysia*, 1963–71, Kuala Lumpur: Oxford University Press.

Salih, K. and Young, M. L. (1985) 'Penang's industrialisation: where do we go from here?', paper presented at Future of Penang Conference, Penang: Malaysian Economic Association (Northern Branch), 6–8 May.

Salih, K. (1988) 'The new economic policy after 1990', *MIER Discussion Paper* 21, Kuala Lumpur: Malaysian Institute of Economic Research.

Salih, K. and Yusof, Z. (1989) 'Overview of the new economic policy and framework for the post-1990 economic policy', *Malaysian Management Review*, 24 (2): 13–61.

Sandhu, K. S. (1969) *Indians in Malaya*, Cambridge: Cambridge University Press.

Shepherd, G. (1980) 'Policies to promote industrial development', in K. Young et al. (eds), op. cit., pp. 182–210

Snodgrass, D. (1995) 'Growth with distribution in a multi-ethnic society', *mimeo*. Harvard Institute of International Development.

Song, O. S. (1923) *One Hundred Years' History of the Chinese in Singapore*, London: John Murray.

Supian, A. (1988) 'Malaysia', in G. Edgren (ed.) *The Growing Sector: Studies of Public Sector Employment in Asia*, New Delhi: ILO-ARTEP.

Thoburn, J. T. (1977) *Primary Commodity Exports and Economic Development: Theory, Evidence and a Study of Malaysia*, London: John Wiley.

Toh, K. W. (1982) 'The state in economic development: a case study of Malaysia's new economic policy', unpublished doctoral thesis, Kuala Lumpur: Universiti Malaya.

Vijaya, Letchumi (1993) 'SMI development programmes', paper presented at MITI/MIDA/FMM Seminar, 'Domestic Investment in Manufacturing', Penang.

Wee, C. H. (1995) *Sabah and Sarawak in the Malaysian Economy*, Kuala Lumpur: S. Abdul Majeed.

Wong, J. (1979) *ASEAN Economies in Perspective*, London: Macmillan.

Wong, L. K. (1965) *The Malayan Tin Industry to 1914*, Tuscon: Arizona University Press.

World Bank (1980) *World Development Report*, New York: Oxford University Press.

World Bank (1990) *World Development Report*, New York: Oxford University Press.

World Bank (1993) *The East Asia Miracle: Economic Growth and Public Policy*, New York: Oxford University Press.

Yoshihara, K. (1988) *The Rise of Ersatz Capitalism in South-East Asia*, Singapore: Oxford University Press.

Young, K., Bussink, W. C. F. and Hasan, P. (eds) (1980) *Malaysia: Growth and Equity in a Multiracial Society*, Baltimore: Johns Hopkins University Press.

6

Singapore: Economic Diversification and Social Divisions

Garry Rodan

Introduction

Whereas the major new dynamic shaping most of South-East Asia is industrialisation, Singapore has already experienced three decades of dramatic and sustained industrial transformation. In the city state, structural pressures and opportunities are now combining to effect a significant diversification of the economic base, involving Singapore as a financial, communications and business centre from which to service the rapidly expanding economies of the region.

Like the decades of industrialisation behind it, the current transformation cannot simply be understood as the unravelling of market processes, important as these are. Rather, the one constant amidst all the flux is the strategic influence of the state in helping to shape economic directions. The form this takes is itself undergoing significant modification as policies like privatisation selectively open up areas of the domestic economy to greater private investment. But this rationalisation and adjustment of role should not be confused with a diminution in the state's economic importance. Indeed, more internationalised accumulation strategies by state-controlled capital is an important element of the current concerted policy push towards greater regional economic integration. It is a process which also involves new alliances between the private sector and the state, as well as heightened grievances from sections of the domestic bourgeoisie.

The maturation and diversification of the Singapore economy are also accompanied by broader social tensions and political challenges. Industrialisation laid the basis for full employment and major social improvements in areas like public housing, education, health and transport. It also generated considerable upward social mobility. In this respect, the World Bank's contention in *The East Asia Miracle* (1993: 188) that governments in East and South-East Asia have 'institutionalised the principle of shared growth' has some resonance with the Singapore experience. However, structural limits to social mobility are now beginning to assert themselves in Singapore as privileged classes and elites consolidate their positions. The current economic transformation has the potential to compound this through sharper differentiation in labour market rewards between the skilled and unskilled. Given the ideological hostility of the ruling People's Action Party (PAP) to welfare-oriented redistributional policies, the government seeks alternative measures to effectively manage the politics of the contemporary economic transition.

In theoretical terms, this chapter demonstrates how the development process is an inherently political one involving dynamic coalitions of interests. In this particular case, the economic and political interests of a bureaucratic state are a critical and prevailing component of this process, significantly shaping accumulation strategies from Singapore and thereby conditioning alliances with and between elements of the private sector. The exact form and impact of political factors may vary from one development experience to another, but they are always present. The search of some political economists for neutral conditions and regulations within which markets might operate is illusory. Furthermore, all market economies, by definition, result in uneven material rewards which also have the potential to manifest in challenges to the established relationships of power defining the economy. The management of these tensions is an increasingly important issue in the political economy of Singapore.

Pre-Industrial Singapore

The incorporation of Singapore under the British colonial umbrella in 1819 was, in the first instance, intended to counter Dutch access to and control over trade in the Malay Archipelago. It subsequently assumed a more general role in fostering trade with the Orient. British and, to a lesser extent, European merchants dominated the ensuing entrepôt trade, facilitated to no small degree by the attraction to Singapore of Chinese and other merchants with established links in the region (see Turnbull 1982: 14; Wong 1960: 74, 84). By the end of the nineteenth century, powerful British and European agency houses had expanded beyond the exportation of raw materials and importation of manufactured goods, into direct control of tin mines and plantations as well as the wholesale and retail trade of manufactures (Hughes 1969: 9). Chinese merchants had established an intricate network of domestic commerce and small-scale collection, distribution and retailing to complement the ascendancy of European capital (see Buchanan 1972: 33). Singapore's role as an entrepôt economy in the service of British colonial capital fundamentally shaped the patterns of capital accumulation by the domestic bourgeoisie, despite some progression from trade and commerce into finance.[1]

It was only really after the Second World War, when Singapore's more stable and rapidly expanding population was accompanied by rising economic nationalism in South-East Asia, that the question of economic diversification began to receive serious consideration by authorities (Turnbull 1982: 234). These trends threatened to undercut the rate of economic growth to Singapore's entrepôt economy at precisely the time when employment growth was increasingly required. The first official investigation into the problem resulted in the commissioning of a report by the International Bank for Reconstruction and Development (IBRD) which was presented in 1955. It recommended an import substitution industrialisation (ISI) strategy to alleviate unemployment. Economic union with Malaya, it was envisaged, would generate a sizeable protected domestic market for manufactured goods which would attract investors. However, given the preoccupation with the issue of Singapore's self-rule, serious action on industrialisation was shelved.

The domestic bourgeoisie's alignment of interests with colonial capital resulted in a detachment by it from the anti-colonial movement. Thus, when self-government came in 1959 the domestic bourgeoisie found itself politically marginalised. The triumphant People's Action Party (PAP) was predominantly led by a small group of English-educated middle-class nationalists who had formed an alliance of convenience with the closely coordinated leaderships of the labour and student movements dominated by the Chinese-educated. The alliance would prove a tempestuous and ultimately unsustainable one, but this was, nevertheless, a crucial historical phase shaping Singapore's economic and political direction. In particular, it kept the government insulated from pressures by established business interests in the formation of a manufacturing strategy. There was no political imperative for a domestic rather than an international industrial bourgeoisie to prevail in any program to attract private investment. In any case, the legacy of colonial Singapore's class structure meant that the state would necessarily play a critical role in industrialisation in Singapore. If this did not take the form of a heavy direct investment role to compensate for the absence of an industrial bourgeoisie, it would at least involve the state in extensive measures to attract or nurture private sector investment in industry.

Industrial Transformation

The PAP government's first initiative towards industrialisation involved an invitation to the World Bank for advice and consequent visits to Singapore in late 1960 from a United Nations Industrial Survey Mission headed by

Dutch economist, Albert Winsemius. What was thus known as the Winsemius Report recommended a program of ISI led by private capital investment, but involving an extensive role for the state in attracting and supporting that investment. This included: control over labour and the holding down of wages; the provision of various industrial estates; upgrading of technical training; tax incentives; and free remittance of profits. In early 1961 the government announced the *Development Plan, 1961–64* (Ministry of Finance 1961), closely reflecting the recommendations of the Winsemius Report. Central to this was the proposed Economic Development Board (EDB) which was entrusted with a range of responsibilities, including the development of industrial estates, provision of industrial loans and technical assistance. Some functions were intended to compensate for the lack of industrial expertise of domestic-based capital, others to lower the establishment and operating costs to capital. However, the government's left-wing critics saw a greater role for the state in direct investment to ensure more favourable conditions for workers.

However, it was the prospect of political merger with the Federation of Malaysia which was to finally precipitate a showdown between the PAP's competing factions. Although the Left had, in principle, supported the idea of political merger with Malaya, when the Federation of Malaysia was proposed in May 1961 by Malayan Prime Minister Tunku Abdul Rahman this raised the prospect of the Left's persecution by a right-wing Federal Government. The consequent intensification of internal PAP friction culminated in a permanent split which led in 1963 to the formation by Prime Minister Lee Kuan Yew's opponents of a new party, the Barisan Sosialis (BS). Support for the newly established BS came not only from the left-wing labour and student movements, but also elements of the domestic bourgeoisie that shared concerns about the status of Chinese language and culture (see Bloodworth 1986: 58–61). This did nothing to endear the PAP leadership to the local business class, instead it engendered further suspicion about it.

The PAP was now not only insulated from political control or influence by business, but also largely spared internal pressures by organised labour. Yet this high degree of relative political autonomy also brought the legitimacy of the government into question. The PAP responded by contending that the national interest could only be represented by a party that sat above the pressures of particular interest groups or classes. This rationale evolved into a comprehensive ideological case for elitist, technocratic government in the years ahead. Meanwhile, the challenge for the PAP was to hold office after having lost the organisational backbone which mobilised mass electoral support for the party. While repression of opponents and state-controlled propaganda were core components of the strategy to maintain office, the PAP understood that only through real and substantive social and economic improvements for the working class could it survive electorally in the medium to long term. This rendered the success of the ISI strategy ever more critical.

Nevertheless, Singapore's membership in the Federation was shortlived. Political differences and mistrust between the governments in Singapore and Kuala Lumpur escalated between 1963 and 1965, culminating in the city state's abrupt separation from the Federation. With this, the vision of industrial expansion through access to a common market seemed an unlikely possibility. Apart from the troubling fact that unemployment was nearly 9 per cent in 1965, the collapse of merger also presented a political opportunity for the PAP's domestic opponents who had opposed merger. To make matters worse, in 1967 the UK Government announced its intention to withdraw its military bases from the island. In that year, spending from the bases accounted for 12 per cent of Singapore's total GNP. The spectre of economic crisis loomed large in policy-makers' minds.

In the re-evaluation of strategy, the PAP's relative political autonomy from both capital and labour was to prove especially important. Having remained committed to private sector-led industrialisation, Singapore's policy-makers looked to the experiences of Hong Kong and Taiwan which had both, by the mid-1960s, successfully embarked on export oriented industrialisation (EOI) programs. Such a direction would necessitate close attention to labour costs. It would also require the attraction of international capital since the domestic bourgeoisie lacked capital and expertise in sufficient quantities to give serious effect to such a program. Despite some expansion in the 1960s, local industrialists lacked the political or economic clout to frustrate this policy direction.

Similarly, the government set about not only further blunting independent labour, but also marshalling the PAP-affiliated National Trades Union Congress (NTUC) in support of its social and economic policies. Against this background, the Employment Act (1968) and the Industrial Relations Act (1969) were introduced, thereby reducing wages and eroding conditions, increasing working hours and severely curtailing union bargaining powers and the capacity for industrial action. The adoption in 1972 of the National Wages Council (NWC) further institutionalised government intervention in the labour market.

Additional measures to promote EOI included a range of specialised institutional initiatives. In this, the EDB assumed a greatly enhanced role in centralising and coordinating the government's investment drive. Among other things, the EDB oversaw a host of generous tax concessions targeting investors in EOI. However, in 1968 the Jurong Town Corporation (JTC) was established to relieve the EDB of responsibility for the rapidly expanding industrial estates which provided centralised infrastructural facilities at low cost. During the same year, the Development Bank of Singapore (DBS) was created to provide below market-rate finance and engage in equity participation to stimulate industrial ventures. In the following year, the government established Neptune Orient Lines (NOL) to expedite foreign trade and ensure lower freight charges for Singapore-manufactured goods. Through the compulsory government-controlled superannuation scheme, the central Provident Fund (CPF), and the Post Office Savings Bank, the government also appropriated a considerable portion of domestic savings. Much of this was channelled into physical and social infrastructure. Direct productive investment was also undertaken by the government as a means of boosting manufacturing (Lee 1978: 138–9).

In a concerted fashion, then, the government set about trying to generate the social, political and economic preconditions for industrial investment in export production. This involved it in conscious efforts to influence the costs of the different factors of production. In effect, the government was helping to shape Singapore's comparative advantage in the production of labour-intensive manufactures.

The EOI strategy proved a spectacular success. Foreign investment in manufacturing rose from S$157 million in 1965 to S$995 million in 1970 and S$3054 million in 1974. In this time, direct manufactured exports from Singapore jumped from a value of S$349 million to S$1523 million and S$7812 million by 1974. The contribution of manufacturing to GDP increased from just over 15 per cent in 1965 to 22.6 per cent in 1974 In this economic transformation assembly work in the electrical and electronics industry was especially important, with international capital adopting Singapore wholeheartedly as an export base for USA and European consumer markets. Other low value-added, labour-intensive industries such as shipbuilding and the textiles, clothing, footwear and leather group of industries underwent significant expansion.

Consequently, Singapore's unemployment rate was brought down to below 4 per cent by 1974. Indeed, by the early 1970s some areas of industry were experiencing labour shortages which soon led to a reliance on imported workers. By 1973 over 100 000 of Singapore's total workforce of 817 400 was comprised of immigrant labour, mainly from Malaysia. In this context, consideration was given to an accelerated policy push into middle-level technologies involving higher value-added processes. However, the effects of the global recession hit Singapore hard in 1974–75, resulting in sizeable investment cutbacks and job losses. This quickly returned the policy priority to employment generation and wage control through the

NWC. Between 1975 and 1978, though, Singapore's economy staged a spirited recovery under the aegis of international capital. Cumulative foreign investment climbed to S$5242 million by 1978 and industrial exports jumped to a value of S$13 633 million in the same year (see table 6.1). In fact, labour shortages now posed a more serious problem. For social and political reasons, the government was at the time disinclined to resolve the problem of labour shortages by simply accelerating the importation of guest workers. Instead, it took the view that a rationalisation of the manufacturing sector was in order: scarce labour should not be hoarded by low value-added, labour intensive industries in which Singapore would be unable to maintain international competitiveness over the longer term. Since some moves were already afoot to introduce more sophisticated technologies to Singapore, the government believed the time was right to accelerate this process and set policy sights on more ambitious economic goals.

Table 6.1 Selected economic indicators, Singapore, 1960–84

Year	GDP growth 1985 market prices	Direct manufacturing exports[a] (S$ million)	Manufacturing value-added per worker[a] (S$)	Cumulative foreign investment in manufacturing[b] (S$ million)	Cumulative local investment in manufacturing[b] (S$ million)	Unemployment rate (%)
1960						
1961	8.5	824	6 100			
1962	7.1	1 158	6 924			
1963	10.5	828	6 900			
1964	−4.3	794	6 845			
1965	6.6	858	7 207	157		8.7
1966	10.6	841	7 732	239		8.7
1967	13.0	912	8 192	303		8.1
1968	14.3	1 014	8 146	454		7.3
1969	13.4	1 971	8 600	600		6.7
1970	13.4	2 044	9 029	995		6.0
1971	12.5	2 362	9 705	1 575		4.8
1972	13.3	2 911	10 388	2 283		4.7
1973	11.3	4 779	12 856	2 659		4.5
1974	6.8	8 520	17 127	3 054		3.9
1975	4.0	7 610	17 763	3 380		4.6
1976	7.2	10 160	19 168	3 739		4.5
1977	7.8	11 412	20 417	4 145		3.9
1978	8.6	13 087	21 179	5 242		3.6
1979	9.3	16 904	23 992	6 349		3.4
1980	9.7	19 875	30 027	7 090	3 471	3.5
1981	9.6	22 894	34 681	8 382	4 060	3.9
1982	6.9	22 227	34 218	9 618	4 910	2.6
1983	8.2	22 922	36 645	10 777	5 646	3.2
1984	8.3	25 058	40 476	12 651	6 798	2.7

Notes: a Data for 1960-83 includes rubber processing and granite quarrying.
         b Figure excludes rubber processing.

Sources: Department of Statistics, Singapore, *Yearbook of Statistics Singapore*, various years; Economic Development Board, Singapore, *Annual Report/Yearbook*, various years; Department of Statistics (1983); Huff (1994).

What transpired was a two-pronged strategy, dubbed the 'Second Industrial Revolution', to both increase technological sophistication and to further raise the contribution of manufacturing to Singapore's economic growth. On the one hand, the strategy involved measures to discourage unskilled, labour-intensive production, but on the other it cushioned the costs for employers attempting to move towards more skilled, higher value-added production. The chief element of the former was the so-called 'corrective wage policy' of the NWC. By the end of 1981, three years of the policy had resulted in wage cost increases to employers of over 50 per cent. Quite intentionally, this hit the most labour-intensive, unskilled operations the hardest. If they were to remain in Singapore, it was expected that technological upgrading would lessen the dependence on labour. The other option was to transfer production outside Singapore to a lower cost site. Meanwhile, tax incentives, below market-rate finance and subsidised training schemes were made available to investors gearing towards higher value-added production. Considerable sums were invested by the government in developing the physical and social infrastructure to support a technological transformation.

The 'Second Industrial Revolution' also witnessed refinements in the role of direct government investment. By 1983 the Singapore government had extensive investments involving fifty-eight companies and S$3 billion paid up capital (Hon 1983). However, they were now used more strategically to promote higher value-added production. In 1983 Singapore Technologies Corporation (STC) was created with a capital base of S$200 million pooled to promote advanced technologies. The government was also a substantial investor in S$2 billion-worth of joint ventures putting together the first fully integrated petrochemical complex in South-East Asia (see Rodan 1989: 153). A further initiative was the establishment of the Government of Singapore Investment Corporation (GIC) which invested some of Singapore's extensive foreign reserves overseas in leading high technology multinational corporations (Rodan 1989: 153–4).

The state's efforts to foster higher valued-added production extended to the social and political realms. In particular, institutional control over organised labour was modified. The country's two largest trade unions, the Singapore Industrial Labour Organisation (SILO) and the Pioneer Industries Employees' Union (PIEU), were broken up into nine industry-based unions. Not only was this new structure more functional for the technical aspects of the new industrial program, it also reduced any potential capacity of organised labour, however constrained by its close relationship with the ruling party, to frustrate the restructuring process. Retrenchments and retraining demands would not necessarily be greeted by union members. The 'Second

Industrial Revolution' was thus a comprehensive and integrated plan of economic, social and political interventions.

Beyond Export Oriented Industrialisation

During the period of the 'Second Industrial Revolution' some significant gains occurred. Foreign investment rose substantially from S$6349 million in 1979 to S$12 651 million in 1984. For the same period, valued-added per worker nearly doubled, up from S$23 992 to S$40 476, attesting to the enhanced productivity of the workforce due to capital investment. Industries like electronics, machinery, chemical and aerospace underwent major technological upgradings and Singapore found some significant production niches, becoming a global centre for the computer disk drive industry, for example (see tables 6.2 and 6.3).

Yet, by the mid-1980s it became apparent that there were structural limits to the expansion of the manufacturing sector in Singapore. From the start of 1980 to the end of 1984 the rate of manufacturing growth was 6.1 per cent compared with Singapore's overall economic growth rate of 8.5 per cent (Ministry of Trade and Industry 1986: 26). Consequently, the relative contribution of manufacturing to GDP dropped from 23.7 per cent in 1979 to 20.6 per cent in 1984 (Department of Statistics, *Yearbook of Statistics Singapore 1984/85*: 78). The opportunities for Singapore as an export manufacturing base were not as extensive in middle level technology as they had been in the earlier phase of low value-added production. The EOI strategy was confronting a paradox: the further up the technological hierarchy Singapore graduated on the basis of lower labour costs—albeit in more sophisticated technologies and in competition with higher-wage countries—the less a proportion labour became to overall production costs. The global recession compounded this problem by generally restricting investment and by exposing the vulnerability of Singapore's heavy dependence on the USA consumer market, particularly in the all-important electronics industry. The economy actually contracted by 2 per cent in 1985, the first experience of negative growth since independence. Manufacturing declined by 6.9 per cent and 3.4 per cent respectively during 1985 and 1986 (Department of Statistics, *Yearbook of Statistics Singapore 1990*: 3). Even with the repatriation of some 60 000 guest workers (Pang 1994: 81), unemployment rose to 6.5 per cent by 1986 (Department of Statistics, *Yearbook of Statistics Singapore 1994*: 16).

Table 6.2 Selected economic indicators, Singapore, 1985–95

Year	GDP growth 1985 market prices	Direct manufacturing exports (S$ million)	Manufacturing value-added per worker (S$)	Cumulative foreign investment in manufacturing[a] (S$ million)	Cumulative local investment in manufacturing[a] (S$ million)	Unemployment rate (%)
1985	−1.6	24 276	42 216	13 160	7 100	4.1
1986	1.9	24 387	48 240	14 120	6 804	6.5
1987	9.4	30 380	52 372	15 830	6 827	4.7
1988	11.3	37 806	55 182	18 131	7 449	3.3
1989	9.4	42 388	58 285	21 524	7 957	2.2
1990	8.8	47 000	61 440	24 133	8 682	2.0
1991	6.7	45 913	65 452	25 831	9 839	1.9
1992	6.0	46 907	69 508	28 565	10 604	2.7
1993	10.1	53 022	79 643			2.7
1994	10.1	62 261	87 743			2.6
1995						2.7

Note: [a] Figure excludes rubber processing.

Sources: Department of Statistics, Singapore, Yearbook of Statistics Singapore, various years; Economic Development Board, Singapore, Annual Report/Yearbook, various years; Department of Statistics, Singapore (1983); Huff (1994).

In response to the mid-1980s economic recession there was a major revision of economic strategy and objectives, enunciated in the report of the Economic Committee (Ministry of Trade and Industry 1986) headed by the then Acting Trade and Industry Minister, Lee Hsien Loong. The immediate priority was to resurrect Singapore's cost competitiveness in manufacturing *vis-à-vis* other NICs and generate employment. Thus, some of the pressures on manufacturers introduced under the 'Second Industrial Revolution', particularly wage costs, were relaxed. With the longer term in mind, though, a shift of emphasis towards the services sector was prescribed. Transport and communications had already increased its share of GDP from 12 per cent in 1980 to 17 per cent in 1984, and financial and business services rose from 20.5 per cent to 24 per cent in the same period (Department of Statistics, *Yearbook of Statistics Singapore 1990*: 87). Singapore's rapid development as a financial centre since the late 1970s had been aided by a host of provisions and incentives by the Singapore government (see Rumbaugh 1995: 39–40). The essence of the new vision was for Singapore to become 'a total business centre'. Accordingly, new tax and other incentives were introduced for companies extending their commercial and manufacturing roles to include financial, marketing, technical and other corporate services to their networks of related companies in the region or worldwide—that is, companies making Singapore their operational headquarters (OHQ).

Table 6.3 Sectoral contributions to Singapore's GDP (%) 1960–94

Year	1960	1970	1980	1985	1990	1994
Total GDP (S$ million, 1985 market prices)	5 058	12 172	28 832	38 924	57 471	78 765
Commerce	26.0	22.1	18.4	17.8	16.5	16.6
Construction	5.6	9.5	6.9	11.4	5.0	6.7
Financial and business services	14.8	17.0	19.9	20.9	26.1	25.0
Manufacturing	17.6	24.9	28.6	22.2	27.2	26.6
Transport and communications	9.3	7.3	11.6	13.7	13.3	13.7
Other services	26.7	19.2	14.6	14.0	11.9	11.4

Sources: Department of Statistics, Singapore, *Yearbook of Statistics Singapore*, various years; Economic Development Board, Singapore, *Annual Report/Yearbook*, various years; Huff (1994).

Compared to previous policy redirections, the views of the domestic bourgeoisie appeared on this occasion to be given more weight. The Economic Committee not only comprised representatives of the Singapore Manufacturers' Association, the Singapore Federation of Chambers of Commerce and Industry, and leading local entrepreneurs, but it also incorporated other local business people through the many sub-committees that were part of the process. The rhetoric in the Economic Committee's report championed entrepreneurship, decried the government's past interventionist economic policies and recommended a program of privatisation.

Fundamentally, though, the new strategy was informed by a reading of emerging trends in international patterns by transnational corporations (TNCs) in the Asia-Pacific region rather than any major re-evaluation of the domestic bourgeoisie's role. This trend included an increasing preparedness by the mid-1980s for TNCs to locate a more comprehensive range of higher value-added processes and services in discrete geographic regions—North America, Europe and Asia—as opposed to adopting a single, hierarchical global division of labour. Known as 'regional focus', this new corporate strategy acknowledged the importance of existing and projected markets within the Asia-Pacific region, including for manufactured products. The Asia-Pacific region was no longer regarded as simply a low-cost production site for export production to the markets of Europe and North America (see Ng and Sudo 1991). Importantly, higher levels of manufacturing investment for regional markets required high value-added services such as accounting, law, training and management services and this opened up a potential niche for Singapore. For the domestic bourgeoisie, this meant both increased competition and opportunity.

Singapore's economic recovery was swift and strong, with GDP growing by an annual average of 9.5 per cent from 1987 to 1990. Although this included the impressive revival of the manufacturing sector, it did not tempt policy-makers to shelve plans for a structural transformation of the economy. The general directions outlined by the Economic Planning

Committee were confirmed and refined in *The Strategic Economic Plan* of 1991. This committee was set up in 1989 and headed by the Minister for Communications, Mah Boh Tan, comprising senior public sector bureaucrats and, again, prominent people from the private sector. It underlined the importance of continued heavy investment in social and physical infrastructure to position the city state as a provider of high value-added services for the region.

The Strategic Economic Plan made explicit reference to the work of American Professor Michael Porter, author of *The Competitive Advantage of Nations* (1990). In actual fact, Singaporean policy-makers had long practised what Porter theorised about the development or creation of competitive advantages. However, they adopted Porter's concept of 'clusters'—the notion that, in industry and business services, vertical and horizontal links to suppliers and customers have a major bearing on overall competitiveness. The EDB took this up with particular fervour in its industrial strategy *Manufacturing 2000*, which aimed at maintaining the sector's contribution to GDP at no less than 25 per cent and employment share at more than 20 per cent in the medium to long run. Moreover, in 1995 the government introduced a S$1 billion Cluster Development Fund for strategic investments (inside and outside Singapore).

Apart from the instances sited above of new forms of consultation with the local private sector, in 1986 the Association of Small to Medium Enterprises was established. In response to this new level of political organisation and activism on the part of small and medium enterprises (SMEs), in 1989 the government committed itself to a SME Master Plan to foster local enterprise, with a complementary boost in the number of schemes assisting the development of local entrepreneurs. However, it would be an exaggeration to interpret this as the emergence of the domestic bourgeoisie as a significant political force. Instead, the post-recession period has been characterised by a general increase in the level and institutionalisation of interaction between capital and state in the policy process. Thus, in 1994, the EDB appointed an International Advisory Council (IAC) comprising chairmen and chief executive officers of the world's leading TNCs. The EDB expects this forum to strengthen its capacity to ascertain the thinking of international capital and its plans and aspirations for the region. Similarly, in 1994 the Regional Business Forum (RBF) was established, bringing together leading local private and public sector figures to coordinate the regionalisation drive. The RBF's first report recommended the formation of private sector business clubs, some of which have already surfaced, such as the Shandong Business Club, the China Business Group and the Myanmar Business Group, as a means of facilitating regional trade and investment.

To a significant extent, wider state consultation with capital bears a functional relationship to the more sophisticated and diversified economy emerging in Singapore and the region. However, private sector participation in the policy process is still largely at the behest of the state. Indeed, through statutory boards, of which there are about seventy, and state-owned companies in manufacturing and commercial enterprises, the Singapore state exerts an enormous influence over the domestic economy. By 1990, through the three holding companies Temasek, Singapore Technology and Health Corporation Holdings, the Singapore state was the sole shareholder of fifty companies with interests in a further 566 subsidiaries. The total assets of these enterprises amounted to S$10.6 billion (see *STWOE* 29 August 1992: 20; Soon and Tan 1993: 23; Vennewald 1994: 27–8).[2] The coordination and control of economic resources is greatly enhanced by the tight interlocking directorships involving a small coterie of civil servants (see Vennewald 1994; Soon and Tan 1993: 28; Chia 1992: 11, 19). This virtual 'class' of public entrepreneurs is closely connected to the ruling political party, the upper echelons of the civil service having been the main recruiting ground for the PAP for some time (see Khong 1993: 12), complemented lately by the armed services (see Huxley 1993). The extensive economic and political reach of the Singapore state necessarily conditions the activities of the locally based private sector. Dependence on the state for contracts and awareness of the political nexus between the civil service and the PAP has promoted cooption rather than forceful interest representation to the government.

As we will see below, the new emphasis on regional economic integration has seen state capital play very much the lead role, with offshore ventures by established local private capital groups often working in cooperation with it. The following account puts into perspective measures adopted since the mid-1980s that might otherwise be interpreted as evidence of a lessening of the economic role of the state. The government has divested its shareholdings in fifty-eight companies (twenty-nine in full and twenty-nine in part) since 1985. Of the statutory boards, only Telecom has been privatised. But this has not generally involved any surrender of control over the economy, nor the likelihood of such. Indeed, as Mukul Asher (1994: 801) argues:

> A strong case can be made that in the Singapore context, its privatisation programme will make the role of government even stronger and more extensive. This is because there is an overall budget surplus, so the divestment proceeds are not intended to either reduce taxes or expand expenditure. Instead, these can be invested at home and abroad.[3]

Regionalisation and Singapore Inc.

One of the Singapore government's most imaginative and symbolic policy initiatives in support of greater integration with regional economies has been the 'Growth Triangle'. First outlined by Goh Chok Tong as Deputy Prime Minister in 1989, the idea of the Growth Triangle was for the Malaysian state of Johor and the nearby Riau Islands of Indonesia to combine with Singapore as a coherent, trans-state economic zone of complementary specialisations. Since the Growth Triangle is the subject of extensive coverage and analysis by James Parsonage in chapter 10 of this volume, only selective observations are necessary here. In particular, the strategy giving substance to the idea of a trans-state Growth Triangle has, paradoxically, involved Singapore government-owned companies in enormous direct investments to establish the necessary industrial infrastructure in Indonesia's Batam and Bintan Islands as well as in the development of heavy industry and tourist resorts in other parts of the Riau Islands (see Rodan 1993a: 239–40). The central role by Singapore's authorities— especially from the EDB—in establishing administrative systems and coordinating projects offshore has been of no less significance. Singapore's reputation for efficient and corruption-free investment processing considerably bolstered business confidence in these projects. The strategy has also involved a range of Singapore government measures, such as tax incentives and training subsidies, to hasten offshore investment in Johor and the Riau Islands (see Rodan 1993a: 240).

Following the Growth Triangle initiative, the Singapore Government was to even more comprehensively promote the city state's regional economic integration by way of offshore investment. In early 1993 a report was handed down by the Committee to Promote Enterprise Overseas, chaired by Teo Chee Hean but again comprising membership drawn from leading local businesses (see Ministry of Finance 1993: 44–8). The government subsequently introduced various incentives to support a bigger regional investment and trade push.[4] This concerted attempt to more fully integrate Singapore-based enterprises with regional economies is an attempt to transcend the limitations of a city state economy and is variously referred to by authorities as the building of an 'external economy', a 'second wing' or 'expanding the economic space of Singapore' (see Mahizhnan 1994; Goh in Kraar 1996: 172). In the development of this 'external economy', the government sees itself playing a pivotal role, freely describing this in terms of a 'Singapore Inc.' According to the EDB (*EDB Yearbook 1994*: 3): 'The Singapore Inc. approach will also be adopted with the private and the public sector forming consortia to undertake regional investment projects on a large scale'.

This has indeed been prosecuted with considerable vigour and resources. As with the Growth Triangle, the Singapore government and affiliated companies have given special emphasis to infrastructural projects throughout Asia that are modelled on Singapore's earlier domestic experience. Huge investments have gone into what one author has described as a series of 'mini-Singapores' (Kraar 1996: 172), involving integrated projects of factories, roads and power plants, established as industrial enclaves within China, India, Indonesia, Burma and Vietnam.[5]

The most ambitious of these projects is the S$28 billion Suzhou Township which is a fifteen to twenty year development involving a twenty-two-member consortium, led by the Singapore government-linked company's (GLC's) Keppel Group, in partnership with the Suzhou municipal council. Some 70 square kilometres will be developed into a township that includes not only industrial parks but also residential estates and commercial facilities (*STWE* 14 August 1993: 24; 3 February 1996: 19).[6] Moreover, in an unprecedented move, Chinese authorities have granted the Singaporeans managerial autonomy in the project. Thus, not only does the project involve the transfer of an extensive range of expertise and capital, it brings with it the networks and, most importantly, the reputation for administrative efficiency and transparent procedures attractive to international investors. Indeed, as Kwok Kian-Woon (1995: 294) explains, this amounts to the export of a 'social technology' involving a 'rational developmentalist mentality' of economic planning that has been institutionalised in Singapore over the last three decades.[7] By early 1996, fifty-nine international companies had committed investments in Suzhou industrial park totalling US$1.8 billion (*FEER* 25 April 1996: 61).

The new emphasis on international accumulation strategies has not only included GLCs such as Singapore Technologies, Keppel Corporation, TDB Holdings, Singapore Telecom, Singapore Power, the Development Bank of Singapore (DBS), and the Sembawang Group, but also the GIC which has invested Singapore's vast foreign reserves—officially and conservatively reported at over S$100 billion in 1996—in as many as twenty-five countries. Private Singapore-based companies have benefited, either as joint-venture partners or independent entities, in the consortia and other projects in which Singapore GLCs have played a leading role, including tourist and property developments, notably in the Growth Triangle and Burma (see Rodan 1993a: 242). However, for the bulk of Singapore's 80 000-plus SMEs in manufacturing, commerce and services, overseas investment is either not appropriate or beyond their realistic capacity in the foreseeable future. In addition to the link-ups with elements of Singapore's private sector, offshore projects have also fostered connections between

Singapore's GLCs and companies connected with major politico-business families in the region such as Indonesia's Salim group.[8]

Apart from direct investments, the Singapore government has engaged in a string of diplomatic initiatives to foster stronger regional economic ties, including the initiation of institutional mechanisms involving government-to-government committees such as the China-Singapore Joint Steering Council, bilateral business councils and memoranda of understanding. Prime Minister Goh, Senior Minister Lee and other senior members of the government have played high profile roles in many of these exercises. In the case of Burma, the Singapore Government's close relations with the military dictatorship of the State Law and Order Restoration Council (SLORC) have been important in opening up new markets for Singapore-based companies (see DeVoss 1995). Singapore companies have been the largest investors in hotel projects in Burma, and in 1995 the second largest investor overall (*Singapore Bulletin*, December 1995: 7–8). When we add to this the Singapore government's budget for the period 1996–2000 of S$4.3 billion for loans and grants to encourage innovation, manpower development and increased trade and investment links with the region, clearly the state is no less involved in shaping economic directions than at any previous stage in post-Independence development.

Table 6.4 Singapore-based overseas investment, 1990–93 (S$ million)

	1990	1991	1992	1993
Total direct investment	16 878	18 608	22 442	28 160
Direct equity investment	13 622	15 184	17 741	21 240

Source: Department of Statistics, Singapore, *Singapore's Investment Abroad 1990–1993*, 1994.

Data in table 6.4 confirm that the 1990s has brought an accelerated investment drive from Singapore into the region. Between 1976 and 1989 total direct investment abroad from Singapore rose from S$1.5 billion to S$8.7 billion. Yet it jumped to S$16.9 billion in 1990 and reached S$28.2 billion in 1993. Significantly, while this reflected foreign-owned TNCs increasingly using the city state as an investment base for regional activities, Singapore-owned companies were equally active. In 1993, for example, 50 per cent of the total direct investment abroad involved either wholly or majority locally owned companies. The stock of total direct equity investment in overseas subsidiaries and associated branches also increased from S$13.6 billion to S$21.2 billion between 1990 and 1993. The importance of this investment surge is underlined by the estimate of government economists that overseas operations accounted for 12 per cent of Singapore's gross national product in 1995 (*FEER* 25 April 1996: 59).

Table 6.5 Singapore's direct equity investment abroad by major regions, 1990–93

	1990		1991		1992		1993	
	S$ M	%	S$ M	%	S$ M	%	S$ M	%
Asia	7 013	51.5	7 401	48.7	9 209	51.9	11 480	54.0
ASEAN	3 567	26.2	3 996	26.3	4 897	27.6	5 934	27.9
Europe	1 095	8.0	1 398	9.2	1 480	8.3	1 550	7.3
North America	716	5.3	1 323	8.7	1 621	9.1	1 867	8.8
Oceania	1 994	14.6	2 068	13.6	2 102	11.8	2 036	9.6
Others	2 802	20.6	2 994	19.7	3 328	18.8	4 308	20.3
Total	13 622	100.0	15 184	100.0	17 741	100.0	21 240	100.0

Source: Department of Statistics, Singapore, *Singapore's Investment Abroad 1990–1993*, 1994.

When we look at the destination of this surge in overseas investment we see a strong bias towards the region (see table 6.5). Between 1990 and the end of 1993, around half of all direct equity invested abroad went to Asia. Data from host governments indicate that, in 1995, Singapore-based investments were the second-largest not only in Burma but also Thailand, and ranked fourth in Vietnam, fifth in China, sixth in Indonesia and eighth in India (*FEER* 25 April 1996: 59).

Table 6.6 Singapore's direct equity investment abroad by sector (S$'000)

	1990	1991	1992	1993
Manufacturing	2 394 955	2 901 110	3 760 053	4 612 735
Construction	69 492	72 301	130 176	199 342
Commerce	1 504 264	1 606 158	1 967 068	2 067 820
Transport	347 237	244 158	289 768	322 035
Financial	7 301 178	8 648 014	9 752 694	11 723 087
Real estate	1 213 058	859 328	962 854	1 294 770
Business services	511 649	605 132	595 253	766 883
Others	279 848	246 882	283 415	253 569
Total	13 621 681	15 183 835	17 741 280	21 240 242

Source: Department of Statistics, Singapore, *Singapore's Investment Abroad 1990–1993*, 1994.

Investment in the finance industry has been the major overseas activity, accounting for S$8.9 billion of all direct equity investment abroad in 1993, followed by manufacturing with S$6.8 billion (see table 6.6). In 1991, 46 per cent of local and 60 per cent of foreign-controlled direct investment abroad was in the financial sector (Department of Statistics 1993: 7). Nearly 60 per cent of overseas investment in finance was in Asia in 1993, while the portion for manufacturing was 44 per cent (Department of Statistics 1994: 26).

The experience in the 1990s thus far seems to suggest a significant increase in Singapore's economic integration with the region and a concomitant diversification of the city state's economic base. More than sixty major TNCs across consumer electronics, construction, transportation, pharmacueticals and other industries have made Singapore their Asia-Pacific headquarters since 1986. Lately this has extended to the use of

Singapore as a regional broadcasting centre. Six international television networks have come to Singapore, among them the USA entertainment and video loan giant Home Box Office, the music channel MTV and USA sports network ESPN. Most recently, the multi-media organisation Walt Disney has joined the list. Various international publishers have also set up operations in Singapore for the purposes of servicing regional demand.

In spite of this progress, questions have been raised by some economists about the extent of structural change taking place. Paul Krugman (1994: 70) contends that Singapore's exceptional economic growth owes most to 'an awesome investment in physical capital' and that increased efficiency in the application of capital and labour, referred to as total factor productivity (TFP), still eludes Singapore. This argument is to some extent premised on a unilinear conception of capitalist development which has its limits for understanding economic trajectories in Asia. It also downplays the ongoing economic importance of the state. On empirical grounds, Krugman's gloomy prognosis has been brought into question too.[9] However, there is some basis to Krugman's thesis and in attempting to raise TFP the Singapore authorities do face some difficulties. Not the least of these is that some 320 000 workers over the age of forty years have less than secondary education, which was mostly through the Chinese medium, and cannot be easily retrained (*ST* 2 May 1996: 1). Singapore's comparatively low percentage of the population in tertiary education will also need to be addressed if a critical mass of researchers is to be generated to support technological innovation.[10] Furthermore, attempts to raise the level of expenditure on research and development (R&D) have been only partially successful, reaching just 1.12 per cent of GDP in 1993 (*STWE* 29 July 1995: 5).

The challenge to effect a more complete structural transformation of the economy is as much a social and political one as it is economic. Many Singaporeans are finding themselves in the wrong half of a dual labour market that separates, with increasing sharpness, the skilled and well-paid from the unskilled and low-paid. The diminished role of the NWC since 1987 and the growth in imported labour, now accounting for approximately one-fifth of the total workforce (Hu 1996), adds to this development. Apart from the moderating of wage pressures at the bottom end of the labour market through the surplus of unskilled guest labourers, there are growing numbers of imported managers, supervisors and professionals whose presence reduces the pressure for a comprehensive upgrading of the skills of the domestic labour force (see *STWE* 11 May 1996: 14). The relaxation since the Second Industrial Revolution of curbs on guest worker numbers has contributed to cost-curtailment and supported increased demand for production, especially in the construction and manufacturing sectors. At the same time, it has boosted government revenue via substantial levies on employers of this labour (see Asher 1994: 798).[11]

Social Divisions in Affluent Singapore

The divergent paths and fortunes of different elements of local business are one feature of the contemporary political economy of Singapore. Other features include the political consequences of an increasingly complex and diverse set of interests embodied in a sizeable middle class, and mounting concerns, especially among the working class and lower middle class, about material inequalities in Singapore.

With the diversification of the economy in the last decade or more, the growth of the middle class has accelerated and its internal composition become more varied. The combined share of the total workforce accounted for by professionals and technicians as well as administrators, executives and managers rose from 18 per cent in 1980 to 24 per cent in 1990 (Department of Statistics 1991: 22). Significantly, professional workers constitute a major component of the expansion, especially from within the finance and business service industry where they accounted for 43 per cent of all workers and were among the most highly paid of all professionals in Singapore in 1990 (Lewis 1993: 26–7).

As elsewhere, the development of the middle class in Singapore has been shaped by the social and economic roles of the state. In the Singapore case, however, the state's activities are heavily oriented towards profit-motivated enterprises, or the support and attraction of private investment through the work of statutory bodies like the EDB and JTC (see Soon and Tan 1993: 22). Comparatively few resources are devoted by the state to areas like welfare and social security which loom so large in the budgets and employment patterns in many advanced industrial economies.[12] Consequently, employment demand is heavily skewed towards professionals with the skills most pertinent to the functioning, administration and accounting of the economy. This, together with the material benefits and high social status the PAP's institutionalised ideology of meritocracy bestows on such state employees and their counterparts in the private sector, produces a set of mutual values and interests between the government and a substantial element of the middle class (Rodan 1996: 31–2).

Broadly, then, the middle class which has emerged over the last few decades has not been one that has any special need or desire to challenge the ruling party's exercise of power. This does not mean that there are not pressures to influence the policy process, as increased social differentiation generates new, discrete interests. But it does mean that the potential for this to be satisfied through extended forms of political cooption is real. The government is thus attempting to incorporate into PAP-controlled institutions a range of different interests including business, professional, ethnic and women's groups.[13] To date, cases of middle-class activism involving the establishment of genuinely independent organisations have been limited (Rodan 1993b, 1996b; Brown and Jones 1995).

Rather than middle-class pressure for a liberal civil society, it is concern about the distribution of material rewards associated with Singapore's economic development that has become the major issue in the 1990s. This first became apparent with the decline in the government's support at the 1991 election from the PAP's traditional social base—the ethnic Chinese working class (Rodan 1993b; Singh 1992). Following the election, the government established the parliamentary Cost Review Committee (Ministry of Trade and Industry 1993) to address public concerns about rising costs of living and inequalities.

Drawing on census data, the committee maintained that the bottom 20 per cent and the middle 60 per cent of households experienced higher rates of income growth than the top 20 per cent in the last decade.[14] However, the claim that income inequality is declining in Singapore is not consistent with the work of academics, who either contend that the statistical evidence is too weak to claim a narrowing of the gap between rich and poor (Rao 1990, 1996a, 1996b), or who actually argue that the disparities are widening (Islam and Kirkpatrick 1986; Krongkaew 1994).[15] In any case, income increases have to be related to changes in costs to fully appreciate any relative shifts in purchasing power. On this point, the Cost Review Committee readily conceded that there had been steep increases in the costs of health care, housing, education and transport. It attributed some of this to Singapore's natural resource constraints which pushed up the cost of land and put a premium on private vehicles. Accordingly, the Committee called on young Singaporeans in particular to modify their expectations.

Despite government attempts through the Cost Review Committee and other officially presented data to reassure Singaporeans that material inequalities are not opening up in contemporary Singapore,[16] this appears to have had little impact on popular perceptions. Furthermore, Prime Minister Goh has effectively conceded that significantly rising absolute inequalities, if not relative inequalities, are engendering a sense of deprivation among lower-income earners: 'They will look at the very rich and find them getting richer while they remain poor. While their income goes up, it will not go up by as much as the incomes of managers, professionals and businessmen' (Goh quoted in *ST* 2 May 1996: 1). Data selected from the 1990 census by the Ministry of Finance show that 20 per cent of households had incomes of less than S$700 per month, a level which involves struggle if not near-poverty for many such people, and a further 50 per cent of households have a monthly income of below S$2100. By contrast, the top 10 per cent of households were receiving a monthly income of S$9700 and above (*ST* 28 August 199: 22). Such absolute variances in income distribution provide a fertile basis for public discord about social divisions. At times, the government has shown a distinct political insensitivity to this. Thus, there was a public outcry in late 1994 and early 1995 in reaction to increases in ministerial salaries, which were already among the highest in the world. The Singapore Prime Minister's annual salary of S$1.1 million (US$812 858), for example, is more than threefold that of the President of the USA (*STWE* 10 February 1996: 24).

However, there have also been a number of public policy measures from the government since the early 1990s to address public anxiety. Amid depictions of the government by opposition parties as mean and lacking compassion for those least benefiting from Singapore's rapid economic development, the PAP has even begun projecting itself as generous in public provisions. The attempt to reconcile this with economic rationality, efficiency and anti-welfare rhetoric has produced an interesting discourse. The fundamental difficulty for the government, however, lies in resolving the contradiction between its deep ideological commitment to elitism on the one hand, and its concern on the other to improve the conditions of the underprivileged. To date, redistribution in favour of the underprivileged has been limited, although there have been quite substantial general increases in government expenditure on social development.

In late 1994 Prime Minister Goh claimed that nine out of ten Singaporeans receive more benefits and subsidies than the taxes they pay, with 70 per cent enjoying a net gain of nearly S$20 000 a year. Included in this figure were subsidies for housing, health care and education services as well as gains from the government's 'asset enhancement programmes' (*STWE* 3 September 1994: 1; *Sunday Times* [Singapore] 28 August 1994: 22). Asset enhancement programs involve transfers of funds from the government via schemes to assist upgrading housing, discounted shares in Telecom (estimated to have involved some S$3 billion), and Edusave— money paid into dedicated accounts for the purposes of education expenditure. These programs are tied to specific expenditure purposes and are generally non-discriminatory—subsidies not subjected to any means testing. The opposition Singapore Democratic Party (SDP) has challenged government claims of huge subsidies, arguing that in the case of housing, for example, revenue collected from levies on the foreign workers who dominate the construction industry and build the flats, car park fees and conservancy charges more than cancel out any subsidies provided by government (*New Democrat*, Issue 1, 1996: 4–5).

Against the backdrop of such opposition arguments, the Prime Minister announced in March 1996 that for the previous year a total of S$1.65 billion of surplus government funds was disbursed through various schemes. This included S$800 million for the Share Ownership Top-Up and Central Provident Fund Top-Up schemes; S$770 million for students, the aged and Housing Development Board (HDB) flat owners, shopkeepers and stallholders; and S$80 million for welfare, charitable and community self-help groups to assist the poor. However, the Prime Minister insisted that these benefits were not to be seen as welfare entitlements, which is why co-payment was usually involved (*STWE* 16 March 1996: 2). Goh has in the past argued that 'the disadvantaged do not expect and cannot demand that they be looked after by the State as a matter of right' (*STWE* 18 September 1993: 1). The PAP's concern is clearly to preserve a paternalistic political relationship more than to avoid public spending. In this relationship, Singaporeans are effectively 'rewarded' by a benevolent government, they are not citizens with intrinsic rights the government is obliged to honour.

Housing is an especially important commodity in Singapore that has become the centre of much public concern about rising costs and social divisions. Escalating prices have not only affected lower income earners, they have also jeopardised the prospects of many acutely status-conscious higher-income earners entering the private property market. The public housing program of the PAP government has been one of the major social and political achievements of the government since the 1960s. Around 90 per cent of Singaporeans now live in public flats and the city state's home ownership rate is among the highest in the world.[17] Yet in contemporary Singapore housing threatens as a source of major disaffection with the government.

The government has responded to public concerns in a variety of ways. It has announced a major upgrading program for older public housing which will cost between S$12 and S$15 billion over the next fifteen to twenty years. A new scheme will also cost the government S$60 million per year to assist low-income people renting HDB flats to buy them. First home buyers are also now eligible for housing grants of up to S$50 000 on the purchase of pre-owned flats. Programs to physically differentiate public housing to a much greater degree and thereby accommodate an increasingly diverse market of consumers are also being introduced, including a new class of housing by the HDB—executive condominiums. While they are expensive, they are expected to be available at between 15–20 per cent below the cost of those in the private market. Grants of S$40 000 are also available to Singaporeans purchasing them.

Prices in the private residential property market in Singapore have more than doubled between 1991 and 1996 (*ST* 27 April 1996: 32). Despite the release by the government of more land for private development, the trend continues. The average price for one 1996 development was S$1200 per square foot (*STWE* 20 April 1996: 13). The considerable attention in the local press to such details must do little for the morale of Singaporeans who are either priced out of the private market or who could never contemplate entering it. Disgruntled Singaporeans of the former category have begun to direct attention to the level of purchases of prime properties by foreigners, an issue enthusiastically taken up in the daily press and prompting the government to introduce new tax and other measures to curb speculation (*STWE* 20 April 1996: 13; *ST* 27 April 1996: 32; *ST* 3 May 1996: 1; *ST* 15 May 1996: 24–5).[18] Yet, the promotion of Singapore as a regional business headquarters necessarily involves increased expatriate executive populations and imported skilled workers, raising competition for limited resources (see Pang 1994: 83).

Conclusion

The economic development of Singapore since the 1960s has involved a considerable guiding hand by the state. As the city state's policy-makers promote economic diversification and regional economic integration following the expansive period of industrialisation, the emphases and forms of the state's economic role are undergoing some change. This includes greater cooperation with the domestic-based bourgeoisie in the development of international capital accumulation strategies and more institutionalised cooperation with the private sector in general. This does not signify a major transformation in the political balance in state-capital relations, but it does reflect growing disparities between different fractions of domestic-based capital. Those capable of embarking on, or consolidating, offshore investment can benefit considerably from state policies, initiatives and resources, and may even be consulted in the policy process. This should be understood, though, as a form of political cooption rather than the legitimation of interest-group politics involving business.

The price of economic success for Singapore's ruling party has been a more complex political economy characterised not just by greater social plurality but also raised material expectations. To consolidate customary levels of political control and effectiveness, this requires more sophisticated management by the PAP so that diverse constituencies and interests are accommodated. The conventional emphasis on discipline based around the idea of Singapore's precarious political and economic existence remains a favoured technique but one likely to produce diminished political returns. Hence, the announcement in early 1996 by the Organisation for Economic Cooperation and Development (OECD) that it had reclassified Singapore from a 'developing country' to a 'more advanced developing country' was received as much with concern as pride by Singapore's leaders. As Deputy Prime Minister Lee Hsien Loong explained: 'If you think that you have arrived and you relax and stop striving, someone is going to overtake you. We have got to get our population to realise that we need to keep moving ahead. For that psychology, it is better to call ourselves not developed' (*STWE*, 10 February 1996: 1). Yet a more fundamental challenge for the government is to ensure that the benefits of economic progress are widely distributed and seen to be so.

In confronting this challenge, the government is grappling with competing political and ideological impulses. The PAP remains ideologically committed to an extreme elitism which it sees as a functional necessity for continued economic success. Yet it is not indifferent to the plight of those least rewarded under this system. Thus, Prime Minister Goh recently stated his government's intention to do more to redistribute wealth in favour of lower and middle-income groups through 'innovative schemes', but not at the expense of rewarding the talented elite on whom he sees the fortunes of these groups ultimately depending (*ST* 2 May 1996: 1). In effect, these 'innovative schemes' will need to perform the same political and social roles the welfare systems historically did in the established liberal democracies, and which have been subjected to so much condemnation from the PAP government. With the dramatic upward social mobility by a virtual generation of Singaporeans benefiting from industrialisation unlikely to be repeated, Singapore's leaders may have to show a great deal of policy innovation if conventional redistributive welfare measures are eschewed.

---

[1] Five domestic banks were set up between 1903 and 1919. The three largest were founded by Hokkien traders and later merged to form the Overseas Chinese Banking Corporation (OCBC), which is today one of the so-called 'Big Four' local banks. Although there were some direct productive investments by local Chinese in the tin and rubber industries, these were also largely circumscribed by European capital (Puthucheary 1960: 83).

² An analysis of the top 500 Singapore-based companies in 1988 also depicted government-linked companies responsible for 60.5 per cent of total realised profits (see Vennewald 1994: 25–8).

³ Asher points out that the arguments for privatisation in Singapore have been quite different from elsewhere. The broadening and deepening of the stock market has been an objective, but this has not necessarily meant a relinquishing of the public sector. Rather, it has opened up opportunities for younger technocrats in the civil service and thereby retained talent that might otherwise have been lost to the private sector. Also, instead of outrightly promoting private firms, the process has more often fostered private/public sector partnerships (see Asher 1994: 800–2).

⁴ These included: co-investment programs drawing on the S$1 billion Cluster Development Fund referred to earlier; tax credits for dividend income remitted from abroad; a double taxation deduction for approved expenses associated with export promotion; the extension to the export of services of double taxation deductions for approved expenses incurred in export promotion; and the introduction of the Overseas Enterprise Incentive scheme providing tax exemptions on qualifying income for up to ten years for approved investments.

⁵ These projects include the S$28 billion Suzhou Industrial Township, the S$250 million Bangalore Technology Park in India and the Wuxi-Singapore Industrial Park in China, of which the first phase alone costs S$196 million.

⁶ In addition to Keppel, the Singapore consortium includes Straits Steamship Land, Wuthelam Holdings, City Developments, Wing Tai Holdings, Temasek Holdings, Singapore Technologies Industrial Corporation, Jurong Environmental Engineering, L and M Group Investments, Lee Kim Tah, and Low Keng Huat (Singapore). It also involves cooperatives such as NTUC Income, NTUC Fair-Price, NTUC Healthcare, NTUC Comfort International and SLF International (see *STWE* 14 August 1993: 24).

⁷ Investment in developmental software will remain a central function of state capital abroad for some time, but in 1996 Singapore Technologies Industrial Corporation announced its intention to substantially extend investments beyond industrial parks into 'hard' infrastructure projects such as expressways, roads, bridges, and power generation and water treatment plants in China. It has committed the sum of S$56.4 million (or US$40 million) for this purpose (*STWE* 27 April 1996: 20).

⁸ The Salim group has been involved in the Wuxi-Singapore Industrial Park, the Singapore-Suzhou Township and the Batam Industrial Park in Indonesia.

⁹ There are studies suggesting a trend towards improvement in the contribution of TFP to Singapore's economic growth, although they do concede a TFP problem (see, for example, van Elkan 1995; Rao and Lee 1995).

¹⁰ Singapore's percentage of the population undertaking tertiary education is currently only about one-third the level for the USA, for example.

11 Gooneratne, Martin and Sazanami (1994: 6) point out that since Singapore has no minimum wage, employers are able to cut wages as a way of absorbing the costs imposed on them by government levies on guest labour.

12 Spending on social security and welfare remains limited, amounting to just 2.15 per cent of total government expenditure in 1993 (see IMF in *Sunday Times* [Singapore] 18 September 1994: 6).

13 This has been attempted through such mechanisms as parliamentary select committees, a nominated members of parliament scheme, the Feedback Unit established by the Ministry of Community Development to take suggestions from the public and explain government policy, and a public policy think tank (see Rodan 1992; 1996: 37–39).

14 The same position was argued in Lau (1992) in a publication by the Department of Statistics, Singapore which compared census data for 1980 and 1990. In this publication, the Department of Statistics insisted that the income gap between rich and poor was narrowing because the bottom 20 per cent and middle 60 per cent of households had increased during 1980–94, from 3.6 per cent to 4.7 per cent of households and from 44 to 46 per cent respectively. However, the income share of the top 20 per cent was reported to have declined from 52 per cent to 49 per cent. It was thus concluded there had been a reduction in the gini coefficient from 0.48 to 0.44 between 1980 and 1994. The gini ratio or coefficient is the standard measure of income distribution used by economists. Economists tend to regard a ratio of between 0.45 and 0.50 as indicative of a high level of income inequality relative to economies experiencing comparable levels of economic development.

15 Rao (1996a) took direct issue with the use and interpretation of 1980 and 1990 census data in Lau (1992), arguing in particular that the gini ratio declined too marginally among the majority ethnic Chinese community to sustain the claims by Lau. Indeed, Rao contends that 'the fact that the income distribution among the majority community has changed little does not look very encouraging' (Rao 1996a: 93). Earlier, Rao's (1990) study drew on data from 1966 to 1989 and concluded that in broad historical terms Singapore had experienced stable inequality over this period around a gini ratio of 0.45 to 0.50 for personal income and 0.41 to 0.42 for household income. Within this general trend there were transitory variations, with income distribution inequality falling from 0.50 in 1966 to 0.42 in 1979 before rising steadily to 0.49 in 1989. Despite the criticisms of the methodology and analysis from Rao, the Department of Statistics, Singapore (1995: 6, 8) continues to make similar claims about a narrowing of the income gap between rich and poor. See Department of Statistics (1995) where it is maintained that the gini coefficient for household income declined between 1980 and 1994 from 0.48 to 0.44 and that the disparity in household income between the top and bottom 20 per cent of households narrowed from 14.4 times in 1980 to

10.5 times in 1990. Krongkaew (1994: 64) compared income distribution in eight East and South-East Asian countries and observed that income inequality in Singapore was the highest of all Asia's newly-industrialising economies.

[16] In 1993, for example, the Department of Statistics maintained that there had been a 3 per cent increase in the 'quality of life' between 1986 and 1992 based on an index of social change incorporating nine categories, see *Straits Times Weekly Overseas Edition* 25 December 1993: 14.

[17] Owning is not freehold but ninety-nine year leasehold.

[18] It is estimated that foreigners make up around 20–30 per cent of buyers of new condominiums in prime areas (see *ST* 27 April 1996: 32).

References

Asher, M. G. (1994) 'Some aspects of role of state in Singapore', *Economic and Political Weekly* 2 April: 795–804.

Bloodworth, D. (1986) *The Tiger and the Trojan Horse*, Singapore: Times Books International.

Brown, D. and Jones, D. M. (1995) 'Democratization and the myth of the liberalizing middle classes', in D. A. Bell, D. Brown, K. Jayasuriya and D. M. Jones (eds) *Towards Illiberal Democracy in Pacific Asia*, pp. 78–106, Oxford: Macmillan, and St Martin's Press.

Buchanan, I. (1972) *Singapore in Southeast Asia*, London: G. Bell and Sons.

Carling, R. C. (1995) 'Fiscal and monetary policies', in K. Bercuson (ed.) *Singapore: A Case Study in Rapid Development*, Occasional Paper 119, pp. 29–33, Washington: International Monetary Fund.

Chalmers, I. (1992) 'Loosening state control in Singapore: the emergence of local capital', *Southeast Asian Journal of Social Science*, 20 (2): 57–84.

Chia Siow Yue (1992) *The Role of Institutions in Singapore's Economic Success*, Paper for the World Bank Project on the High Performing Asian Economies, November.

Department of Statistics, (various years) *Yearbook of Statistics Singapore*, Singapore.

Department of Statistics (1983) *Economic and Social Statistics 1960–1982*, Singapore.

Department of Statistics (1991) *Census of Population 1990*, Advance Data Release, Singapore.

Department of Statistics (1993) *Direct Investment Abroad of Local Companies, 1991*, Occasional Paper on Financial Statistics, Singapore.

Department of Statistics, (1994) *Singapore's Investment Abroad 1990–1993*, Singapore.

Department of Statistics (1995) *Income Growth and Distribution*, Occasional Paper on Social Statistics, Singapore.

DeVoss, D. (1995) 'Doing business with the Generals', *Asia Inc.*, November: 54–61.

Deyo, F. (1981) *Dependent Development and Industrial Order: An Asian Case Study*, New York: Praeger.

Economic Development Board (various years) *Yearbook*, Singapore.

Economic Planning Committee (1991) *The Strategic Economic Plan: Towards a Developed Nation*, Singapore: Ministry of Trade and Industry.

Gooneratne, W., Martin, P., and Sazanami, H. (1994) 'Labour migration within Asia: an introduction', in W. Gooneratne, P. Martin and H. Sazanami (eds) *Regional Development Impacts of Labour Migration in Asia*, UNCRD Research Report Series, no. 2, pp. 1–24, New York: United Nations Centre for Regional Development.

Haas, M. (1989) 'The politics of Singapore in the 1980s', *Journal of Contemporary Asia*, 19 (1): 48–76.

Hon Sui Sen (1983) 'Statutory bodies and government-owned companies', a response from the Minister for Finance to a question in parliament from Toh Chin Chye, *Parliamentary Debates Singapore*, 43 (1), 30 August, columns 12–15 and 89–96.

Hu, R. (1996) *1996 Budget Statement*, Business Times Online, 28 February.

Huff, W. G. (1994) *The Economic Growth of Singapore*, Cambridge: Cambridge University Press.

Hughes, H. (1969) 'From entrepôt trade to manufacturing', in H. Hughes and You Poh Seng (eds) *Foreign Investment and Industrialisation in Singapore*, Canberra: Australian National University Press.

Huxley, T. (1993) *The Political Role of the Singapore Armed Forces' Officer Corps: Towards a Military-Administrative State?*, Strategic and Defence Studies Centre, Working Paper No. 279, Canberra: Australian National University.

International Bank for Reconstruction and Development (1995) *The Report of the International Bank for Reconstruction and Development on the Economic Development of Malaya*, Baltimore: Johns Hopkins University Press.

Islam, I. and Kirkpatrick, C. (1986) 'Export-led development, labour-market conditions and the distribution of income: the case of Singapore', *Cambridge Journal of Economics*, (10): 113–27.

Khong Cho Oon (1993) 'Managing conformity: the political authority and legitimacy of a bureaucratic elite', paper presented at the workshop on Political Legitimacy in South-East Asia, 23–26 February, Changmai, Thailand.

Kraar, L. (1996) 'Need a friend in Asia? Try the Singapore connection', *Fortune*, 4 March: 172.

Krongkaew, M. (1994) 'Income distribution in East Asian developing countries: an update', *Asian-Pacific Economic Literature*, 8 (2): 58–73.

Krugman, P. (1994) 'The myth of Asia's miracle', *Foreign Affairs*, 73 (6): 62–78.

Kwok Kian-Woon (1995) 'Singapore: consolidating the new political economy', *South-East Asian Affairs 1995*, pp. 291–308, Singapore: Institute of Southeast Asian Studies.

Lee Sheng-Yi (1978) *Public Finance and Public Investment in Singapore*, Singapore: Kong Brothers Press for the Institute of Banking and Finance.

Lau Kak En (1992) *Singapore Census of Population 1990: Housholds and Housing*, Statistical Release 2, Singapore: Department of Statistics.

Lewis, P. (1993) *On the Move: The Changing Structure of Singapore's Labour Market*, Asia Research Centre Paper 1, Perth: Asia Research Centre, Murdoch University.

Mak Lau-Fong (1993) 'The rise of the Singapore middle class: an analytical framework', in Hsin-Huang, Michael Hsiao (eds) *Discovery of the Middle Classes in East Asia*, Taipei: Institute of Ethnology, Academia Sinica.

Mahizhnan, A. (1994) 'Developing Singapore's external economy', *Southeast Asian Affairs 1994*, pp. 285–301, Singapore: Institute of Southeast Asian Studies.

Ministry of Finance Republic of Singapore (1961) *State of Singapore, Development Plan, 1961–1964*, Singapore: Government Printer.

Ministry of Finance Republic of Singapore (1993) *Final Report of the Committee to Promote Enterprise Overseas*, Singapore: Singapore National Publishers.

Ministry of Trade and Industry, Republic of Singapore (1986) *The Singapore Economy: New Directions*, Report of the Economic Committee, Ministry of Trade and Industry, Singapore.

Ministry of Trade and Industry, Republic of Singapore (1993) *Report of the Cost Review Committee*, Singapore: Singapore National Publishers.

Ng Chee Yuen and Sudo Sueo (1991) *Development Trends in the Asia-Pacific*, Singapore: Institute of Southeast Asian Studies.

Pang Eng Fong (1994) 'Foreign workers in Singapore', in W. Gooneratne, P. Martin and H. Sazanami (eds) *Regional Development Impacts of Labour Migration in Asia*, UNCRD Research Report Series, no. 2, pp. 79–94, New York: United Nations Centre for Regional Development.

Paix, C. (1993) 'The domestic bourgeoisie: how entrepreneurial? how international?', in G. Rodan (ed.) *Singapore Changes Guard: Social, Political and Economic Directions in the 1990s*, Melbourne: Longman Cheshire.

Porter, M. (1990) *The Competitive Advantage of Nations*, London: Macmillan.

Puthucheary, J. J. (1960) *Ownership and Control in the Malayan Economy*, Singapore: Eastern Universities Press.

Rao, V.V.B. (1990) 'Income distribution in Singapore: trends and issues', *The Singapore Economic Review*, XXXV(1): 143–60.

Rao, V. V. B. and Lee, C. (1995) 'Sources of growth in the Singapore economy and its manufacturing and service sectors', Federation of ASEAN Economic Associations Conference, Singapore, December.

Rao, V. V. B. (1996a) 'Singapore household income distribution data from the 1990 census: analysis, results and implications', in B. Kapur et al. (eds) *Development, Trade and Asia-Pacific: Essays in Honour of Professor Lim Chong Yah*, Singapore: Prentice-Hall.

Rao, V. V. B. (1996b) 'Income inequality in Singapore: facts and policies', in Lim Chong Yah (ed.) *Economic Policy Management in Singapore*, Singapore: Addison-Wesley.

Rodan, G. (1989) *The Political Economy of Singapore's Industrialization*, London: Macmillan.

Rodan, G. (1992) 'Singapore's leadership transition: erosion or refinement of authoritarian rule?', *Bulletin of Concerned Asian Scholars*, 24 (1): 3–17.

Rodan, G. (1993a) 'Preserving the one-party state in contemporary Singapore', in K. Hewison, R. Robison and G. Rodan (eds) *Southeast Asia in the 1990s: Authoritarianism, Democracy and Capitalism*, Sydney: Allen & Unwin.

Rodan, G. (1993b)'The growth of Singapore's middle class and its political significance', in G. Rodan (ed.) *Singapore Changes Guard: Social, Political and Economic Directions in the 1990s*, pp. 52–71, New York: St. Martins Press.

Rodan, G. (1993a) 'Reconstructing divisions of labour: Singapore's new regional emphasis', in R. Higgott, R. Leaver and J. Ravenhill (eds) *Pacific Economic Relations in the 1990s: Cooperation or conflict?*, pp. 223–49, St Leonards: Allen & Unwin.

Rodan, G. (1996a) 'Class transformations and political tensions in Singapore's development', in R. Robison and S. G. Goodman (eds) *The New Rich in Asia: Mobile Phones, McDonalds and Middle Class Revolution*, pp. 19–45, London and New York: Routledge.

Rodan, G. (1996b) 'State-society relations and political opposition in Singapore', in Garry Rodan (ed.) *Political Oppositions in Industrialising Asia*, pp. 95–127, London and New York: Routledge.

Rumbaugh, T. (1995) 'Singapore's experience as an open economy', in K. Bercuson (ed.) *Singapore: A Case Study in Rapid Development*, Occasional Paper 119, Washington: International Monetary Fund.

Seah Chee Meow, (1973) *Community Centres in Singapore*, Singapore: Singapore University Press.

Singh, B. (1992) *Whither PAP's Dominance? An Analysis of Singapore's 1991 General Elections*, Petaling Jaya: Pelanduk Publications.

Soon Teck-Wong and Tan C. Suan (1993) *The Lessons of East Asia: Singapore, Public Policy and Economic Development*, Washington: World Bank.

Turnbull, C. M. (1982) *A History of Singapore 1819–1975*, Kuala Lumpur: Oxford University Press.

van Elkan, R. (1995) 'Accounting for growth in Singapore', in K. Bercuson (ed.) *Singapore: A Case Study in Rapid Development*, Occasional Paper 119, pp. 4–10, Washington: International Monetary Fund.

Vennewald, W. (1994) *Technocrats in the State Enterprise System in Singapore*, Working Paper No. 32, Asia Research Centre, Perth: Murdoch University.

Wong Lin Ken (1960) 'The trade of Singapore, 1819–1869', *Journal of Malayan Branch of the Royal Asiatic Society*, XXXIII: 5–315.

World Bank, (1993) *The East Asian Miracle: Economic Growth and Public Policy*, New York: Oxford University Press.

Newspapers and Journals

Far Eastern Economic Review (FEER)

New Democrat

Straits Times Weekly Overseas Edition (STWOE) (recently renamed Straits Times Weekly Edition)

Straits Times Weekly Edition (STWE)

Straits Times (ST)

7

Vietnam: The Transition From Central Planning

Melanie Beresford

Introduction

Vietnam is unique among the countries covered in this book in that it is in the process of transition from central planning to a market economy. In order to understand the special characteristics of this transition we need first to comprehend the nature of the socialist economy as it developed between 1954 and 1980—the social basis of the state and the way in which interests of different groupings in society were represented within the state apparatus. This is what gave Vietnamese socialism its particular institutions and economic structure. During the 1980s the latter began to undergo dramatic changes, partly as a result of spontaneous developments at the grassroots level and partly as a result of state efforts to manage the process. The structure and institutions which Vietnam has today can only be understood as the outcome of a protracted struggle between various interests over the shape of the socialist state, a struggle which necessitated constant reform and has ultimately led to the demise of the planned economy. As one Vietnamese intellectual put it to the present author: 'The system was patched and repaired so many times that it was eventually repaired out of existence. It became something quite different'. The relations of power which developed throughout this process of struggle and reform are evident in the existing institutions and structure which give the Vietnamese market economy its peculiar characteristics today. While the ultimate result of the current market-oriented transformation may be some form of capitalism, Vietnam is still ruled by a Communist Party, the distinction between public and private property remains blurred and there is continuing pressure from parts of the state to defend the rights of workers and poor peasants. This means that the path is still unclear and likely to be characterised by considerable social and political tensions in future.

This chapter first gives a brief outline of the division of the country between north and south during 1954–75 and the attempts to construct rival socialist and capitalist economies in the two halves. (More detail can be found in Beresford [1988].) There follows a discussion of the socialist reality in northern Vietnam focusing on the role that economic interest played in creating the north's economic structure and institutions and the forces generating the transition process, particularly after reunification of the country in 1976. The third section looks at developments since the 6th Communist Party Congress in 1986 at which the historic decision to institute *Doi Moi* (Renovation) was taken. It focuses on the interaction between interests and policy in promoting (or obstructing) change and discusses the ways in which contending social forces organise and find expression within the framework of a Communist Party state. The final section discusses the major issues in contention today.

Historical Background

For two decades, from 1954 to 1975, two separate regimes claimed sovereignty over the whole of Vietnam. The Democratic Republic of Vietnam (DRV) was based in the northern half of the country with its capital at Hanoi. It had been established by an indigenous nationalist movement (the Viet Minh) led by the Vietnamese Communist Party on 2 September 1945 following the surrender of Japan. It had fought and won a nine-year war against France's attempt to retake its former colony, a war which culminated in the humiliating French defeat at Dien Bien Phu in 1954. During this war, the Viet Minh had extended its control over nearly three-quarters of the country, although the French and some anti-communist groups had managed to retain the main cities and some of the rural areas, particularly in the far south. The inability of the communist-led movement to win decisively in the south provided the basis for the formation of an alternative regime, initially led by the Emperor Bao Dai who was favoured by the French to promote their own interests. As the French war effort faltered, however, the USA increasingly took over the role of preserving Western interests in Indochina and, by 1955, had installed in Saigon a political leader (Ngo Dinh Diem) less subservient to the French. From that year the southern regime became known as the Republic of Vietnam (RVN).

During 1954 an international conference at Geneva had temporarily divided the country at the 17th parallel with an agreement to hold reunifying elections in 1956. Knowing that the Viet Minh would have won easily, the Diem regime and its US supporters refused to hold these elections. Between 1956 and 1959 the struggle between the regime and the southern Viet Minh gradually escalated until, in 1960, the DRV officially formed the

National Front for the Liberation of Southern Vietnam (NLF) and began a war to reunify the country. US combat troops were introduced in 1964, systematic bombing of the DRV by the US Air Force began in 1965 and lasted until 1968. By 1969 the USA had over half a million troops in the country and yet it had become clear by then that no end was in sight. For all the expense and effort the USA had put into defending the RVN, the communists' Tet (Vietnamese New Year) offensive in early 1968 had driven home the lesson that US war strategies were not working. In the USA itself the war became increasingly unacceptable politically and by 1973 the last US troops were withdrawn. Within two years the RVN collapsed completely and the DRV armed forces entered Saigon on 30 April 1975.

The economic policies pursued in the two halves of Vietnam during 1954–80 reflected the different social bases of the regimes. In the case of the DRV its economic structure was initially developed in response to the needs of the peasantry who constituted over 80 per cent of the population, although their interests and those of the state were later to diverge in ways which, as we shall see, led to pressure for reform. The RVN, on the other hand, drew its support from a much narrower base: large landowners, the small indigenous capitalist class and the bureaucrats.

Popular support for the DRV in the northern half of the country was consolidated during the 1950s by agrarian reform policies, beginning with rent reduction campaigns from 1951 and culminating in land reform during 1954–56 which saw some 810 000 hectares redistributed from landlords and rich peasants to poor and middle peasants (White 1983, Moise 1976). However, a push from the poorest farmers and more radical sections of the Party spilled over into the collectivisation of agriculture, commencing as a spontaneous movement during 1958 and established as official policy during 1959. Collectivisation had advantages for poor peasants since, even after land distribution, plot sizes in the crowded northern river deltas were often too small to provide an adequate subsistence, let alone surpluses. Moreover, poor peasants lacked draft animals and equipment or access to irrigation water and therefore faced potential economic insecurity. Collectivisation, on the other hand, guaranteed access to these as well as providing minimum standards of nutrition, health, education and welfare. From the point of view of the state, collectives were seen as a means of increasing agricultural productivity (through consolidation of fields, rehabilitation and construction of large-scale irrigation works, more equal distribution of equipment and animals and improvements in social security and living standards). This would, in turn, provide agricultural surpluses to finance industrialisation. Support for collectivisation also came from other socialist countries, notably China.

Cooperatives were therefore established in the rural economy in three stages (basically following the Chinese model of the mid-1950s), beginning with the formation of production responsibility teams, then graduating to lower level cooperatives, in which land and equipment were collectivised and a small rent paid to their former owners, and finally higher level cooperatives in which all income was distributed according to a system of workpoints (Beresford 1988: 130; Gordon 1981). By the mid-1960s virtually the whole rural population had been incorporated into these higher level cooperatives.

Industrial development in the north was to be carried out through a series of five-year plans (FYP). Following nationalisation of the remaining French-owned plants, the first FYP was instituted for 1961–65 and resulted in the establishment of much new industrial capacity which, in turn, was brought progressively under the control of Soviet-type planning institutions. Output targets and prices, input supplies, domestic wholesale and retail trade, and imports and exports were determined by central government (usually via the intermediary of provincial and district authorities). The planners intended to establish a vertically integrated economy in which individual production units had no commercial contact with each other, exchanging goods only with higher levels which in turn distributed them according to plan. The State Bank of Vietnam, was responsible for recording the value of transactions against the accounts of individual units and transferring revenue between these accounts and the state budget, thus eliminating the need for cash.

Renewed warfare against the USA-backed southern regime created many difficulties for this newly created socialist economy. However, as we shall see below, the economic system itself was responsible for many of the problems besetting the DRV during the 1960s and 1970s. An overwhelming focus of the authorities on war fighting as well as strong popular support for the objective of national reunification meant that many of these economic problems were not tackled, or even regarded seriously until much later. Indeed most in authority genuinely believed that the wartime emergency called for stronger centralisation and planning, rather than less. In the meantime, there occurred numerous (mostly illegal) experiments by individual households and production units aimed at overcoming the day-to-day difficulties imposed by the system. These experiments would ultimately provide the foundations of the reformed economy.

Meanwhile the southern half of the country continued to develop a capitalist economy. While the narrow political base of the RVN regime ensured that the demands of the peasantry were largely ignored in areas under its control during the 1960s, land reforms were carried out by the NLF. Land ownership remained a contested issue until 1970–73 when the RVN decided to distribute 'land to the tiller', effectively legitimating occupancy under NLF reforms and finally resolving a problem which had led the majority of farmers to give their support to the communists. This reform came too late to save the RVN from military defeat, but it did have the effect of boosting agricultural production in the fertile Mekong River delta region as, with the ending of landlord domination, farmers gained control over their surplus output (Beresford 1989: 66–7, 101–3; Ngo Vinh Long 1984).

During the war, US economic aid flowed into the RVN, allowing high levels of consumption of imported goods. Under normal circumstances, these might have boosted economic growth, however, speculative investments were encouraged by insecurity and inflation while productive investment suffered. Commerce, services and bureaucracy expanded while the economy as a whole stagnated. Corruption became a serious problem in both civilian and military life. Refugees flooded into the main cities to escape the devastation of US bombing and chemical warfare, leading to the growth of slums and a large unemployed or semi-employed population. The complete withdrawal of US aid in 1975 thus provided the basis for a serious crisis of the southern economy, although the reformed agricultural sector also implied the possibility of a fairly rapid recovery. This failed to materialise, largely because after the official reunification in 1976 the DRV attempted to impose its own economic model on the south.

Vietnamese Socialist Economy and Pressure for Reform

Tensions over agrarian policy were already evident during the land reform of the 1950s, as many in the Party leadership would have preferred a reform less damaging to the interests of rich and upper 'middle-class' peasants whom they regarded as more productive and likely to foster rapid agricultural development. A 'Rectification Campaign' carried out in 1956 went some way to redressing this problem, but tensions between poor and rich peasants and between Party factions remained. While subsequent collectivisation appears to have had an impact in reducing tension (Elliott 1974; White 1983), it proceeded throughout the 1960s against some resistance from local forces. During 1960–63 the number of peasant households belonging to cooperatives actually fell, while there was apparent lack of interest from upper 'middle' and rich peasants,[1] who provided a high proportion of both Party membership and cooperative management boards (Gordon 1981: 30–5). The move towards establishing higher level cooperatives was renewed in 1964, but Adam Fforde's detailed study of the cooperative management system during the 1970s shows that few actually functioned as the government intended (Fforde 1989). Instead a wide variety of arrangements existed, including extension of household plots beyond the 5 per cent of land officially permitted (often in favour of cadres and their families and friends) and refusal to issue plan targets by cooperative managers. Like other socialist economies, Vietnam developed an 'outside' (in the Vietnamese terminology) or second economy which, as we shall see, both fed on the socialist structures and tended to undermine them in the long run.

The basic problem with cooperatives, not only in the north but in the south after 1976, lay in the lack of incentives for increased output and productivity. At the root of this problem was the official pricing system which offered low procurement prices to cooperatives in order to maintain low wages in the urban sector and boost industrial profits of state-owned enterprises. Market prices for both agricultural produce and industrial goods, on the other hand, tended to be above the official prices from 1960 onwards, encouraging the diversion of state-supplied goods on to black markets and leading farmers to devote more time and effort to production on their household plots, the output from which could be legally sold on the free market. Estimates of peasant income earned from these plots rose from about 40 per cent in the 1960s to 60–70 per cent by the late 1970s. Since those with priority access to state-supplied goods were generally Party members and officials, the system created opportunities for capital accumulation and new class divisions within the countryside. It is important to note here that it was their position within the cooperative structures which gave cadres such opportunities to profit from outside activities and this is one of the key factors leading to widespread perception of the unfairness of the system.

Distribution of income within the cooperatives also created problems for increasing output and productivity. It was, in a formal sense, highly egalitarian since minimum supplies of food and security of land tenure were guaranteed, while health, education and other welfare facilities (such as child-care) were also provided to all members. The advantages to poor peasants and to women were enormous: life expectancy, infant mortality, literacy levels and other social indicators were all better than for other countries at similar per capita income levels (Beresford 1995b). But these benefits came at a high cost, as cooperative production stagnated in the long term and egalitarianism became a case of 'shared poverty' rather than rising overall living standards. Moreover, as demands from the government for procurement to feed urban cadres and industry rose, the residual output available for distribution to cooperative members declined, creating further incentives to reduce labour in cooperative fields and increase that applied to household plots and to commercial activities outside the plan (Beresford 1989; Fforde 1989).

Conflict of interest was, therefore, very apparent in the actual functioning of the cooperative system in Vietnamese agriculture. The needs of the poorest families for a more equal distribution and those of households most able to benefit from outside activities differed, as did those of cooperative farmers as a whole and of the government. Pressure on lower- and middle-level cadres tended to increase as, on one side, farmers' demands to reform the cooperative system increased and, on the other side, the centre's procurement requirements also rose.[2] These contradictions only sharpened after reunification, particularly in the Mekong delta–Ho Chi Minh City region, where the market economy was firmly entrenched. The so-called 'middle' peasants, the chief beneficiaries of land reforms in the 1960s and 1970s, stood to lose substantially through the collectivisation campaign of 1977–78. Their passive resistance meant that available surpluses dropped sharply (Beresford 1989: 113–15) and black market activities intensified, despite the massive government clampdown on mainly ethnic Chinese businesses during 1978.

As Vietnam entered the 1980s the scope of 'outside' economic activity in the rural areas continued to increase and a powerful coalition of rural interests in favour of system reform began to emerge. Southern provincial politicians, in particular, fearing the loss of their peasant support base, became vocal spokespersons for change within the Party and government apparatus. They found allies among the generation who had earlier argued that collectivisation in the north had gone too far.

A similar state of affairs existed within the state-owned industrial enterprise system. This had expanded rapidly following implementation of the First FYP (1961–65), but problems were already emerging (Fforde and Paine 1987) before USA wartime operations destroyed much of the capacity created. High investment rates in industrial capacity, particularly in heavy industry projects with long lead times, created demand for resources which could not be met from the existing stream of output (Beresford 1989). While the resulting shortages were to some extent alleviated by foreign aid, they were never eliminated altogether and the Vietnamese economy, like other socialist economies, became characterised by chronic disequilibrium between supply and demand. Moreover, shortages within the planned sector were aggravated by the difference between official and market prices which diverted resources to the outside economy (Fforde and Paine 1987).

In the debates over reform of the economic system which began in the 1970s, two basic positions can be identified. The first essentially accepted the ability of planners to achieve an equilibrium growth path within the framework of central planning and in general opposed high rates of capital construction and insufficient attention to the real financial and resource capacity of the economy. The second located the real problem of the planned economy precisely in the attempt to achieve equilibrium by *administrative* means and pointed to a built in tendency to disequilibrium which cannot be overcome, due to the existence of real conflicts of interest. Once established, the shortage economy provides incentives to managers and planners which ensure that the tendency to disequilibrium is exacerbated rather than overcome (see Beresford 1989; de Vylder and Fforde 1988, for a fuller discussion).

In reality, large capital construction projects involve vested interests among the ministries, managers and employees engaged on them, so that simply scrapping them becomes politically difficult. While the use of market forces and 'loose' planning norms to allow more local initiative can alleviate the shortages, they threaten the positions of those in the central ministries whose control over resources is threatened and who exert pressure to recentralise control. Market forces, on the other hand, allow for the accumulation of capital independently of the centre. This may be private capital, especially in the farm and retail sectors, but is more likely to take the form of 'corporatisation' (that is, the separation of legal ownership and control) in the state-owned enterprise (SOE) sector (Beresford 1993). The ability of the planning authorities to reimpose centralisation after a period of decentralisation, then, depends crucially on its power *vis-à-vis* such relatively independent *loci* of capital accumulation.[3] In my view, the emergence of such capital accumulation in the periphery in Vietnam introduced not only a systemic *tendency* towards breakdown of the planned economy, but introduced structural changes which laid the foundations for a relatively successful transition to a market economy.

In the political sphere, the ability and willingness of the state to reimpose centralisation determines whether the planned economy continues to exist in a recognisable form, or whether an irreversible transition takes place. The planned economy in Vietnam was not a straightforward 'top down' system, but one in which the plan emerged as the outcome of a process of negotiation between groups with differing, often conflicting, interests. The most important interests which can be identified here are firstly, that of central authorities whose power derived from the planning system and whose control was threatened by outside activity; secondly, that of local authorities and enterprise managers (including cooperative managers) who strove to minimise surplus extraction to higher levels and accumulate capital within their own sphere; thirdly, those who drew little benefit from the planning system due to lack of access to the kinship and/or Party networks which held local power.

The process of negotiation between the contending factions and interests led to a balance of power which began to shift in the late 1970s and early 1980s. During 1979–81 a number of key reforms were pushed through, essentially legalising some outside activity by individuals and production units. The earliest reforms, in the second half of 1979, allowed limited autonomy to farmers and enterprises. More important were the introduction in 1981 of output contracts in agriculture, permitting allocation of cooperative land to individual households and sale of surplus product on the free market, and the so-called 'Three Plan System' in industry, for the first time sanctioning production for the market by SOEs. In the same year, prices and wages were brought closer to market prices. By early 1982 a recovery was underway and Party fears of the re-emergence of capitalism were renewed, causing it to take measures against further private sector development, including a renewed attempt to collectivise southern agriculture. However, the structural changes taking place since the second half of the 1970s meant that these were only partially successful, if at all. The outside economy continued to expand while the difficulties of the planned sector continued, not least because in reality the bulk of state investment still flowed into large-scale heavy industry projects. Shortages in the latter sector were exacerbated by the attempted recentralisation, leading to higher inflation in the increasingly monetised outside economy (176 per cent in 1984 [World Bank 1994: 131]). Moreover, the necessity to subsidise unprofitable SOEs contributed to high budget deficits, inability to finance the external debt and renewed stagnation in the agricultural sector.

In face of the new crisis, in September 1985 the government introduced a further round of reforms aimed at price liberalisation, ending the ration system and monetisation of wages for state employees. By that time almost all the southern and central provinces plus a number in the north were refusing to implement plan directives and/or pressuring the government for wider reform. However, the impact of this reform was blunted by several different political pressures on the government at this time. The southern provinces were already at a more advanced stage of commercialisation, with a correspondingly higher cost of living, and demanded big wage increases for government workers which resulted in large budget deficits. SOE directors, especially those in the older, less efficient northern sector, demanded retention of the two-price system for industrial inputs.[4] Within three months of the reform's implementation Vietnam had galloping inflation,[5] and the reform was partially reversed. Nevertheless, a further substantial shift towards commercialisation of the economy had been achieved.

Doi Moi (Renovation)

By 1986 the Vietnamese socialist economy had already undergone major structural and institutional changes towards the formation of a market economy. Official statistics on this period do not reflect these changes as they continued to measure only the planned economic sector. However, other evidence indicates strongly that transactions taking place outside the scope of the plan were of increasing importance (de Vylder and Fforde 1988). Not only was a large proportion of agricultural output being marketed, but there were also signs of a rudimentary capital market as SOEs sought ways to utilise capital accumulated through non-plan activities. Under the impact of these developments and also the fact that the rationing system had largely disappeared after the 1985 reform, political pressure for further changes to the system increased.

Not enough research is available for a full understanding of how the Vietnamese political system works and how these pressures from interests associated with increased commercialisation of the economy are translated into policy changes at the top. The major reason for this has been the essentially closed character of this system to outsiders and the concern of Party leaders from earliest times to present a unified front to the rest of the world. We therefore have little understanding of how political positions are arrived at or, in other words, how the leadership gains its political constituency. What we do know is that a number of the top leaders in Vietnam have held different political positions in relation to key issues and that alliances within the leadership have often been quite fluid (Beresford 1988: 88–90). This suggests that, far from being communist ideologues, the Vietnamese leaders have behaved much like politicians elsewhere and that their power is contingent upon building coalitions of interest which provide support for particular sets of policies. It follows from this that if the structure of interests changes substantially, the position of political leaders on questions like reform must change accordingly. Members of key Party organisations who continue to reflect the dominant interests of the old system are likely to be replaced by representatives of the new dominant interest groups.

This appears to be what happened at the Sixth Party Congress in December 1986. Membership changed in favour of a younger generation with less experience of revolutionary activity and more versed in the problems of managing the socialist economy, particularly at the local level. There was also increased representation in the highest positions of those with long experience in the south (Beresford 1988: 113), a change which seems to reflect recognition of the relative success of the more commercialised southern region in overcoming the constraints of the planning system. Moreover, the emergence of reformers at the top in turn opened the way for many more junior cadres to express their views freely.

Doi Moi in the Economic Sphere

The 1986 Congress was a turning point for the reform process in the sense that it took a conscious decision to build a market economy in Vietnam, although this was (and still is) portrayed by the Party leadership as 'a socialist market' or 'a market economy under state guidance'.

By the late 1980s the planned economy had disintegrated to the extent that there was no longer sufficient strength of interest in attempting to make it work. Even within the SOE sector, which had strongly resisted the proposal to implement market prices in 1985, there were differences of interest between those enterprises which continued to depend heavily on price subsidies and those for whom the majority of their activities were by now market oriented. The upsurge of inflation during 1986 forced state sector workers, unable to survive on their official wages, to devote more and more time and effort to outside activities. Productivity in the SOEs declined, requiring higher levels of subsidy and generating yet more inflation through rising budget deficits. While Soviet aid continued to be available in the late 1980s, the rise of Gorbachev to power and the introduction of *Perestroika* (restructuring) in the Soviet Union in 1986 meant that aid would no longer be provided on the same generous terms as before (it was finally cut in 1991). The Vietnamese state was therefore faced with a looming fiscal crisis that convinced even the highest levels that further reform was their only option. Whereas reform prior to 1986 was largely spontaneous in character, with the state attempting to manage (sometimes even reverse) the process of change, it is only after 1986 that we can speak of the state playing a role in pushing the pace of change.

The reforms of the late 1980s are widely regarded in the Western literature as the only ones of importance. Indeed a number of authors have described the reform package of 1989, in particular, as constituting a 'Big Bang' or 'shock therapy' type of reform which introduced the market economy for the first time in Vietnam. What we have seen in this chapter, however, is that the market economy had been in the process of forming in Vietnam since the late 1970s. If we want to understand the relative success of these reforms in Vietnam (as compared, for example, to the results of 'shock therapy' in some East European countries or the Soviet Union), then it is important to understand that, to all intents and purposes, the Vietnamese economy by the late 1980s was already a market economy. Policies designed for a market economy are more likely to work in a market economy than in a planned economy.

The first major reform of *Doi Moi* was the Foreign Investment Law, passed in December 1987 and implemented in the following year. By South-East Asian standards, this was a very liberal law, allowing for 100 per cent foreign ownership and profit repatriation, significant tax holidays and concessions for enterprises investing in a number of priority areas (including exports, consumer goods, technology transfer and processing of local raw materials). The resulting inflow was small at first, but by 1994 it amounted to $US1.5 billion per annum and constituted about a third of Vietnam's total investment effort. The majority was in joint ventures with Vietnamese partners, mostly SOEs. Moreover, in the 1990s Western aid to Vietnam has begun to increase rapidly and this has brought with it a stream of foreign contractors. Foreign companies and aid donors now constitute an important interest group applying pressure for further market oriented reforms and for changes to the legal system to bring it into line with international norms.

In March 1988 Resolution 10 of the Politburo effectively led to decollectivisation in the Vietnamese countryside. Although many cooperatives continued to exist, their functions were severely curtailed and, most importantly, they were no longer given a role in production. Instead land was distributed for a period of fifteen years (or fifty years in cases of perennial crops) to households who were put in charge of all decisions relating to production and investment. Government procurement contracts were abolished and all output could be sold, whether to the government or in the open market, at market prices. This reform had very positive effects on agricultural output and in 1989 Vietnam shifted from being a net importer of rice to being the world's third largest exporter (though as we shall see below, the immediate increase in supply was also related to the success of anti-inflation measures in that year). However, a number of the more useful functions provided by the cooperatives were adversely affected. In particular, householders were reluctant to make contributions to cooperative health and welfare facilities and the cooperative maintenance of irrigation schemes also tended to suffer. Responsibility for provision of these, as well as other local infrastructure, has shifted to the commune (village) and district levels of government, yet these also suffered severe cuts in revenue after 1989. The impact of these changes has inevitably fallen hardest on the poor. However, as overall living standards have risen, the potential for rural class conflict has been blunted at the same time.

During 1989 the government abolished official prices (except for a handful of government monopolies), floated the exchange rate and the introduced positive real interest rates in the banking system. The first two of these mean that all prices, including the exchange rate against foreign currencies are now basically market prices, although the State Bank does control the rate of devaluation through its market operations. Direct subsidies from the state budget to SOEs were effectively ended. Positive real interest rates[6] (temporarily reversed in 1990–91) were used to encourage savings, halt the 'dollarisation' of the economy,[7] eliminate SOE subsidies via cheap credit provision, halt the growth of the budget deficit and bring inflation under control.

The impact of this package on the economy was dramatic. Inflation was reduced from 308 per cent p.a. in 1988 to only 36 per cent in 1990 (World Bank 1994: 131) and the domestic savings ratio improved. With the reduction in inflation, much of the incentive to hoard goods for speculative purposes disappeared and there was an increase in supply (particularly of rice) so that shortages were eliminated. Industrial output has also risen rapidly since the reform. At the same time, many of the least profitable SOEs closed their doors—the number of still operating enterprises had been halved by 1993—and in industry there was a significant reduction in employment as formerly subsidised SOEs sought to become profitable. Within the state-owned sector, the total number of employees has been reduced from 3.86 million in 1985 (about 15 per cent of the total labour force) to 2.92 million in 1993 (or 9 per cent). Employment in the cooperative industrial sector declined even more precipitously, from 1.2 million in 1988 to 287 600 in 1992 (GSO 1994). Women workers were the majority affected. Unemployment rose to some 13 per cent of the workforce, but many of those laid off were able to find work in the burgeoning household sector or in the informal sector, particularly in farming, trade and services. There was therefore a substantial shift in the employment structure.

Control over inflation has seen a significant return of confidence in the Vietnamese currency, increased savings ratios and rapid growth of GDP (at over 8 per cent p.a. in 1992–96). These sustained high growth rates in a labour-intensive economy mean that unemployment has begun to diminish, although given the irregular character of much informal sector employment, underemployment remains a serious problem affecting women most of all.

One of the implications of the above analysis is that the high growth rates of the early 1990s were brought about chiefly through improvements in efficiency. The elimination of hyperinflation reduced speculation and waste of resources, while the abolition of subsidies to SOEs provided incentives for improved productivity and a shift of resources into more profitable areas of production. A major issue in the economic debates has therefore been how to improve rates of investment in order to sustain growth over the long run. This has a parallel in government concerns to improve revenue collection in order to be able to finance much needed infrastructure projects and restore funding to education and health without resorting to inflationary deficit financing. Reform of the financial sector and taxation system have thus been key priorities (see below).

Social and Political Impact of *Doi Moi*

The impact of this series of reforms on the politics and culture of Vietnam was just as dramatic as that on the economy. As Vietnam has opened to the world market there have been significant gains in the ability of individuals to become involved in decisions concerning their livelihood and in freedom of speech (although not without setbacks). National Assembly debates have become much more lively and, for the first time in 1992, candidates not 'approved' by the Party were permitted to run for election. The demise of the cooperative system in the rural areas has also allowed the revival of traditional Vietnamese cultural and religious practices. Kinship networks have become more important in village life. New types of cooperative organisation, such as the numerous credit groups, have also spontaneously arisen to fulfil some of the social functions previously carried out by the cooperatives. Independent trade unions are permitted and some have already emerged in private enterprises, although so far mainly of the 'house union' type. There have been a number of strikes in foreign invested enterprises where unfamiliar, and to Vietnamese workers harsh, labour management systems have been introduced. The government has so far shown no inclination to interfere in the independent settlement of these disputes.

Vietnam is now a much more open society than at any stage in the so-called socialist period (Thayer 1991: 26–33). The Party's concern for political stability has ensured that such freedoms have not extended beyond the limits of what it considers acceptable at any given time, but these limits are still expanding. As in the 1980s, when policy tended to attempt to manage change but was unable to control it, the desire of Party leaders to preserve legitimacy means that they have been forced to react to shifts in public opinion.

This is evident, for example, in the movement to create a society 'ruled by law' in the first half of the 1990s. 'Rule by law' (often mistranslated in English language sources as 'rule of law') represents the attempt to codify

Vietnamese laws and regulations into a coherent and internally consistent system in order to increase the transparency of bureaucratic processes and eliminate opportunities for corruption, nepotism, favouritism and arbitrariness from their implementation. As such, it implies reform of the bureaucracy as well as rewriting of many laws to make them applicable to a market economy. Unlike the Western concept of 'rule of law', however, law does not represent a higher authority than the Party-state. It does attempt to create a level playing field in which the popular perceptions of unfairness in the application of law, which characterised the socialist and earlier transitional society, will be eliminated.

Several important tests of 'rule by law' have already taken place. In one case which received widespread publicity, a well-connected policeman who had murdered a young man for a petty amount of cash appealed against his death sentence. A huge crowd gathered outside the court in Hanoi on the day of the appeal to reinforce public opinion that the sentence should be upheld. Whether because of this demonstration or not, the appeal failed and the policeman was subsequently executed. Other cases include the demolition of houses illegally built on the dyke protecting Hanoi, the successful ban on fireworks at the Vietnamese New Year in 1995 and a recent campaign to enforce traffic rules.

Current Issues

Sustainable Growth

For the Party and government the main question is how to sustain economic growth at a sufficient rate to catch up to its South-East Asian neighbours in ASEAN (CPV 1994: 24).[8] This is seen as the surest way to avoid the potential for social conflict inherent in the reform process and maintain Communist Party rule in the medium term.

As mentioned above, high growth rates of GDP in the early 1990s largely came from improvements in the efficiency with which resources were used. Gross investment in the state sector in 1992 was around 8 per cent of GDP, which is far too low to sustain rapid economic growth. (This did not, however, include investment in the private sector and in own-capital of SOEs.) Raising the rate of savings and investment has therefore had a high priority. In 1994 the estimate for all investment had risen to about 20 per cent of GDP, of which foreign investment takes up a large share.

In order to sustain these increases, reform of the financial sector has been a major area of concern in the 1990s. In 1990 the government separated commercial banking from the central bank functions of the State Bank of Vietnam and established four commercial banks which, in principle, should lend only according to commercial criteria. However, the process of commercialisation of the banking system has been slow. During the period of negative real interest rates, the demand for funds by SOEs remained high, while savings deposits were low and many banks were technically insolvent. When interest rates were raised again to positive levels in 1992, many SOEs were unable to repay their loans and the State Bank was forced to provide funds to cover these non-performing debts which meant that the system of subsidising SOEs continued by an indirect method. More recently the Agricultural Bank, in particular, has been under pressure to shift its lending portfolio away from SOEs towards private farmers and it has made some progress. But given the lack of adequate accounting practices in all Vietnamese enterprises, banks have little genuine ability to assess commercial risk and therefore continue to prefer SOEs because of the greater certainty that they will be bailed out by the government. Private ventures have had to rely on informal borrowing which is both more risky and limits the capital available. This is one of the factors restricting the growth of large-scale private capitalists in Vietnam—a situation with which many in the Party may be content, if they see an emerging bourgeoisie as a threat to Party rule.

The second area of concern in relation to sustained economic growth has been the fiscal system. As explained above, the main sources of finance for state-led industrialisation were the profits of industrial enterprises and foreign aid. However, by the 1980s the majority of SOEs were clearly unprofitable and, of over 12 000 then existing, the government was forced to rely on only a few hundred to provide the bulk of revenue. The flow of aid from the West remained slow, while Soviet aid was withdrawn in 1991. Expenditure needs, on the other hand, continued to grow and the resulting large budget deficits could only be financed in an inflationary way. The need to stabilise the economy by reducing the budget deficit provided a strong impetus to reform of the fiscal system, both to increase revenue collection and to widen the tax base to include the private sector.

Revenues have increased substantially, from a low of 11 per cent of GDP in 1988 to 26 per cent in 1994 and the deficit has been greatly reduced. Since 1986 some of the revenue burden has shifted from SOEs to taxes on foreign trade and oil production. However, attempts to increase the private sector's tax share have been less successful. Personal income tax provides only 1 per cent of revenue (World Bank 1994: 127; IMF 1994: 58), and other taxes have not been increased. It is for this reason that the government intends to introduce a value-added tax within the next few years. One of the main factors leading to popular resistance to tax payments is that, in recent times, the severe compression of local budgets during the fiscal crisis led local authorities to extract 'contributions' from the population. This practice is often illegal and, rightly or wrongly, has been tarred with the brush of corruption. People now question why they should pay twice, to the central government and to the localities.

A third issue of concern to the sustainability of growth is the environment. The socialist system was certainly not free of environmental problems (Beresford and Fraser 1992), but given the priorities of the regime and the lack of countervailing power, these did not become an important issue. The emergence in the government's agenda of environmental issues in the 1990s is testimony both to its greater openness and to the increasing ability of the people to organise themselves. As the growth rate has accelerated, the socialist economy's legacy of poorly developed infrastructure and dirty industrial plants has proven highly inadequate. In rapidly expanding urban centres like Hanoi and Ho Chi Minh City, the colonial-period drainage system can no longer cope with the volume of effluent; old factories, new construction sites and growing numbers of motorbikes pour dust and chemical pollution into the atmosphere; public transport systems are run down and traffic clogs the multitude of narrow streets. In the country as a whole, only 17 per cent of households have access to a reticulated water supply or deep drilled wells with a pump, while the rest rely on hand-dug wells, rivers, lakes, natural springs, ponds or rainwater (SPC/GSO 1994). The latter group are most vulnerable to pollution and are concentrated in the Mekong River delta.

Institutional frameworks for dealing with environmental problems remain weakly developed and when conflicts of interest arise, solutions may be the result of an *ad hoc* public protest, as in the case of the successful campaign against construction of a coal washing plant in a residential area (Beresford 1995a). The government has also acted firmly in cases where public safety is threatened (see above). However, it seems more likely to act in favour of those with market power; that is, when pollution and congestion become evident bottlenecks to growth. For the time being, environmental guidelines are often flouted.

Equity

The distributional impacts of growth are very much on the agenda and, in particular, the question of corruption and lack of transparency at all levels of the state apparatus. These issues are seen as central to maintenance of the regime's legitimacy and its credentials as a leader of the transition to socialism. During the crisis period of the late 1970s and 1980s the Party suffered a loss of legitimacy and its attempt to rebuild is largely based on the simultaneous attempt to sustain economic freedoms and promote growth while providing a social safety net for the poor. However, the Party and government contain a diversity of views on the relationship of growth and equity, leading to a continuing struggle over relative priorities.

Some of these concerns are reflected, for example, in the labour and land laws passed during 1993–94. The Labour Law contained many provisions aimed at protecting the rights of workers, but it is by no means clear that these will be fully implemented if, as is also the intention, foreign direct investment and privated domestic investment are also to be encouraged. The Land Law also prominently addressed the security needs of farmers and consequently contains a number of provisions inimical to the development of markets in land and to the development of capitalism in agriculture. This too may turn out to be impossible to implement in many areas.

The future of the health and education sectors are also key equity issues. The legacy of the socialist system in this area was relatively high levels of literacy and public health indicators (Beresford 1995b) as well as high popular expectations, particularly in relation to education. In the fiscal crisis of the late 1980s and early 1990s, however, public spending in both these areas dropped significantly. Teachers and health workers found it impossible to live on their state salaries and most took up second occupations or left the system completely. Schools began to demand contributions from parents in order to maintain classes, but with staff frequently demoralised by their poor wages and conditions, standards fell and well-off parents found it necessary to pay for private tuition outside school hours. In health care, there has been a shift away from use of clinics and even private practitioners towards self-prescribed drugs which are freely available in the new market environment. As state revenues have begun to recover, there are signs of some improvement in health sector funding, especially at the local level, and foreign non-governmental organisations are also now active. Education remains more problematical as it not only requires more funds (private sector education can take a small portion of the burden for this), but extensive restructuring of the curriculum to adapt to the needs of a market economy.

Centre-local Relations

Centre-local relations have been one of the most difficult areas of reform within the state apparatus. From the point of view of the centre, its capacity to manage the economy is affected by the distribution of power existing at the outset of the reform process, by uneven regional development and by the degree of integration of local levels into the national administrative apparatus. Problems can emerge in particular from:

(a) regional autarchy arising from the operation of the socialist system with incomplete market reform giving rise to a situation in which regions are able to defy central policy directions (this also refers to the rise of local networks using family, Party and business connections to build quasi-independent control over a protected local economy);

(b) the emergence of wide regional disparities between provinces and regions together with essentially contrived attempts to equalise income differentials which can lead to resentment (by wealthier regions at having to subsidise the poorer ones and by the poor regions at perceived privileges meted out to the more prosperous).

Where local authorities are vested with significant power, reforms aimed at centralising control over the levers of economic management, especially fiscal policy, can be seriously undermined.

Under the impact of market-oriented reforms, competition between centre and localities over budget resources became a major issue. Under the traditional socialist system, local budgets of the poorer regions or those with a large number of 'political' production units were heavily subsidised by the centre. After reform, these subventions were reduced since chronic fiscal deficits could no longer be tolerated because they fuelled inflationary tendencies. The problem was exacerbated by low revenue raising ability of the centre. However, increased buoyancy of revenues since 1993 and corresponding deficit reduction could reduce pressure on the localities from the centre. Nevertheless, most provinces continue to require subsidies which leads to tensions between these and the wealthier provinces.

The purpose of reforms decentralising economic decision-making powers initiated in 1979 was to unleash productive energies of individual units and the new policies did not at first contain any well articulated regional dimension. However, the emergence of a *de facto* regional 'policy' can be seen in Vietnam where the autarkic policies of the 1970s have been replaced by a more market-oriented system with promotion of trade and investment links with the international economy. In the 1980s regional differences in growth rates began to emerge more strongly, with the southern region around Ho Chi Minh City and the Mekong delta seen as a model to be emulated by other areas. The northern industrial centres around Hanoi and Haiphong have more recently experienced similar rapid rates of growth, while some rural provinces, especially in the highlands, have lagged behind. The two industrial regions have not received differential treatment in fiscal relations as in China, but the relative independence of the southern region has nevertheless been bolstered by its greater wealth, corresponding lower dependence on central subventions and higher levels of rents.

The normal expectation is that as the level of market influence extends throughout Vietnam there will be a gradual erosion of the strength of regional political autonomy and of the ability of local authorities to allocate resources independently of or in contradiction to the direction of central policy. Higher levels of regional integration will ensure that it is in the interests of localities to contribute to central budgets for maintenance of inter-locality infrastructure and for macroeconomic stabilisation and growth policies. However, this will not occur if localities are at the same time denied resources needed to sustain local growth and social welfare systems. The answers will have implications for the distribution of accumulation and growth, the efficiency of resource allocation and the longer term ability of the centre to manage economy. These questions are also closely related to the equity issues discussed above. At the present time, capital construction expenditures (including those financed by aid) by the central authorities are tending to be spent on large infrastructure projects which will have benefits primarily for the high growth regions and for the centrally managed SOEs and foreign investments concentrated in those regions.

State Enterprise Reform

Financial autonomy of SOEs has been progressively increased since 1981, allowing the process of capital accumulation and cross shareholdings. Since 1989 direct budget subsidies have ended and cheap credit has been progressively eliminated with the result that enterprises must restructure and become profitable or close down. Although many assets of these enterprises are still owned by the state and a capital 'fee' is charged, the growth of relatively autonomous capital accumulation has implied an increase in the proportion of 'own-capital' which is regarded by the enterprises themselves (though not by the bureaucracy) as their collective property. Workers are entitled to a share in the enterprise profits in the form of bonus and welfare funds which grow in size as productivity rises and serve to reinforce the collective ownership ethos. In at least a few SOEs bonus funds have been distributed in the form of shares (although these are not yet legally recognised). Labour recruitment practices also reflect this ethos, especially in the north, where many enterprises try to provide employment security and to recruit new workers from among family members of the existing workforce. The question of ownership of the SOEs is therefore very unclear. Increased autonomy from government control suggests a process of 'corporatisation' of public enterprises (Beresford 1993), while there also appears to be developing a corporatist, rather than a confrontational, model of labour relations.

Corporatisation of the state-owned sector does not necessarily represent the formation of a private sector. Administrative interference in the operation of enterprises still exists, both directly via the line ministries and indirectly via the banking system. Soft budget constraints on enterprises, although considerably reduced since the beginning of the reform, are also still in existence. The political difficulties associated with imposing genuine bankruptcy on unprofitable SOEs mean that, in common with other reforming planned economies, Vietnam has been subjected to a stop-go cycle in relation to financial discipline. There still exists a lot of confusion between the macroeconomic management role of the *government* in a market economy (as the reformers see it) and the productive function of the *state* ministries and enterprises. Moreover, there is now considerable anecdotal evidence pointing to the formation of a state 'business interest' (Fforde 1993: 310). This is not 'nomenklatura capitalism' as it does not yet imply the privatisation of assets—ownership of capital assets by production units and ministries remains dependent on their positions within the state apparatus. In fact, attempts to use state positions in order to accumulate *private* capital have been clearly identified as corrupt and, where possible, severely punished. The state 'business interest' forms a powerful, sectional interest within the state preventing the establishment of a capitalist economy based on private property, most clearly by creating barriers to entry for new capital. Moreover, given the power of the 'state business interest', the continuing low rate of investment by this sector must throw some doubt on sustainability of growth.

State enterprise reform remains one of the most sensitive areas in the mid-1990s. As market discipline on enterprises becomes greater, pressure on workers seems likely to increase, both for increased productivity and to end the sharing of profits between worker funds and investment funds. Moreover, foreign investors and international competition are likely to generate pressure to halt the real wage increases which state sector workers have experienced in recent years. Development of a stronger labour market and the movement of new, inexperienced workers into the waged labour force is also likely to see dilution of the culture of social or collective enterprise ownership. While some discussion currently exists about worker participation in privatisation or equitisation schemes for SOEs, little has been achieved.

Equitisation of enterprises is presently considered a more appropriate model of ownership reform than privatisation and implies that the state will retain a considerable share in enterprise ownership, along with managers, workers, other SOEs and private capitalists. Some pilot projects had been carried out in small enterprises by the mid-1990s, but there were as yet no concrete plans for large scale equitisation schemes. Steps had also been taken to increase corporatisation through measures to remove the control of line ministries over production-related decisions and there were plans to amalgamate or abolish these ministries. In several industries, enterprises were to be incorporated into large vertically integrated corporations within which most enterprises would retain their autonomy in investment, production and marketing decisions. However, unprofitable enterprises and those producing inputs for the whole sector would be under the control of the corporation. If successful, this should go some way towards reducing the duplication of capacity which existed when SOEs were under the control of different ministries and provinces. Given the interest of ministries and provinces in retaining control over these enterprises, however, this change is likely to be slow.

Conclusion

The planned economy in Vietnam by no means corresponded to the 'command economy' model favoured by Western commentators on 'actually existing socialism'. Rather it was a system in which the demands of conflicting interest groups within society were negotiated through the mediation of the state apparatus. These interests included sections of the state apparatus itself which derived power from their control over the planned allocation of resources; local authorities, enterprise and cooperative managers, and their dependants who strove to minimise surplus extraction to higher levels and accumulate capital within their own sphere; and lastly, those who lacked access to the networks holding local power. The fundamental legitimacy of the regime, due to its nationalist credentials and reliance on popular mobilisation during wartime, not only blunted any social conflict inherent in the system, but created a tradition in which negotiated solutions were possible. After unification of the country in 1976 the shift in the balance of interests in favour of relatively independent capital accumulation led to an increasing erosion of the power of interests associated with the planning system as inflation tended to aggravate shortages generated within the plan. Serious economic crises in 1978–80 and in the mid-1980s progressively destroyed the remaining legitimacy of the planning system within the central apparatus itself and led to the political consensus around *Doi Moi*. Moreover, by the end of the decade, structural and institutional change had proceeded far enough that abolition of the remaining vestiges of planning produced little serious dislocation of the economy.

State-led marketisation of the economy under *Doi Moi* has very much reflected the interests of those groups which rose to dominance during the early part of the transition: SOE managers (often in alliance with the new group of foreign investors) as well as local and provincial kinship and political networks linked to relatively autonomous capital accumulation processes. However, within the increasingly plural Vietnamese society new groups are emerging which are not necessarily linked to the above networks. These include private domestic capitalists and may also include wealthy farmers. Until now, their potential for expansion has been limited by discrimination in policy implementation and lack of capital, and they are unlikely to become a dominant class in the foreseeable future. Those SOE workers who remain in employment after the shakeout of the early 1990s have also benefited through their ability to share in the profits of enterprises which most still regard as collectively owned. However, as market discipline on enterprises increases over time and a labour market becomes more established, these benefits are likely to diminish.

At the bottom end of the scale are those who have not participated in capital accumulation and are largely excluded from state power: poor peasants, the unemployed and underemployed. Women are the majority of all these categories. For many, their livelihood has become increasingly precarious, although there is so far no evidence that this is true for the great majority. Nevertheless, the crisis in health and education, increasing regional differentiation and increasing income differentials mean that there is potential for social conflict. Therefore, not only the sustainability of growth, but equitable distribution of the benefits are likely to remain important issues for some time to come. Within the Party-state apparatus the perceived need to maintain the legitimacy of Communist Party rule is also likely to ensure, however, that the interests of the poor are not completely ignored. Although what we have called the 'state-business bloc' remains powerful, there are also many in the Party and bureaucracy who see a continuing need for compromise with the interests of the poor in order to block the rise of opposition parties based on interests associated with private capital.

---

[1] The category of 'middle' peasants, defined as those producing 'on average' sufficient for household subsistence and not requiring hired labour, is problematical as it includes many above-average farms which produced a marketable surplus. In the southern region, in the 1970s it was the 'middle' peasants, accounting for 72 per cent of the rural population, who produced the bulk of the south's large marketed grain surplus (Beresford 1988: 149).

[2] In the late 1970s, for example, there were reports that procurement officials were sometimes arrested by cooperative members and their own bicycles were appropriated because the centre was unable to supply the contracted amounts of industrial goods in exchange for procurement quotas.

[3] The term 'relative independence' is used here since 'capital accumulation in the periphery' remains in many cases contingent upon access to resources, supplied at low official prices within the plan, which can then be diverted on to parallel markets.

[4] The system of subsidised input prices continued at around 60–70 per cent of the market price, while the government's wages bill rose by 220 per cent. The budget deficit rose from 24 per cent of expenditure in 1984 to 31 per cent in 1986 and 39 per cent in 1988 (Spoor 1989: 125; World Bank 1994: 126).

[5] Inflation during 1986 was 487 per cent p.a. (World Bank 1994: 131).

[6] Meaning that the rate was pegged a few points above the inflation rate.

[7] One of the characteristics of high inflation in all the former socialist countries has been increasing use by the population of the US dollar as the preferred currency for domestic transactions, since it holds its value better than the local currency.

[8] Party Secretary Do Muoi, in delivering the political report to the Mid-term National Conference of the Party held in January 1994, enumerated four basic challenges facing the country today: 'The challenges lie in: the danger of our economy falling further behind those of other countries in the region and the world due to our low starting point, our still low and unstable growth rate and the fact that we have to develop in an environment of tough competition; the possibility of our going astray from the socialist orientation if we fail to correct deviations from the path laid down for its implementation; corruption and other social evils; "peaceful evolution" schemes and activities undertaken by hostile forces'.

References

Beresford, M. (1987) 'Vietnam: "northernizing" the south or "southernizing" the north?', *Contemporary Southeast Asia* 8 (4): 261–75.

Beresford, M. (1988) *Vietnam: Politics, Economics and Society*, London: Pinter.

Beresford, M. (1989) *National Unification and Economic Development in Vietnam*, London: Macmillan.

Beresford, M. (1991) 'The impact of economic reforms on the south', in Dean Forbes et al. (eds) *Doi Moi: Vietnam's Renovation and Policy Performance*, pp. 118–35, Canberra: Department of Political and Social Change, RSPAS, Australian National University.

Beresford, M. (1993) 'The political economy of dismantling the "bureaucratic centralism and subsidy system" in Vietnam', in K. Hewison, R. Robison and G. Rodan (eds) *Southeast Asia in the 1990s*, pp. 213–36, Sydney: Allen & Unwin.

Beresford, M. (1995a) 'Economy and environment', in Ben Kerkvliet (ed.) *Dilemmas of Development: Vietnam Update 1994*, Political and Social Change Monograph 22, pp. 69–88, Canberra: Department of Political and Social Change, RSPAS, Australian National University.

Beresford, M. (1995b) 'Political economy of primary health care in Vietnam', in Paul Cohen and John Purcal (eds) *Health and Development in Southeast Asia*, pp. 104–19, Australian Development Studies Network.

Beresford, M. and Fraser, L. (1992) 'Political economy of the environment in Vietnam', *Journal of Contemporary Asia*, 22 (1): 3–19.

Beresford, M. and McFarlane, B. (1995) 'Regional inequality and regionalism in Vietnam and China', *Journal of Contemporary Asia*, 25 (1): 50–72.

CPV (1994) Communist Party of Vietnam, Political Report of the Central Committee (7th Tenure) Mid Term National Conference, Hanoi: The Gioi Publishers.

de Vylder, S. and Fforde, A. (1988) *Vietnam: An Economy in Transition*, Stockholm: SIDA.

de Vylder, S. and Fforde, A. (1995) *From Plan to Market*, Boulder: Westview.

Elliott, D. (1974) 'Revolutionary reintegration', PhD thesis, University of Michigan, Ann Arbor.

Fforde, Adam (1989) *The Agrarian Question in North Vietnam 1974–78*, New York: M. E. Sharpe.

Fforde, A. (1993) 'The political economy of reform in Vietnam—some reflections', in Borje Ljunggren (ed.) *The Challenge of Reform in Indochina*, pp. 293–326, Cambridge MA: Harvard University Press.

Fforde, A. and Paine, S. H. (1987) *The Limits to National Liberation*, London: Croom Helm.

Forbes D. K., Hull, T. H., Marr, D. G. and Brogan B. (eds) (1991) *Doi Moi: Vietnam's Renovation Policy and Performance*, Political and Social Change Monograph no. 14, Canberra: Australian National University.

Gordon, Alec (1981) 'North Vietnam's collectivisation campaigns: class struggle, production and the "middle peasant" problem', *Journal of Contemporary Asia*, 11 (1): 19–43.

GSO (1994) *Nien Giam Thong Ke 1993*, Hanoi: General Statistical Office.

IMF (1994) *Vietnam: Recent Economic Developments*, Washington, June.

Kalecki, M. (1969) *An Introduction to the Theory of Growth in Socialist Economy*, Oxford, Basil Blackwell.

Kerkvliet, B. J. T. (1994) 'Decollectivising the land: everyday politics, policy changes and village–state relations in Vietnam', unpublished ms.

Kornai, J. (1982) *Growth, Shortage and Efficiency*, Oxford: Basil Blackwell.

Luong, H. V. (1992) *Revolution in the Village: Tradition and Transformation in North Vietnam 1925–1988*, Honolulu: University of Hawaii Press.

Marr, David G. and White, Christine P. (eds), (1988) *Postwar Vietnam: Dilemmas in Socialist Development*, Ithaca: Cornell University, Southeast Asia Program.

Moise, E. E. (1976) 'Land reform and land reform errors in North Vietnam', *Pacific Affairs*, 49.

Ngo Vinh Long (1984) 'Agrarian differentiation in the southern region of Vietnam', *Journal of Contemporary Asia*, 14 (3): 283–305.

Shimakura, T. (1982) 'Cycles in the Chinese economy and their politico-economic implications', *Developing Economies*, XX (4).

Spoor, Max (1987) 'Finance in a socialist transition: the case of the Democratic Republic of Vietnam (1955–64)', *Journal of Contemporary Asia*, 17 (3): 339–65.

Spoor, Max (1988) 'State finance in the Socialist Republic of Vietnam', in David G. Marr and Christine P. White (eds) *Postwar Vietnam: Dilemmas in Socialist Development*, pp. 111–32, Ithaca: Cornell University, Southeast Asia Program.

Thayer, Carlyle (1991) 'Renovation and Vietnamese society: the changing role of government and administration', in D. K. Forbes, T. H. Hull, D. G. Marr and B. Brogan (eds) *Doi Moi: Vietnam's Renovation Policy and Performance*, pp. 21–33, Political and Social Change Monograph no. 14, Canberra: Australian National University.

SPC/GSO (1994) *Vietnam Living Standards Survey*, Hanoi: State Planning Committee and General Statistical Office, September.

White, Christine P. (1983) 'Mass mobilisation and ideological transformation in the Vietnamese land reform campaign', *Journal of Contemporary Asia*, 13 (1): 74–90.

World Bank (1994), *Vietnam: Public Sector Management and Private Sector Incentives Economic Report*, 20 September.

8

Labour and Industrial Restructuring in South-East Asia

Frederic Deyo

Economic development comprises a broadly based transformation of economic and social institutions, one of whose outcomes is sustained economic growth. The socio-political process through which this transformation occurs normally produces winners and losers as it alters patterns of economic claims and income flows. For this reason, economic development is typically a contested process, one in which shifting and emergent groups and coalitions contend for favourable economic positions in a changing and uncertain social order and in which the very nature and extent of development is an outcome of social and class contention.

Such a political-economy view of economic development may usefully be applied to an understanding of the role of organised labour in the rapid industrialisation of South-East Asia during recent years. Of particular importance here are changes in the 'labour systems' through which labour is socially reproduced, mobilised for economic ends, utilised in production, and controlled and motivated in support of economic goals. These changes are joint products of the economically driven labour strategies of government and business elites, of global political and economic pressures and constraints, of the process of industrialisation itself, and in some cases of the individual and collective responses of workers to elite strategies and industrial pressures.

This chapter explores the labour implications of recent economic development in the capitalist countries of South-East Asia, a region where rapid industrial development is being powerfully shaped and conditioned by two global transformations in economic ideology and policy. The first of these is the growing world influence of neo-liberalism, a broad economic approach emphasising increased reliance on market forces to direct international and national economic processes. The second involves adoption of more flexible post-Fordist production systems as a condition for success under intensified global economic competition. Neo-liberalism and the need for increased production flexibility have powerfully influenced the nature of industrial change in the region, and, by consequence, the emergent characteristics of labour systems as well. Thus, the global context in which industrialisation is occurring is closely related to the issues of contestation between workers and trade unions on the one hand, and on the other, governmental and business elites during the process of economic development.

The chapter begins with a brief overview of labour and development in capitalist South-East Asia during past years. It explores the way in which changing economic strategies of governments and firms over recent years have been reflected in corresponding labour policies, and the response of workers to those policies. It is seen that, with some exceptions, labour has been unable effectively to challenge elite strategies, and that labour's weakness derives in large measure from political constraint, rooted initially in the imperatives of regime survival itself, and subsequently in the perceived requirements of the economic strategies of both firms and governments. Equally important, however, have been the labour impact of economic structural changes themselves, the temporal sequencing of political and economic change, and the world system timing of recent industrial deepening in a volatile post-Fordist global economy.

Labour and Development in South-East Asia

Until relatively recently, the economies of South-East Asia were based largely on agriculture and service industries. As late as 1970 services alone accounted for two-thirds of GDP in Singapore, and 46 per cent in Malaysia. Services and agriculture together accounted for 75 per cent of GDP in Thailand and Malaysia, 67 per cent in Indonesia, and 69 per cent in the Philippines. These two sectors together accounted for even larger shares of the workforce, ranging from 73 per cent in Singapore to 95 per cent in Thailand (World Bank 1995).

Much of the large service sector consisted at that time of small commercial establishments, government services, and a heterogeneous grouping of informal-sector jobs, self-employment, and unpaid family labour. In agriculture, the overwhelming majority of the workforce owned or worked on small family farms, the remainder serving as plantation workers.

Trade unions and other types of labour organisations in these countries were confined largely to plantation workers (especially in Malaysia, Indonesia and the Philippines), dockworkers, miners, workers in large manufacturing companies, transportation workers, mill workers and, where legal, state enterprise employees (especially public utility workers) and civil servants. Despite low levels of unionisation and confinement of unionism to these few occupational sectors, organised labour played a significant role in national politics through to the mid-twentieth century. In Malaysia and the Philippines, labour militancy during the 1930s economic crisis strengthened calls for a realignment of development policy away from reliance on exports of primary products and toward protection and encouragement of domestic industry. In Indonesia, the Philippines and Malaysia, labour participated actively in anti-colonial independence struggles, especially after the Second World War (Ingelson 1981; Arudsothy and Littler 1993; Wurfel 1959). During the 1950s militant labour groups in the Philippines pushed successfully for enactment of what was, by regional standards, highly progressive labour legislation (Ofreneo 1995). And in Singapore, leftist labour mobilisation came close to creating a socialist government in the early 1960s, and clearly shaped the social agenda of the government that was established after the defeat of the Left (Deyo 1981).

By the late 1960s, however, restrictive state controls over trade unionism and labour activism had substantially tamed and depoliticised organised labour. In Malaysia ethnic and communal conflict which threatened both state structures and conservative ruling groups was met by tight controls over organised labour. More generally across the region, a political assault on the independence and power of organised labour, associated with Western-supported efforts to contain communism, initiated a downward spiral of union influence (cf. Hewison and Rodan 1994). In Singapore, Chinese communalism precipitated a violent confrontation with Anglicised moderates in the ruling People's Action Party (PAP) eventuating in decimation of the Chinese Left, consolidation of moderate PAP leadership, and the instituting of tight pre-emptive controls over organised labour through a government-dominated National Trades Union Congress (NTUC). While labour organisation and agitation in the Philippines initially elicited a response of political accommodation and establishment of a liberal, USA-modelled labour relations system, it was met in later years by ever harsher restrictions, culminating in 1972 in martial law repression justified, in part, by the need to contain left radicalism and popular sector insurgency (Hutchison 1993).

Similarly, organised labour in Thailand was effectively suppressed under intermittent anti-communist military rule over much of the 1940s–1960s period, especially after the 1957 coup (Hewison and Brown 1994), while in Indonesia, following the 1965 military coup which sought to eliminate growing leftist influence in national politics, the powerful, communist-aligned, national labour federation SOBSI was banned and later replaced by a government-sponsored federation.

But this was not to be the end of labour influence in the region. Even under tight political controls, labour continued in some countries to wield moderate influence in policy-making and implementation, in many cases through community-based political mobilisation only loosely linked to trade unionism. Labour's oppositional potential became especially evident during interludes of political crisis and government transition. Following the collapse of military rule in Thailand in 1973, for example, labour militancy frequently paralysed business in Bangkok and elsewhere. During the subsequent three years of open politics, labour pressed successfully for enactment of new legislation which provided for union recognition and collective bargaining rights, established a minimum wage and required employers to provide a number of new benefits, including workers' compensation. These labour victories were to provide a foundation for re-establishment of worker rights under subsequent democratic governments in the late 1980s and 1990s. Similarly, it will be seen below that during the 1980s economic crisis in the Philippines, a seemingly moribund labour movement mounted strong opposition to structural reforms, as well as to the Marcos regime itself, in the context of a growing debt and economic crisis. In Singapore and Malaysia a greater degree of political continuity precluded such periods of dramatic labour mobilisation. But even here, workers and unions continued to press employers and government for improved wages and benefits during the 1960s and into the 1970s.

If by the late 1960s, South-East Asian labour movements had already been partially tamed under restrictive political regimes, subsequent decades were to see a further diminution in labour influence at enterprise and national levels. This continuing decline may be documented by reference to changes in union density, collective bargaining and the political role of trade unions. Regional union densities are now relatively low, ranging from a high of 17 per cent in Singapore to roughly 5 per cent in Thailand. In many cases, densities have persistently declined over recent years: in Singapore from 24.5 per cent in 1979 to the current level of 17 per cent by 1988, and in Malaysia from 11.2 per cent in 1985 to 9.4 per cent in 1990 (Kuruvilla 1995). Union density in Malaysia's private sector is especially low, at only about 7 per cent (Arudsothy and Littler 1993). During the 1980s unionisation rates in Thailand's private sector grew slowly from a very low base. However, the forced dissolution of unions in the highly organised state enterprises following the 1991 military coup brought a sharp overall union decline and subsequent stagnation in the private sector extending, significantly, into the post-coup restoration of democratic rule. And in Indonesia, despite intensive organisational recruitment efforts during the early 1990s, no more than a tiny fraction of workers were successfully enrolled in plant-level affiliates of the official national trade union federation, SPSI (Lambert 1993).

The weakness of organised labour in the region is further reflected in the restriction of collective bargaining agreements to a very small percentage of employed workers (roughly 5 per cent in Thailand, the Philippines and Malaysia) (Arudsothy and Littler 1993; Brown and Frenkel 1993; Kuruvilla 1995).

At national levels, the weak political position of organised labour is similarly evident across the region. In Singapore and Indonesia, the dominant national trade union federations, the NTUC and SPSI respectively, are so tightly integrated into ruling party structures that they cannot provide an independent channel of representation for workers (Deyo 1989; MacIntyre 1994). In Malaysia the largest national federation (the Malaysian Trades Union Congress) is losing ground to a more conservative federation (the Malaysian Labour Organisation) promoted by and closely aligned with government. While Thai state enterprise workers achieved some limited political goals in 1980s, including, most notably, passage of the comprehensive Social Security Act of 1990, the trade union movement has more generally been marked by divisiveness and factionalism, seen most dramatically in the existence of five competing major union federations, each supported by high-level military and/or bureaucratic elites. Labour movements in the Philippines and Malaysia suffer from similar problems of disunity and internal competition (Hutchison 1993; Crouch 1993; Arudsothy and Littler 1993).

The weak political position of labour is perhaps most dramatically seen in the relative inability of workers to seize new opportunities flowing from democratic reforms in Thailand and the Philippines. In both cases, workers joined with middle-class groups in opposing military rule and in instituting more open electoral regimes. But in neither case has the organisational strength or collective bargaining position of trade unions been subsequently enhanced. To the contrary, labour movements under the new democratic regimes have remained divided and stagnant.

In this context, it is not surprising that elite developmental and labour policies have not been significantly challenged by organised labour. And to the limited extent labour opposition has been politically consequential it has taken other forms, typically involving spontaneous action on the part of small groups of workers or, alternatively, community groups protesting government policies or inaction in matters of local concern.

It is noteworthy that stagnation in South-East Asian labour movements has occurred in the context of a number of social transformations which have historically been associated elsewhere with the empowerment of labour and other popular sector groups. At the base of these transformations is the continuing course of industrial development itself. Singapore,

Malaysia, the Philippines and Thailand all have substantial manufacturing sectors and provide industrial employment for a sizeable portion of their workforces (see table 8.1).

Table 8.1 Industrial and social indicators in South-East Asia (%)

	Manufacturing as % of GDP[a] (1993)	Heavy industry* as % of total manufacturing value-added[a] (1992)	% Industrial[t] employment[b] (1988–93)	% Urban population[a] (1993)	% Literacy[c] (1994)	Wage-employment as % of non-agricultural total (1988–93)[d]
Singapore	28	63	34	100	87	86
Malaysia	na	45	28	52	80	80
Thailand	28	45	14	19**	89	63
Philippines	24	23	16	52	88	65
Indonesia	22	21	14	33	86	82

Notes:
* Chemicals, machinery, and transport equipment.
[t] Mining, manufacturing, construction, utilities.
** Based on household registrations. Since many urban dwellers maintain official residence in rural areas, the actual percentage of 'urban' is higher.
[a] World Bank (1995).
[b] ILO, *WLR* (1995).
[c] *World Almanac* (1994).
[d] Frenkel (1993): 18.

In addition, these countries boast high levels of literacy and education, usually considered useful predictors of social awareness and political participation. Industrial development in Singapore, Malaysia and Thailand has been associated with industrial deepening into more sophisticated, technology- and capital-intensive production. Important too is the emergence across the region of an ever more settled and thus organisable urban proletariat (see Hadiz 1994), an important basis for early unionisation in Western countries.

But despite these seemingly propitious socio-economic changes, rapid industrialisation has nowhere spawned effective trade unionism or enhanced worker participation in political or economic arenas. In part, the weak collective bargaining and political role of South-East Asian labour during recent years derives from a continuation of earlier forms of political constraint. But unlike the experience in earlier periods, these controls have been oriented less to the political imperatives of regime stability than to the requirements of economic strategies and initiatives. Because of this increasingly close relationship between labour regimes and developmental initiatives (see Deyo 1989; Kuruvilla 1994) we consider them together in the discussion that follows, emphasising throughout the ways in which the successful implementation of each of these strategies has been rooted in, while subsequently reinforcing, the weakness of South-East Asian labour movements. The strategies of particular regional interest during recent years include the shift to light, export oriented industrialisation (EOI), economic liberalisation and structural adjustment, industrial deepening under 'second-stage' EOI, and most recently, adoption of post-Fordist flexible production systems in manufacturing. While light EOI and neo-liberal economic reforms are common to all the capitalist countries in the region, the implications of industrial deepening and post-Fordist flexibility are greatest in the industrially most advanced countries (Singapore, Malaysia, Thailand and, arguably, the Philippines). For this reason, much of the remaining discussion centres on these countries.

Labour Under Early Export Oriented Industrialisation

Light industry based EOI in Singapore and the Philippines in the 1970s and in Malaysia, Thailand and Indonesia in the 1980s, was premised on the successful mobilisation of low-cost semi-skilled labour for the assembly and manufacture of products for world markets. In Singapore, Thailand and Malaysia, and in the export processing zones of the Philippines and Indonesia, light EOI relied heavily on direct foreign investment. The twin pressures of cost-containment on the one hand, and multinational locational requirements for both political stability and low labour costs on the other, encouraged governments to impose or enhance stringent constraints on labour.

In Singapore these pressures elicited new labour legislation in 1968 which reduced permissible retrenchment benefits, overtime work, bonuses, maternity leave and fringe benefits. Thenceforth, unions could not demand, nor could managers offer, benefits greater than those stipulated under law. In addition, the National Wages Council, a tripartite body, established wage guidelines which held industrial wages down through the 1970s. And a second statute gave management full discretionary power in matters of promotion, transfer, recruitment, dismissal, reinstatement, assignment or allocation of duties and termination. These topics were now removed by law from the range of legally negotiable issues. This array of legislation, along with a parallel set of new investment incentives, was followed by a wave of foreign investment that continued until the mid-1970s recession (Deyo 1989; Begin 1995).

In Malaysia, where legislation similar to that in Singapore also removed from collective bargaining a broad range of personnel matters, union organisation and collective bargaining was banned throughout the early 1980s in electronics, the single most important manufacturing export industry. Union militancy was repressed as well in the export-processing zones of the Philippines.

In the wake of declining oil prices in the early 1980s, the Indonesian government intensified labour controls under a more centralised national labour federation and gave special encouragement to light, export orien- ted manufacturing in order to replace diminished oil export earnings (Hadiz 1994: 195). In Thailand, even under democratisation, exclusion of state enterprise workers from union coverage continues to undermine the national labour movement insofar as the state enterprise unions have tradi- tionally played the lead union role in this country. And in the Philippines the new Aquino regime took a hardline stance on strikes defined as illegal under existing Marcos-era labour legislation (Ofreneo 1994).

The association between early EOI and cost-containment labour regimes has been noted frequently in the literature on East Asian industrialisation (Kuruvilla 1994). A recent International Labour Office (ILO) study of world employment trends explicitly links the globalisation of capital, production and markets with reduced labour protection in developing countries, arguing that global integration has adversely affected the ability of governments to enact and enforce labour legislation. Globalisation, notes the ILO, gives 'governments an incentive to dilute, or fail to enact, measures intended to protect the welfare of workers, or to turn a blind eye to infringements of legislation with this in mind' (ILO 1995a: 72–3). This same report notes, as well, an increasing erosion of the quality of formal- sector employment through reduced job security, the diminished signifi- cance of local or national-level collective bargaining, new policies of 'firm-centric cooperation,' union-avoidance strategies, promotion of com- pany unions, and reduced access to information and diminished bargaining leverage on the part of local unions.

Less well documented has been the extent of labour opposition to such labour practices and the cost-focused labour systems they have created. Throughout the region groups of young women, newly hired in export- manufacturing industries, have protested against low wages and benefits, harsh working conditions and lack of enforcement of labour standards leg- islation. Such protest was especially pronounced in the export processing zones of the Philippines and Indonesia (Lambert 1993).

But having said this, it remains the case that such protests have rarely brought enduring gains for workers. This follows, in part, from the char- acteristics of employment in light export sectors, with their relatively transitory labour force comprising large numbers of young women in unstable low-skill jobs with little career opportunity. The difficulties of organising such a workforce into effective unions along with the associated difficulty of mounting well-organised pressure on employers continue to undercut labour movements among workers in the large export manufacturing sectors of South-East Asia (Deyo 1989).

Economic Liberalisation and Structural Adjustment

Compounding this EOI-linked structural demobilisation of labour are continuing international pressures, often associated with ongoing regional and global trade agreements to further open domestic markets to imports. Trade liberalisation has, in turn, subjected firms to intensified competition in both domestic and international markets. In developing countries, with their relatively labour-intensive, export oriented industrial structures, managers have sought to meet these new competitive pressures through cost-cutting measures directed in large part at reducing labour costs. Such measures have, in turn, both reflected and reinforced labour's already weakened bargaining position, for competitive pressures have created a credible threat of shutdowns, retrenchments and relocation of production to cheaper labour sites in the absence of effective labour cost containment. In addition, some cost-cutting measures, including the increased use of temporary and contract labour and greater out-sourcing of production, have directly undercut organised labour while at the same time addressing a second set of competitive requirements, discussed below, stemming from the globalisation of post-Fordist production systems.

A broader set of structural adjustments, including privatisation of state enterprises, reduced state regulation of the economy and reduced public expenditures, has been urged by international agencies and lenders, including, most prominently, the World Bank, the Asian Development Bank and the International Monetary Fund in response to high levels of indebtedness among some countries in the region (Hutchison 1993). Privatisation, most prevalent in Thailand, Malaysia and the Philippines, in turn subjects large firms in public transport, communications, and other sectors to heightened competition, and thus to the possibility of workforce retrenchment, and of wage and benefit reduction.

These and other neo-liberal economic reforms were often met by labour opposition and public demonstrations. In Malaysia, Thailand and the Philippines trade unions fought vigorously for legislation to restrict the use of temporary workers. In Thailand public sector unions successfully slowed privatisation during the late 1980s through mass demonstrations in central Bangkok. And in the Philippines, a wave of strikes and public demonstrations opposing structural adjustment measures during the early 1980s contributed to the political crisis which ultimately brought down the Marcos regime.

These labour campaigns against economic liberalisation slowed, but did not stop, ongoing economic reforms. In Thailand the use of temporary workers continued to expand despite legislation restricting the practice. Efforts to stop privatisation met with somewhat greater success, particularly in Thailand. There, as elsewhere, effective opposition drew strength from a cross-class defensive coalition of labour and bureaucratic/military elites whose economic interests and power base were threatened by privatisation of the enterprises they controlled. In addition, state enterprise workers were politically insulated from many of those competitive pressures which progressively undercut organised labour in the private sector. Thus, they continued to struggle from a position of relative strength. But in the end, larger state interests dictated a silencing of opposition from this 'un-restructured' Thai labour sector through a banning of all state enterprise unions following the 1991 military coup.

The more general undermining of labour movements under these various economic reforms has been furthered by a third strategic initiative, this time at the enterprise level, involving the incremental introduction of flexible production systems in manufacturing sectors during the 1990s. The following discussion assesses the labour impact of this most recent industrial transition in the context of industrial deepening.

Labour Under Recent Industrial Restructuring

In response to heightened economic competition and uncertainty, firms in South-East Asia, as elsewhere, are seeking to enhance manufacturing flexibility and adaptability. Flexibility here refers to the ability quickly, efficiently and continuously to introduce changes in product and process. Such flexibility yields a superior capacity to respond to the intensified pressures of liberalised trade, world market volatility, market fragmentation, heightened demand for just-in-time production and continuous improvements in productivity and quality, and rapid technological change.

A useful distinction is often made between 'static' and 'dynamic' forms of flexibility. Static flexibility, which focuses on short-term adaptability and cost-cutting, is the predominant managerial approach in the labour-intensive export sectors which have figured so prominently in many of Asia's developing economies. The reasons for this strategic choice are clear. Intensified global competition under trade liberalisation places firms under extreme pressure to cut costs in the short term. Risky long-term investments in training, research and organisational development may be eschewed where they seem to place a firm at a short-term disadvantage *vis-à-vis* other firms which do not make these investments. In some countries, lack of adequate public investment in collective goods (for example, training, R&D, physical infrastructure) further discourages such long-term investments, while lack of effective government support for minimal labour standards, adequate wages and benefits, and fair employment practices encourages firms to compete through labour-cost reduction and union avoidance.

That such static flexibility, with its negative consequences for labour, predominates in many countries and industrial sectors in the region is clear. There is evidence of increased use of subcontracting, casualisation and contract labour ('numerical flexibility') in the large export sectors of the Philippines (Ofreneo 1994) and Thailand (Deyo 1995). Guy Standing (1989) similarly documents increasing numerical flexibility and casualisation in electronics and other export sectors in Malaysia. Indeed, even in industrially advanced Singapore, James Begin (1995) reports continued reliance on static numerical flexibility. In many cases, employers adopt such labour strategies, in part, explicitly to undercut unions or unionisation drives, for these strategies have the known effects of creating an insecure, floating workforce and of encouraging a further dispersal of production to small, contracted firms and households. As a result, unions throughout the region have fought strenuously to institute legislative restrictions on the employment of temporary workers and similar practices (Charoenloet 1993; Ofreneo 1994).

Dynamic flexibility strategies, while less prevalent, are pursued in product niches requiring high levels of quality, batch vs. mass production, and continuing innovations and improvements in process and product technologies. Such strategies are encouraged where states underwrite a supportive social infrastructure of training, education and R&D, where they enforce adequate labour standards and where they provide incentives to firms to invest in training and organisational development. Finally, such strategies are most likely to be undertaken by large, resourceful firms which are able, in part, to create their own support infrastructure and which operate in relatively protected or oligopolistic markets characterised by moderate, rather than extreme, competitive pressure. For these various reasons, dynamic flexibility strategies tend to occur in the upper-tier NICs with developmentally active states (for example, in Singapore, South Korea and Taiwan) and among dominant firms in semi-protected industrial sectors across the region.

Most research on dynamic flexibility has focused on the experience of innovative industrial firms in the developed countries of Japan, Europe and North America. In these settings, dynamic flexibility has generally been associated with enhanced worker welfare and security, as well as with increased worker participation in organisational decision-making as firms have sought both to increase worker commitment and loyalty, and to encourage workers to assume increased responsibility for enterprise success. A distinction has been drawn in this regard between 'bargained' forms of flexibility which are associated with strong, independent unions and high levels of participation in instituting and operating new production systems, and 'participative' flexibility, characterised by captive, enterprise unions and more circumscribed forms of worker participation confined largely to shop-floor problem-solving (see Turner/Aur, Herzenberg, and other chapters in Deyo 1996). This body of research would seem generally to suggest that the instituting of dynamic flexibility, whether of bargained or participative forms, should have a similarly salutary effect for workers and perhaps unions in developing Asia. In fact, it has not.

In the higher value-added market niches where technological and product quality requirements preclude continued reliance on low-skill temporary workers and static flexibility, Asian firms are pressed to make long-term investments in worker training, product development, organisational restructuring and other programs supportive of enhanced dynamic flexibility. In addition, such firms may sometimes institute suggestion systems, modified quality circles, labour-management councils and other means of mobilising worker involvement in quality and productivity improvements (on Malaysia, see Rasiah 1994). But even such instances of dynamic flexibility, typically accompanied by improved wages and benefits and other measures to enhance the stability and commitment of workers, rarely permit the level of worker decision-making involvement found even under participative flexibility. Indeed, improvements in compensation levels and working conditions are as often introduced to avoid unions as to foster long-term organisational improvements (see, for example, Deyo 1995). In general, flexibility-enhancing organisational reforms are overwhelmingly attentive to managerial agendas driven by competitive economic pressures, to the exclusion of the social agendas of workers and unions. And in many cases, such strategies, along with their relatively benign labour welfare policies, are confined to a few critical production processes, thus fostering internal labour market dualism between core, stable workers on the one hand, and casual or contract workers on the other (Deyo 1995).

The reasons for the more autocratic forms of dynamic flexibility found across the region are not hard to discern. First is the vicious circle defined by initially weak labour movements, the subsequent reorganisation of industry exclusively around managerial goals and the resultant institutionalisation of autocratic forms of industrial flexibility. Here we see most clearly the way in which the political resources and effectiveness of labour determine institutional outcomes which reinforce existing power inequalities. And in the cases of Singapore and Malaysia, as noted earlier, state labour regimes have further encouraged the instituting of autocratic flexibility by excluding from collective bargaining such matters as job assignment and work transfers, which are important elements of 'labour flexibility'. Correspondingly, multinational corporations often insist on operating in a union-free environment (Rasiah 1994; Kuruvilla 1994).

Second, it may be that worker participation and empowerment is less critical to the success of programs of dynamic flexibility in developing Asian countries than in industrially mature economies. Following Amsden (1989), we may distinguish between innovative and learning-based industrialisation. Innovative industrialisation relies on development of a stream of new products and technologies for changing markets. Learning-based industrialisation, by contrast, relies on local adoption of technologies and products developed elsewhere. In so far as newly developing countries pursue technology-dependent, learning-based industrialisation, employers may seek to institute forms of flexibility which minimise worker participation in favour of unchallenged managerial control over production. This is so because learning-based industrialisation depends mainly on local adaptation and implementation of already debugged production processes and products, thus minimising need for an extensive involvement of workers in dealing with shop-floor production problems. In such a context, engineers and production managers assume the primary role in reorganising production around imported technologies. Thus, more autocratic forms of flexibility are adequate to the demands of industrialisation. Given that multinational firms, whose investments provide a major conduit for technology transfer and diffusion to Asian firms, are reluctant to relocate major R&D functions to foreign subsidiaries in developing countries, the perpetuation of learning-based industrialisation into future years may imply a long-term stability of such autocratic forms of flexibility and a corresponding discouragement of union or worker empowerment at the workplace level. In such a context, human resource mobilisation efforts will continue to confine collective forms of shop-floor participation to cooptive, officially sanctioned and closely circumscribed deliberative fora such as quality circles and labour-management councils, and more generally to eschew collective participation in favour of suggestion systems, informal consultation, merit-based incentives, job ladders and other individualised modalities of worker participation and involvement in organisational development.

In this context, state labour regimes have been substantially transformed. As employers have increasingly gained the upper hand in their dealings with

workers, labour market 'deregulation,' the counterpart of economic liberalisation and marketisation, has proven a more effective policy than has
continued repression. Such deregulation has not typically been accompanied by proactive labour protection measures which might provide a level
playing field for unions in their bilateral dealings with employees. The
Malaysian and Thai governments have ratified only eleven of the ILO's
labour conventions, and only minimally enforce existing labour standards
legislation, thus effectively subjecting workers to capricious managerial
domination, attacks on unions and non-compliance with minimum wage,
health and safety legislation (Brown and Frenkel 1993). In Thailand union
organisers receive no legal protection during organisational drives up until
the actual date of official union registration, thus impeding organisation
drives. In both Thailand and the Philippines labour-market deregulation
under democratic reforms, unaccompanied by corresponding measures
to strengthen and institutionalise trade unionism, has thus resulted in
increased employer domination at the enterprise level along with heightened union factionalism and conflict (Brown and Frenkel 1993).

Industrial Deepening: The Impact of Timing

It has been noted that, despite the political and economic constraints facing
organised Asian labour, there have been recent cases of dramatic labour acti-
vation. In the 1980s, resistance to structural reforms in the Philippines
provides the most striking exception to the broader pattern of decline among
organised labour. In order to understand this exception, as well as the more
general context of union weakness and decline, it is necessary to consider the
temporal context of the economic structural changes discussed earlier. In
particular, how have the domestic developmental sequencing of political and
industrial change and the historical timing of regional industrialisation influenced the consequences of development for labour? In this section, we
explore the impact of developmental sequencing. Later discussion examines
the effects of historical timing across the region more generally.
  The Philippines differs from the other countries in this discussion in its
continuing economic and industrial stagnation and relatively higher levels
of unemployment, which are powerful impediments to strong labour movements. In such an inhospitable setting, it is not surprising to find low and
declining overall union densities, negligible unionism in many industries,
and collective agreement coverage for only a small proportion of wage
workers. It is important to recognise, however, that the growing economic
crisis of the 1980s precipitated what Rene Ofreneo (1995: 3) refers to as a

'rise in militant unionism whose depth and breadth has no parallels in the country's history' (see also Hutchison 1993). In part, this militancy, which preceded the 'democratic coup' of 1986, can be attributed to a sustained and successful process of domestic market-oriented industrialisation which sets the Philippines off from other countries in the region and which parallels more closely the experience of several Latin American countries which pursued similar development strategies during the 1950s and 1960s. The Philippine 'exception' underscores the importance for labour movements of differences in developmental sequencing.

It is often noted that early Latin American industrialisation in the 1940s and 1950s was based on import substitution, while that occurring during subsequent decades in Asia was more strongly rooted in export manufacturing. Latin American import substituting industrialisation (ISI) sought initially to defuse a growing political crisis occasioned by the collapse of primary export-based development by fostering industrial growth, employment, and labour peace under policies of economic nationalism and the building of corporatist political coalitions which encompassed strong if dependent trade union federations. Protection of local companies from foreign competition permitted sustained industrial development along with ever higher wages and social benefits for politically supported trade unions in key economic sectors. Indeed, rising industrial wages were seen as supportive of continued industrial growth by increasing consumer demand for local products.

The Philippines, like these ISI-based Latin American countries, entered a sustained ISI phase during the 1940s and 1950s which combined successful import substitution with labour regimes, in this case more liberal than corporatist, which encouraged unionisation and collective bargaining. In response to a balance of payments crisis associated with the collapse of the primary-commodity export strategy of the early decades of the twentieth century, protective tariffs and foreign exchange controls marked a shift toward ISI-led development. This shift was, in turn, associated with enactment of new labour legislation which greatly enhanced worker welfare and union security in the formal sector of manufacturing. Especially important was new minimum wage legislation in 1951 and the Industrial Peace Act of 1953, subsequently dubbed the 'Magna Carta of Labour', which was patterned after the USA's National Labour Relations Act of 1935 in providing protection for trade unions and encouraging effective collective bargaining at the enterprise level (Ofreneo 1995). This 'misplaced' Latin American experience, stemming, in part, from USA political influence, eventuated in the emergence of strong local unions which, under subsequent years of martial law and state repression (1972–86), sustained a latent, community-based opposition movement. This movement provided the essential foundation for labour mobilisation and militancy during the economic and political crisis of the mid-1980s.

The Philippine case contrasts strongly with the developmental sequencing of political and economic change elsewhere in the region. Early industrialisation in most countries of the region was accompanied by authoritarian state controls which sought either to repress or to coopt organised labour. It was noted that these political controls, encouraged and supported by Western governments, comprised part of a larger global strategy of communist containment. Pre-emptively demobilised or coopted at the outset of sustained industrialisation, most Asian labour movements lacked the political capacity to shape new labour relations institutions in the early years of industrial development. The developmental labour systems subsequently created were to ensure continued labour subordination during later years of industrial deepening.

The Impact of Historical Timing

The historical timing of industrial change was as important as developmental sequencing for Asian labour movements. It was noted that labour-intensive EOI under post-Fordist production regimes differs from earlier standard-production EOI in shifting the locus of labour control from state to enterprise and in fragmenting and dispersing the workforce to a greater extent than in the earlier period. More generally, the global-temporal context of industrial deepening into more capital and technology-intensive production has shifted appreciably. Despite unfavourable developmental sequencing, earlier Fordist industrial deepening gave somewhat greater encouragement and scope to emergent labour movements in South Korea, Taiwan and elsewhere. Under flexible production regimes, by contrast, incipient deepening into high value-added industrial production in Thailand and Malaysia has encouraged pre-emptive enterprise participative structures which often displace or coopt unions. In Thailand, such resourceful and market-dominating firms as Toyota, Siam Cement and Yamaha have instituted skill-based job ladders, suggestion systems, enterprise unions, carefully circumscribed quality circles, dualistic internal labour markets and other measures which mobilise worker ideas and involvement while at the same time discouraging independent or oppositional collective action. In Malaysia, state encouragement for Japanese-style enterprise unionism contributes further to such an effect. More advanced technology-deepening in Singapore has been associated with even greater official sponsorship and encouragement of labour management councils, sponsored enterprise unions and other pre-emptive participatory fora.

Conclusion

To return to the central question guiding this chapter, it is clear that organised labour has not played a forceful role in the socio-political construction of the developmental labour systems through which enterprise and state elites have sought to further their economic strategies. In addition, organised labour's role has largely diminished over recent decades in response not only to political constraint, but also under economic structural changes associated with liberalising economic reforms and the introduction of new more flexible production systems in manufacturing. Finally, the developmental sequencing and world-system timing of development have further contributed to labour's continuing decline.

A number of political and developmental ramifications flow from the diminished role of South-East Asian labour. First, economic liberal reforms and recent post-Fordist industrial restructuring have been associated with a growing irrelevance and anachronism of state labour controls, increasingly supplanted by enterprise controls rooted in market discipline or cooptive participation, except in a few 'unrestructured' sectors (for example, Thai state enterprises) where repressive controls remain in place (cf. Hutchison 1993: 208). Second, the growing power of employers over workers at enterprise levels provides an opportunity for national governments to respond more fully to international demands for improved recognition of labour and human rights without threatening economic growth or political stability. Third, it also provides a solution to the problem posed by the internal instability of authoritarian regimes under sustained economic development and the rapid expansion of a new middle class, in part by enhancing the prospects for democracy by reducing the likelihood that labour can exploit new political opportunities offered by parliamentary reforms. Alternatively stated, the structurally rooted elimination of labour from democratic politics under regimes of 'exclusionary democracy' has enhanced the usefulness to business of parliamentary institutions, thus creating a critical political base for those institutions.

But what are the implications of a labour-exclusionary path of development for the economic future of the region? In the short term, a relatively weak popular sector role in economic policy processes (increasingly dominated by private firms) speeds development inasmuch as private firms and government agencies need attend but minimally to the social implications of their strategies. The longer term is less certain, however. Developmental sustainability requires, *inter alia*, a political base. Popular sector exclusion fosters inequities and alienation among large segments of the population, thus eroding this essential base and in the long term bringing increased resistance even to the democratic institutions increasingly favoured by elites.

Similar considerations apply to the sustainability of firm-level economic strategies as well. The absence of strong, politically protected unions along with human resource policies that are oriented mainly towards cost-cutting and static flexibility, have encouraged short-term opportunism at the expense of long-run organisational development. Where training and human resource investments have been undertaken, they have not been accompanied by increased worker participation in production decisions, so essential to higher levels of organisational adaptability and innovation. Nor have they generally been accompanied by the commitment-enhancing labour policies which provide a motivational foundation for the success of such participation.

States can play an essential role in this regard by providing the collective goods individual firms are unlikely to attempt, and by providing incentives and inducements for longer-term R&D, training, and organisational development on the part of firms. A good example of such an inducement is the use, in Singapore and Malaysia, of a skill development levy under which employers contribute to a general training fund, receiving back portions of that fund to be used only for approved employee training programs.

Without such external pressures and support from unions or states, short-term cost-cutting and static flexibility will characterise firm level strategies in much of the region, while consigning even the more advanced industrial sectors of these countries to learning-based, rather than innovative, global production and market niches. Only in Singapore and, arguably, Malaysia where strong states have played a more forceful role in encouraging training, research and development and other investments in organisational adaptability, are the long-term prospects brighter.

References

Amsden, Alice (1989) *Asia's New Giant: South Korea and Late Industrialisation*, New York: Oxford University Press.

Arudsothy, Ponniah and Littler, Craig R. (1993) 'State regulation and union fragmentation in Malaysia', in Stephen Frenkel (ed.) *Organised Labour in the Asia-Pacific Region*, pp. 107–32, Ithaca, NY: ILR Press.

Begin, James P. (1995) 'Singapore's industrial relations system: is it congruent with its second phase of industrialisation?', in Stephen Frenkel and Jeffrey Harrod (eds) *Industrialisation and Labour Relations*, pp. 64–87, Ithaca, NY: ILR Press.

Brown, Andrew, and Stephen Frenkel (1993) 'Union unevenness and insecurity in Thailand', in Stephen Frenkel (ed.) *Organised Labour in the Asia-Pacific Region*, pp. 82–106, Ithaca, NY: ILR Press.

Charoenloet, Voravidh (1993) 'Export-oriented industry in Thailand—implications for employment and labour', in Arnold Wehmhoerner (ed.) *NIC's in Asia: A Challenge to Trade Unions*, pp. 7–14, Singapore: Friedrich-Ebert Stiftung Foundation.

Crouch, Harold (1993) 'Malaysia: neither authoritarian nor democratic', in Kevin Hewison, Richard Robison, and Garry Rodan (eds) *Southeast Asia in the 1990s*, pp. 133–58, Sydney: Allen & Unwin.

Deyo, Frederic (1981) *Dependent Development and Industrial Order*, New York: Praeger.

Deyo, Frederic (1989) *Beneath the Miracle: Labour Subordination in the New Asian Industrialism*, Berkeley: University of California Press.

Deyo, Frederic (1995) 'Human resource strategies and industrial restructuring in Thailand', in Stephen Frenkel and Jeffrey Harrod (eds) *Industrialisation and Labour Relations*, pp. 23–36, Ithaca, NY: ILR Press.

Deyo, Frederic (ed.) (1996) *Social Reconstructions of the World Automobile Industry*, Baskingstoke: Macmillan.

Europa World Yearbook (1995) London: Europa Publications.

Frenkel, Stephen (1993) 'Variations in patterns of trade unionism: a synthesis', in Stephen Frenkel (ed.) *Organized Labour in the Asia-Pacific Region*, pp. 309–46, Ithaca, NY: ILR Press.

Hadiz, Vedi R. (1994) 'Challenging state corporatism on the labour front: working class politics in the 1990s', in David Bourchier and John Legge (eds) *Democracy in Indonesia, 1950s and 1990s*, pp. 190–203, Monash papers on Southeast Asia no. 31, CSEAS, Monash University.

Hewison, Kevin, and Rodan, Garry (1994) 'The decline of the Left in Southeast Asia', in *The Socialist Register 1994*, pp. 235–62, London; Merlin Press.

Hewison, Kevin and Brown, Andrew (1994) 'Labour and unions in an industrialising Thailand: a brief history' *Journal of Contemporary Asia*, 4 (4): 483–514.

Hutchison, Jane (1993) 'Class and state power in the Philippines', in Kevin Hewison, Richard Robison and Garry Rodan (eds) *Southeast Asia in the 1990s*, Sydney: Allen & Unwin.

Ingleson, John (1981) 'Worker consciousness and labour unions in colonial Java' *Pacific Affairs*, 54 (31): 485–501.

International Labour Office (1995a) *World Employment 1995: An ILO Report*, Geneva: ILO Press.

International Labour Office (1995b) *World Labour Report*, Geneva: ILO Press.

Kuruvilla, Sarosh (1994) 'Industrialisation strategy and industrial relations policy in Malaysia and the Philippines', Proceedings of the Forty-Sixth Annual Meeting of the Industrial Relations Research Association, Boston, 3–5 January.

Kuruvilla, Sarosh (1995) 'Industrialisation strategy and industrial relations policy in Malaysia', in Stephen Frenkel and Jeffrey Harrod (eds) *Industrialisation and Labour Relations*, pp. 37–63, Ithaca, NY: ILR Press.

Lambert, Rob (1993) *Authoritarian State Unionism in New Order Indonesia*, Working Paper no. 25, Asia Research Centre, Murdoch University (October).

Leggett, Chris (1993) 'Corporatist trade unionism in Singapore', in Stephen Frenkel (ed.) *Organised Labour in the Asia-Pacific Region*, pp. 223–48, Ithaca, NY: ILR Press.

MacIntyre, Andrew (1994) *Organising Interests: Corporatism in Indonesian Politics*, Working Paper no. 43, Asia Research Centre, Murdoch University (August).

Ofreneo, Rene E. (1994) 'The labour market, protective labour institutions, and economic growth in the Philippines', in Gerry Rodgers (ed.) *Workers, Institutions, and Economic Growth in Asia*, pp. 255–301, Geneva: International Institute for Labour Studies.

Ofreneo, Rene E. (1995) 'The changing terrains for trade union organising' unpublished manuscript, School of Labour and Industrial Relations, University of the Philippines.

Piriyarangsan, Sungsidh and Kanchada Poonpanich (1994) 'Labour institutions in an export-oriented country: a case study of Thailand', in Gerry Rodgers (ed.) *Workers, Institutions, and Economic Growth in Asia*, pp. 211–54, Geneva: International Institute for Labour Studies.

Rasiah, Raja (1994) 'Flexible production systems and local machine tool subcontracting: the case of electronics components transnationals in Malaysia' *Cambridge Journal of Economics*, 18 (3): 279–98.

Standing, Guy (1989) 'The growth of external labour flexibility in a nascent NIC: Malaysian Labour Flexibility Survey (MLFS)' World Employment Programme, Research Working Paper no. 35 (November) Geneva: ILO.

World Bank (1995) *World Development Report 1995: Workers in an Integrating World*, Washington, DC: Oxford University Press.

Wurfel, David (1959) 'Trade union development and labour relations policy in the Philippines' *Industrial and Labour Relations Review*, 12 July.

9

South-East Asia and the Political Economy of APEC

Andrew MacIntyre[1]

Thirty years of very good economic growth among nearly all of the ASEAN countries has had profound consequences. Not only are these countries now much wealthier, they are also generally less gripped by fundamental problems of political stability and nation-building. To be sure, there are still many serious domestic challenges—most notably the need for further progress with poverty reduction and the improvement of income distribution. But it is also clearly the case that the transformations of the domestic political economies of these countries that are underway have permitted governments to lift their sights from the purely domestic and to give greater attention to international policy issues. In addition, changes in global economic and security conditions have given South-East Asian governments powerful incentives to devote more resources to international challenges.

Other chapters in this volume focus primarily on domestic dimensions of political and economic change in particular South-East Asian countries. The concerns of this chapter lie primarily at the international level, centring on the development of APEC. The purpose of this chapter is to outline the key empirical features of the APEC story, focusing on the competing interests at stake, and then to go to the major debates in the scholarly literature about APEC. The structure of this essay is straightforward. I begin by tracing the evolution of APEC and South-East Asian responses to it,[2] and then introduce the issue of the alternative conception for regional cooperation championed by Malaysia and built around the idea of an 'Asia for Asians'. Attention is then turned to questions of APEC's actual significance and the intellectual and policy debates over the preconditions for successful regional cooperation.

The analytical focus of the literature on APEC is noticeably different to that of the literatures on the domestic political economy of the various South-East Asian countries. The central concern is less with the advantages secured by any particular class, sector, industry or group, and more with the advantages secured by particular nation states. And while we can, indeed, discern two broad categories of writers which dominate this literature—

policy-oriented trade economists, and international political economists—as will be seen, the lines of debate between the two do not divide as neatly as might be expected. In terms of the three theoretical and ideological camps described by Richard Robison, Garry Rodan and Kevin Hewison in chapter 1 of this volume, we can see most (but not all) of the intellectual protagonists in debates about APEC as drawing primarily on the neo-classical and new institutionalist literatures.

The Evolution of APEC

APEC is the first regional institution to bring the governments of the Asia-Pacific region together in one forum. It is a loosely structured consultative body, with a modest organisational base in the form of a small secretariat in Singapore. As is now well known, APEC was launched as an intergovernmental ministerial forum in Canberra in 1989 at the initiative of the Australian government.[3] The original group was made up of twelve countries: the six ASEAN countries, Japan, South Korea, Australia, New Zealand, Canada and the USA. Motives for participation varied among countries and subregions, but all were united by a defensive instinct in the face of the chronic difficulty then being experienced in the negotiation of the Uruguay Round of the General Agreement on Tariffs and Trade, and growing concern about the rise of protectionism and of economic regionalism elsewhere with the strengthening of the European Community and (for those APEC participants on the western side of the Pacific) the likely negotiation of the North American Free Trade Association (NAFTA). APEC did not suddenly come into existence fully formed in 1989. It grew out of a lengthy process of institutional experimentation in support of the concept of economic cooperation across the Asia-Pacific region stretching back nearly three decades (Funabashi 1995; Soesastro 1994b; Woods 1993; Drysdale 1988). Working in tandem for much of the time, Japan and Australia played the leading roles in exploring possibilities for institutionalising economic cooperation among varying combinations of leading non-communist countries around the Pacific Rim. Although there was government support in the background, much of the pioneering work in the development of different institutional frameworks was undertaken by academic economists (Saburo Okita and Kiyoshi Kojima from Japan, and Sir John Crawford and Peter Drysdale from Australia). The essential idea underlying much of this work was that the complementarity between the advanced industrial economies and the developing economies of the region meant that there were large gains to be secured through increased trade.

In the late 1960s two notable institutions were established. The Pacific Basin Economic Council (PBEC) was founded in 1969 out of a meeting of the Australia–Japan Business Cooperation Committee and comprised the peak business associations from the five industrial democracies in the region: Japan, Australia, the USA, Canada and New Zealand. Meanwhile the first Pacific Trade and Development (PAFTAD) conference was held in Japan and grew into a regular conference bringing together the leading academic economists around the Asia-Pacific region concerned with regional trade issues. Over time, a growing and reasonably coherent group of economists from around the Asia-Pacific region was brought together on a regular basis and united by a common conviction as to the economic benefits to be had from freer regional trade arrangements. However, as H. Soesastro (1994b: 83–4) notes, relatively little attention was given to the political practicalities of achieving significant and broadly based regional economic cooperation. This began with the creation of an umbrella organisation in 1978 which became known as the Pacific Economic Cooperation Council (PECC). PECC brought together tripartite teams of academic economists, business people and government officials (participating in an unofficial capacity) from twenty countries around the Pacific in which representatives of both PAFTAD and PBEC were included.[4]

The formation of PECC was very important for at least two reasons. First, the involvement of government officials, even if on an informal basis, brought national and subregional political differences to the foreground. Secondly, it brought the developing countries of the region, notably the ASEAN countries, to centre stage. PECC was, in effect, the forerunner to APEC. The key change that came with the launching of APEC was the agreement among participating countries to elevate regional economic cooperation to a fully fledged and exclusively intergovernmental institution. PECC continues to function as an important source of ideas for APEC, but the creation of APEC as an institution bringing government ministers and now heads of state together was indicative of a desire for greater policy action.

In spite of this history of gradual institutional evolution, the final move to establish APEC was not without problems. The ASEAN states were—and, as will be seen, in some respects remain—ambivalent about the creation of APEC. Although South-East Asian economists, business people and officials had been enthusiastic participants in PECC, and although South-East Asian governments had taken an important preliminary step towards regional intergovernmental cooperation with the establishment of the ASEAN Post-Ministerial conferences in 1984,[5] the creation of APEC was worrying. The principal concern at this stage was not so much the possibility of facing external demands for trade liberalisation (substantive agendas of this sort were still several years away), but more fundamentally that the individual and collective voices of ASEAN states would be lost in the larger institution, and that APEC would be dominated by the West.

The USA (which was not included in the conception of APEC initially floated by Australian Prime Minister Bob Hawke) at this point remained wary of a multilateral framework for regional economic cooperation. Not surprisingly, Washington's preference was to deal with the countries of the region on a bilateral basis, thereby maximising the bargaining advantages deriving from its size. Although the ASEAN states and the USA both concluded that the potential benefits of supporting APEC outweighed the risks, their respective concerns were not unfounded, and accordingly, have lingered.

Map 9.1   The APEC economies

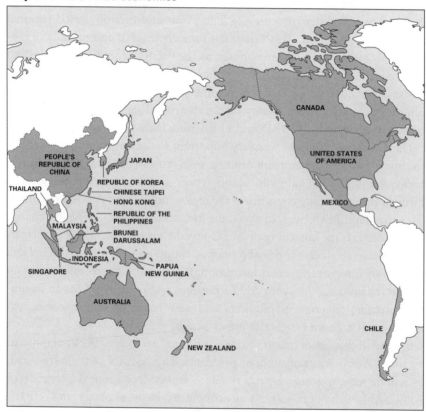

In the period since 1989 APEC has developed from being an uncertain seminar to an institution of some substance. Its membership has grown from twelve to eighteen with the addition of China, Taiwan, Hong Kong, Papua New Guinea, Mexico and Chile (see map 9.1). Bringing about a situation in which China, Taiwan and Hong Kong were all officially represented under the same roof was in itself a pioneering achievement. APEC is no longer just a tentative annual meeting of regional foreign ministers; there are now regular heads of governments meetings, finance ministers meetings, various senior officials meetings, more than a dozen 'working groups' dealing with different sectoral issues,[6] a small secretariat, a private sector advisory organisation (the APEC Business Advisory Council), and independent research capabilities through designated APEC-studies centres in universities around the region. In terms of promoting serious multilateral economic policy coordination, APEC's most notable achievement to date is the agreement at the Bogor Summit in 1994 to establish a free trade and investment region in the Pacific by 2010 for high income countries, and 2020 for the others. From a global perspective, these developments are potentially of great significance given that APEC countries account for roughly half of the global economy and that APEC is the largest single regional trading agreement. (Table 9.1 shows that of the 61 per cent of world trade covered by regional trading agreements, APEC is the largest even when the subregional groupings subsumed within it—AFTA, CER, and NAFTA—are excluded.)

Table 9.1 Regional free trade agreements' share of world trade (1994)

Region	Agreement	Share of world trade
Asia	AFTA	1.3
	CER	0.1
	APEC	23.7[a]
The Americas	NAFTA	7.9
	MERCOSUR	0.3
	FTAA	2.6[a]
Europe	EU	22.8
	EUROMED	2.3
Total		61.0

Notes:

[a] Excluding subregional groupings

AFTA   Indonesia, Singapore, Malaysia, the Philippines and Brunei.

CER   Australia and New Zealand.

NAFTA   USA, Canada, and Mexico.

MERCOSUR   Argentina, Brazil, Paraguay and Uruguay.

FTAA   NAFTA and MERCOSUR and twenty-seven other western hemisphere countries.

EUROMED   Twelve Mediterranean countries.

Source: Bergsten (1996): 106.

An East Asian Economic Grouping?

Precisely because ASEAN was then the only relevant and coherent sub-regional association in the Pacific, South-East Asia enjoyed an enhanced bargaining position in the early years of APEC's existence. For example, it was clear at the outset that unless the ASEAN states collectively were satisfied with the proposed framework for APEC, it could not proceed because of their potential to impose a bloc vote veto. The other middle-sized participants—South Korea, Canada, Australia or New Zealand—did not enjoy such bargaining advantages. South-East Asian concerns to guard against the possibility of APEC becoming a forceful and strictly rules-based organisation dominated by the advanced industrial economies were strongly reflected in the original agreement reached in Canberra in 1989, and were codified shortly after in early 1990 with the adoption of a formal ASEAN policy document, the so-called Kuching Consensus (ASEAN 1990).

While these concerns were shared in some degree by all ASEAN states, it is clear that Malaysia had much deeper reservations about APEC than the others. In December 1990 the Malaysian Prime Minister, Dr Mahathir Mohamad, floated a proposal for an alternative regional organisation for Asians only: an East Asian Economic Group (EAEG). In the context of the apparent slide towards protectionism and defensive regionalism in Europe and the Americas at the time, this move was generally interpreted as an alarming challenge by the western countries of the region and encouraged images of a discriminatory Asian trading bloc centred around Japan. Although Malaysian observers insist that Dr Mahathir's conception was widely misunderstood and *never* conceived in economically discriminatory terms (Ariff 1994: 115), and while it was indeed NAFTA rather than APEC which was the principal inspiration for the East Asian regional initiative, there can be little doubt that it was promoted by the Malaysian government in a strong and challenging way as a rival organisation to APEC. Strong opposition to the proposal came from a number of quarters. Much attention has been given to Washington's hostility and Japan's unenthusiastic response. Less well noted was Indonesia's unusually blunt rejection of the idea, together with the opposition of Singapore and South Korea. In response, a face-saving compromise was negotiated within ASEAN whereby the concept would be softened to become an East Asian Economic Caucus (EAEC)—a discussion forum for Asian economies *within* APEC (as distinct from outside APEC, Malaysia's preferred fallback position).

Mahathir's EAEC strategy is a fascinating and important puzzle. There has been almost no good analysis of the subject. Because of political sensitivity it has not been subject to much serious public scrutiny in Malaysia, or indeed, elsewhere in South-East Asia. And in the Caucasian countries of the Pacific (and Europe) it has been dismissed much too quickly and gleefully as an economically misguided strategy of an idiosyncratic, publicity-hungry politician.

In addition to any personal disdain Dr Mahathir may feel towards Western societies, there are grounds for believing that domestic political considerations have been an important motive for him. The image of a plucky Malaysian Prime Minister standing up for Malaysian and Asian interests in the face of Western brow-beating and reaching for a leadership role within South-East Asia may have been helpful in Malaysia's strongly nationalist political climate.[7] But is this an adequate explanation for the pursuit of a regionally divisive strategy leading to what would appear to be a decidedly second-best option in economic terms?

Notwithstanding the marked increase in intra-East Asian investment, trade and development assistance flows, trans-Pacific ties remain very important for both South-East Asia and North-East Asia. Table 9.2 illustrates this with simple trade flow data for South-East Asia: trade with APEC as a whole remains decidedly larger than trade just within East Asia. Interestingly, Malaysia is no exception to the general pattern. For 1989–93, East Asia accounted for 55 per cent of Malaysia's exports and 58 per cent of its imports, whereas the corresponding figures for APEC were 77 per cent and 79 per cent.

Table 9.2 ASEAN direction of trade patterns (1989–93 average, %)

		ASEAN	EAEC	APEC
Malaysia	exports	28	55	77
	imports	20	58	79
Indonesia	exports	11	63	80
	imports	9	46	68
Thailand	exports	13	40	65
	imports	12	55	70
Philippines	exports	7	36	77
	imports	10	47	73
Singapore	exports	22	45	71
	imports	18	53	72
Brunei	exports	20	87	91
	imports	39	54	74
ASEAN	exports	23	60	89
	imports	19	64	87

Notes: ASEAN = ASEAN 6.

EAEC = ASEAN plus Japan, South Korea, China, Taiwan and Hong Kong.

APEC = EAEC plus USA, Canada, Mexico, Chile, New Zealand, Australia and Papua New Guinea.

Source: International Monetary Fund, Direction of Trade Statistics, various issues.

Using more or less elaborate statistical techniques, various studies (Panagariya 1993; Frankel 1993; Cohen 1995) have emphasised two main claims. First, if the purpose of regional economic cooperation is to maximise net welfare gains by lowering regulatory barriers and transaction costs, it makes little sense for East Asian countries to opt for an East Asia-only trade area, because of the ongoing importance of the western countries in the Pacific (principally the USA) as markets, sources of capital and technology, and providers of education. Second, quite apart from the hypothetical question of whether it would be rational to create an East Asian trade area, the empirical reality is that there is little evidence of a credible 'Yen bloc' emerging in East Asia. The idea of an economically meaningful East Asian regional trade grouping is dealt a further blow by the awkward fact that the current sole candidate for economic leadership, Japan, shows no sign of being willing to accept such a role. Many studies have now presented strong evidence to support the argument that while Japan's tariff barriers may indeed be low, it imports a disproportionately small share of manufactured goods relative to other industrialised economies (Lincoln 1990; Ravenhill 1993; Doherty 1995; Bergsten and Noland 1993). Much attention has been given to the tensions this has created in relations between the USA and Japan. But there is also real potential for friction between Japan and most East Asian countries because of the meagre level of manufactured goods Japan imports from East Asian producers. Indeed, the public call by Singapore's Lee Kuan Yew for Japan to yield ground in the 1995 trade dispute over automobiles with the USA may indicate increasing recognition of and frustration with Japan's aberrant import profile in South-East Asia (Richardson 1995).[8]

In the face of these economic obstacles, how are we to explain Dr Mahathir's crusade for the EAEC? It could, of course, simply be a case of an obdurate politician whose pursuit of electoral advantage and a place in diplomatic history painted him into a diplomatic corner from which he could not emerge without severe loss of face. There is insufficient primary research on the political economy of Malaysian trade policy to permit a confident answer to this puzzle. However, there are several grounds for concluding that whatever Dr Mahathir's reasons for embarking on this strategy, it should not be disregarded. First, regardless of the economic optimisation of trans-Pacific cooperation, it would be foolish to overlook the extent to which Dr Mahathir's message resonates in East Asia (Smith 1994: 18). While this is particularly true in Japan, one readily encounters similar sentiment in other parts of the region as well. More important than high profile statements of eager support from the likes of Japanese polemicist, Shintaro Ishihara, is the extent to which Dr Mahathir seems to have picked up on what to some will appear as pan-Asian pride and self-confidence arising from the region's extraordinary economic achievements, and what to others appears as chauvinistic Asian triumphalism. Pronouncements such as the following capture the tenor of his message:

Some of the talk about Asia's economic success is motivated by less than good intentions. Fearing that one day they will have to face Asian countries as competitors, some Western nations are doing their utmost to keep us at bay. They constantly wag accusing fingers in Asia's direction, claiming that it has benefited from unacceptable practices, such as the denial of human rights and workers' rights, undemocratic government, and disregard for the environment ...

At the same time that the West demands that Asian countries open their markets and stresses the undesirability of economic blocs, it has strengthened its own blocs ... Western leaders talk of open regionalism while manning the barricades to keep others out. (Mahathir and Ishihara 1995: 39, 45)

In short, regardless of the economics of the situation, there may well be a sizeable popular constituency for this message. A second reason for not dismissing the idea of an 'Asia for Asians' is that the economic basis for an East Asian grouping may not be as weak as is often supposed. As John Ravenhill (1995c: 2–7) has argued, the widely cited study by Jeffrey Frankel (1993, noted above), failed to capture the pace at which the pattern of regional trade has shifted in the last five years, and thus, notwithstanding the sophistication of his statistical techniques, probably understates the plausibility of an East Asian framework for regional economic cooperation. Ravenhill's claim is not that Frankel is in error, simply that the level of economic interdependence has grown more quickly than Frankel suggested. A third reason for paying attention is that whatever the merits of Dr Mahathir's original proposal, the fact that there *is* now an EAEC which holds periodic meetings may help to prevent the USA—or more generally, Western—domination of APEC. Moreover, the ASEAN-sponsored Asia-Europe Meeting (ASEM) in Bangkok in 1996 can be viewed as a further step towards the development of the EAEC idea, for this was the first time the heads of the major countries in South-East Asia and North-East Asia had met together in the absence of the USA. In the event that Washington was to seek to pursue an agenda unpalatable to East Asians or pursued particularly aggressive unilateral trade measures outside APEC, the existence of the EAEC does serve as a political 'safety net' (Ariff 1994: 115).[9] And, as noted, Ravenhill's argument suggests that it is a more substantial safety net than previously thought. Certainly this would appear to be the principal message South-East Asian and North-East Asian governments were hoping to send to the USA through the ASEM. Rather than developing newly discovered bonds between Europe and Asia, it would seem that for the Asian countries the purpose of the meeting was to signal to Washington that they were now willing to explore all trade coalition options (particularly as preparations for WTO negotiations take shape) and not simply rely on Washington's goodwill.

Does APEC Really Matter?

In both academic and media discussion, APEC generates strongly diverging reactions. Some see it as a very valuable initiative promoting regional economic cooperation and long-run economic growth, while others view it as weak, ineffectual and even wrong-headed. If we leave aside government officials, the most enthusiastic supporters of APEC are to be found among policy-oriented trade economists. There have been many conferences and publications around the Asia-Pacific region arguing the benefits of APEC from this perspective (Elek 1992; Elek 1995; Garnaut and Drysdale 1994; Bergsten 1994; Soesastro 1994a; Yamazawa 1992). Distilling this literature, the principal benefits attributed to APEC may be summarised as follows:

- APEC has the potential to make an important contribution to the economic welfare of the region through promoting multilateral cooperation to liberalise regulatory barriers hindering trade and investment flows.
- Extending this point, APEC can facilitate regional economic growth through various low profile functional initiatives to reduce transaction costs in areas such as standardisation of packaging and labelling requirements, the harmonisation of customs procedures and the expediting of trade and investment-related document processing.
- APEC could help to contain trade and investment disputes among member countries, with the possible creation of a dispute mediation mechanism.
- Paralleling the critical role APEC allegedly played in inducing European concessions in the Uruguay round of GATT negotiations, APEC could provide valuable leverage in global trade negotiations under the WTO.[10]

In short, supporters of APEC typically view it in very enthusiastic terms, as the institutional capstone of regional economic integration which will serve to keep member countries on the path to liberalisation and sustained economic prosperity. In contrast, those sceptical of APEC see it as a weak and ineffectual institution that is productive of little except encouraging statements of intent. The main current of criticism in the academic literature comes from international political economists, though there are also significant contributions from orthodox economists. Critiques of APEC can be sorted into three rough groups.

Perhaps the most fundamental criticism is one made several years ago and which has not received the attention it deserves. In what can perhaps best be termed an 'ultra-market' critique, economist Helen Hughes (1991) argued that APEC (and, by extension, other regional institutions) are incapable of doing much to strengthen regional trade and investment flows as these are determined by the decisions of individual business people and consumers, not governments. According to this very astringent view, APEC is the creation of initiative-hungry officials and policy-oriented academics in search of their next international conference. While APEC may indeed grow to be a substantial (and thus expensive) institution, there will be very little it can do to actually expand sustainable trans-border economic activity.

A second and more general critique of APEC is that it rests on a doubtful foundation, given the low propensity for serious cooperation in the Asia-Pacific region displayed thus far. By comparison with the substantial histories of regional cooperation and institution building in both Europe and the Americas, economic cooperation around the Pacific appears quite paltry. As A. Fishlow and S. Haggard (1992: 30) put it, in comparing the experiences of economic cooperation in different regions of the world 'the puzzle with reference to the Pacific is not to explain the progress of regional initiatives, but their relative weakness'. A number of writers have noted that the Pacific is so divergent that there is ultimately no shared sense of community among its members (Manning and Stern 1994; Katzenstein forthcoming; Fishlow and Haggard 1992; Higgott 1994). The extraordinary diversity of the region in terms of population, size, wealth, political system and socio-cultural traditions is held to be so great as to preclude the possibility of serious cooperation and integration. As the Malaysian commentator (and, now, supporter of Dr Mahathir's critique of APEC) Chandra Muzaffar said: 'As a concept, "Asia Pacific" makes little sense. Unlike East Asia or South Asia or Southeast Asia, it has no shared history or common cultural traits. Asia Pacific is not even an accepted geographical entity' (quoted in Higgott and Stubbs 1995: 526).

The third and most elaborate critique of APEC is the one that has commanded greatest attention in recent years. It focuses on APEC's institutional attributes and argues that quite apart from any of the circumstantial hurdles just mentioned, APEC's organisational rules will ensure that it never develops into a forceful regional institution. At the insistence of the ASEAN countries (and with the support of all the western Pacific members) two of the core organising principles of APEC were the notions of 'open regionalism' and governance by consensus. In essence, open regionalism refers to a pattern of regional economic cooperation which is open to all comers and does not discriminate against other countries. Thus the benefits of trade liberalisation in one APEC country are available not just to all other APEC countries, but also to any non-APEC country as well. Moreover, the benefits are available regardless of whether the other APEC or non-APEC countries have undertaken reciprocal reforms. This is in effect a pure, or unqualified form of the Most Favoured Nation (MFN) principle. As such, it stands in striking contrast to the organising principles of the other major frameworks for international economic cooperation. Not only are all the other major regional economic groupings built around a discriminatory or preferential form of MFN (*viz.* the EU, AFTA, NAFTA, MERCOSUR, or the mooted Free Trade Area of the Americas), even the General Agreement on Tariffs and Trade (GATT) and its successor organisation, the World Trade Organisation (WTO) limits the benefits of its negotiations to member countries.

The other key defining principle is the notion that APEC should operate on a voluntary and consensual basis, rather than in a formalised manner governed by binding rulings. As a result, APEC tends to move only as fast as its slowest members, allows dissenters to abstain from implementing aspects of an agreement which they do not like, and avoids detailed and specific commitments to action. In this regard, it very much resembles ASEAN. The logic behind the insistence on the consensus principle by the ASEAN countries and the other East Asian members of APEC was, of course, to guard against the possibility of unwanted commitments being forced upon them.

A number of writers (Aggarwal 1994; Kahler 1994; Fane 1995; Raven-hill 1995a) have argued that these organising principles are politically naive and effectively preclude the possibility of APEC making significant headway with economic policy coordination. In varying degrees, these writers draw on the logic of game theory, and in particular, on the 'Prisoner's Dilemma' model of the problem of achieving cooperation among actors with limited information and no way of obtaining a mutual binding agreement. They argue that without specific reciprocity requirements the incentives for individual countries to free ride on the back of the efforts of those who volunteer to liberalise first will be such as to undermine the whole process of cooperation. Quite simply, countries which do liberalise will be unwilling to move further if their concessions are not reciprocated. This problem of incentives for cooperation is further compounded by the fact that APEC rules extend benefits not just to APEC countries, but to any country at all.

In the absence of rules requiring all members to commit to specific reforms and to limiting access to these benefits only to members, critics argue little reform action at all is likely. Exporters from other countries (most notably the Europeans) will be swift to take advantage of the situation, causing producers in APEC countries to lobby their governments to obtain reciprocal concessions, provide appropriate compensation or abandon the endeavour. In short, adopting a more 'virtuous' set of rules than the GATT/WTO is held to be both silly, and ultimately, counterproductive.

In response to these criticisms P. Drysdale and R. Garnaut (1993), two policy-oriented trade economists closely associated with APEC's development and the idea of open regionalism, have put forward an alternative theoretical account which, they argue, better captures the dynamics of economic liberalisation in the Pacific. Simply put, they argue that APEC sceptics have mistakenly emphasised the Prisoner's Dilemma game and its logic of non-cooperation, for the reality is that for the last decade or so most countries in the Pacific *have* undergone major economic liberalisation drives. Accordingly, rather than the Prisoner's Dilemma framework, the appropriate model is what they call 'Prisoner's Delight'—a situation in which enlightened self-interest leads participants to liberalise unilaterally. This argument has considerable appeal, for it does at least offer a theoretical explanation for the reality of widespread unilateral trade liberalisation in Asia over the last decade. However, in cogent responses to the Drysdale and Garnaut position, G. Fane (1995) and J. Ravenhill (1995a) have argued that if the true nature of trade cooperation in the Pacific does, indeed, correspond to the Prisoner's Delight model, then the inescapable implication of this is that APEC is fundamentally *irrelevant* to the whole process. In other words, if participants are willing to liberalise unilaterally regardless of the behaviour of others, international cooperation is superfluous. Fane and Ravenhill go on to argue persuasively that in order to correctly theorise the dynamics of economic cooperation in the Pacific it must be recognised that there is a real cooperation problem and that the Prisoner's Dilemma is indeed the appropriate framework for analysis. I find the critiques of Fane and Ravenhill persuasive but, as I will argue later, not entirely satisfactory.

APEC and South-East Asia: Ideas and Interests

Theoretically-based scholarly arguments about APEC's weakness come primarily from North America and, to some extent, Australia. In South-East Asia, APEC has not been viewed as a weak and ineffectual institution. Indeed the underlying concern in most of South-East Asia has been that APEC will be too strong and intrusive and subject to USA manipulation. All the ASEAN countries (except Brunei), like most other countries in the western Pacific, have been singled out for bilateral pressure from the office of the US Special Trade Representative on issues ranging from market access in particular industries, through intellectual property rights issues, labour rights, and environmental and health standards (Ariff 1994; USTR 1994; USTR 1995). Regardless of any merits attaching to the USA position on these bilateral issues, one of the by-products of such action has been to encourage concerns in South-East Asia that APEC is an attempt to extend Western (particularly USA) influence into Asia and to tap into the economic dynamism of Asia in an effort to avoid being overtaken economically. These concerns are held to some degree by all South-East Asian governments, but it is the Malaysian government that has articulated them most strongly. To quote Dr Mahathir again:

> The United States created NAFTA and it wants to use APEC to hold back the East Asian countries. American support for NAFTA but not EAEC is quite illogical. If Washington favors one, then it ought to be in favor of both. You cannot justify NAFTA and APEC and reject the EAEC. That is hypocrisy. It is like saying, 'I can do what I like, but you may only do what I let you'. (Mahathir and Ishihara 1995: 46–7)

Beyond the nationalist-cum-anti-Western 'EAEC critique' of APEC, other scholarly discussion of APEC in South-East Asia is remarkably positive and couched in very much the same fashion as the arguments of APEC supporters elsewhere in the Pacific region. Most of the South-East Asian academic literature on APEC appears in the form of conference papers, presented at the innumerable conferences on APEC that are held around the region. Although many of the conference proceedings are not widely available, a flavour of much of the discussion can be gained from major South-East Asian scholarly publications, such as the *ASEAN Economic Bulletin*.

One of the striking features about this literature is that there is remarkably little *debate* about the costs and benefits of APEC in terms of particular classes, sectors, or industries. Why is there is so little disagreement about APEC among South-East Asian intellectuals, commentators, and indeed, economic interests? There is, after all, vigorous debate about many other aspects of economic policy among academics, journalists, and NGOs in most of South-East Asia. Why is there no sharp critique from either the Left or the Right?[11] An important part of the explanation is, of course, that to date APEC has done very little that directly affects the interests of particular economic groups. For instance, as yet, apart from supporting the obligatory conferences and dialogue processes that have long existed in PBEC and have now grown around APEC, few, if any, big South-East Asian firms have taken a serious interest in APEC—either to support its expansion or to resist possible trade reforms. Certainly there have been major adjustments in economic policy throughout South-East Asia since the mid-1980s, but it is very difficult to attribute them to APEC. In short, in contrast to the EU, NAFTA or even AFTA, APEC has not become a site for the negotiation of industry-specific trade deals. Accordingly, the domestic politics of APEC are much more relaxed.[12] More generally, however, the scarcity of debate about APEC reflects the fact that even more so than in most other parts of the Pacific, in South-East Asia intellectual discussion of APEC is overwhelmingly conducted by policy-oriented economists. As R. Higgott and others have argued, there does indeed appear to be something approaching a self-contained and self-referential epistemic community, or network of policy experts that dominates discussion of APEC.[13]

But this gives rise to a seeming paradox which returns us to the theoretical debates about the organisational preconditions for successful multilateral economic cooperation and brings us closer to understanding the real significance of APEC for South-East Asia. Given that most South-East Asian scholars working in this area are convinced of the benefits of freer trade, why is there not more interest among them in reciprocity and rules, rather than open regionalism and consensus? Why do South-East Asian economists (and policy-makers) not favour a set of organisational arrangements which should *hasten* the liberalisation of trade regimes?

One response to this question is to build on the logic of the Prisoner's Dilemma and to point out that the South-East Asian countries have particularly powerful incentives to preserve the option of free riding as their trade and investment regulatory environments are much less open than those of APEC countries in the Americas or the southwest Pacific. While the South-East Asian countries have made notable unilateral tariff reductions, such reductions have typically been 'unbound', that is, not-registered with the GATT/WTO and thus not strictly binding. Indeed, with the exception of Singapore, the bound tariff levels of the ASEAN countries are among the very highest in the world. Under a strict rules-based system of trade reciprocity, in order to bring themselves into line with the more open economies of the region the South-East Asians (other than Singapore) would have to bear very high initial adjustment costs, and would have to do so according to a schedule agreed to by all other members, rather than at their own pace and on the basis of their own economic and political priorities.

One can thus make a very plausible argument that although South-East Asian economists and policy-makers may indeed have a general preference for freer trade, they are not oblivious to the fact that their countries would bear some of the highest adjustment burdens if APEC were to have a rules-based regime featuring specific reciprocity requirements. To this extent, the earlier mentioned arguments drawing on the Prisoner's Dilemma model provide us with a powerful insight into the strategic calculations of the ASEAN countries. But such analysis is not complete, for the potential gains from collaborative trade liberalisation are only one element of the wider calculation about APEC. There are two propositions to be emphasised here: first, that the contribution of multilateral cooperation to liberalisation and national economic growth is modest, and second, APEC is only partly concerned with economics. Taking the first of these points, while it does indeed seem that coordinated multilateral liberalisation is most unlikely to be achieved on a sustained basis in the absence of reciprocity requirements, this is not to say that liberalisation and growth cannot take place in the absence of this condition. There are *many* factors that bear upon the decisions of governments to liberalise, and APEC policy intellectuals such as C. Fred Bergsten and Ippei Yamazawa surely exaggerate when they say: 'Interstate arrangements are usually necessary to implement such liberalization ...' or that economic growth in the region 'needs a free trade regime to sustain it'.[14] A survey of *detailed* studies of the political economy liberalisation in the industrialising countries of Asia reveals that multilateral collaboration has been far from the top of the list of contributing factors (Haggard 1990; Timberman 1992; Shirk 1994; MacIntyre 1994). This is not to preclude the possibility that collaboration may become increasingly necessary in the future, but simply to note that it is hard to link collaboration with the reform processes underway in much of Asia.

The second and more important point to be emphasised here is that while all of the free trade rhetoric about APEC does indeed invite the conclusion that liberalisation is the *raison d'être* for the organisation, it would be a serious mistake to settle on such a conclusion. As Paul Krugman (1993) said in response to the confused debate about NAFTA's real significance: 'It's foreign policy, stupid!'. His point was that although NAFTA would be productive of some small net increase in national economic welfare in the USA, contrary to the wildly overblown rhetoric about its economic impact the real reasons for pushing for its establishment were foreign policy driven. Much the same is true of APEC.

The reasons for supporting APEC and insisting on open regionalism and consensual governance go way beyond collaborative liberalisation. At the most fundamental level, the core concern is quite simply that APEC should exist as a forum which brings together nearly all of the countries in what is widely recognised to be a diverse, volatile, and fast changing region. More specifically, it provides a wider setting in which difficulties in the bilateral relationships among the three major players—the USA, Japan and China—can be massaged and perhaps eased. To use the language of strategic studies, APEC plays a vital role as a 'confidence-building' institution in the Pacific. It is not a coincidence that APEC was developed and managed through foreign ministries. Moreover, even as APEC has evolved and technical economic ministries have been drawn in to handle functional issues, foreign ministries (and the foreign policy advisers of heads of state) continue to be central.

Loose organisational rules have been an essential precondition for holding the organisation together and, in particular, of reducing widely felt anxieties about the possibility of USA dominance. Particularly in the new security environment of the post-Cold War period, none of the South-East Asian countries favours a contractual or rules-based framework in which the USA might use its leverage to impose unwelcome burdens. Loose organisational rules enabled both the establishment of APEC and have minimised the risk that it would collapse as result of unacceptable demands being made on it by any party. The risk for the South-East Asians (and indeed all of the countries in the western Pacific) is that without the pay-offs that might be expected to flow from a rules-based regime emphasising specific reciprocity requirements, the USA may well lose all interest in APEC.

Once we put aside the excessively narrow view that APEC is principally about collaborative economic liberalisation and recognise the wider foreign policy dimensions, not only does the insistence on open regionalism and consensual governance make more sense, but APEC ceases to look like a puny bureaucratic plaything, and begins to look much more like a truly remarkable achievement.

Conclusion

There has been a great deal of discussion about increasing economic interaction among varying sets of countries in the Asia-Pacific region in recent years. APEC is the first framework for regional economic cooperation which promises to bring most of the major economies on both sides of the Pacific together. Although launched officially in 1989, the idea of creating an institution of some sort to serve this function had, in fact, been under development in policy-oriented academic circles for many years. The catalyst for action in the late 1980s was the widespread concern that the global trading system was in danger of fragmenting because of chronic problems in the Uruguay Round of GATT negotiations and the apparent slide towards defensive regional trading groups.

The ASEAN countries have faced a difficult choice from the outset. On the one hand, they feared being left out in the cold should there have been a breakdown in GATT negotiations and a scramble to erect more inward-looking and exclusionary trading arrangements in Europe and the Americas. On the other hand, they have also feared that APEC might become an arena in which the interests of the leading economies—Japan and, in particular, the USA—would predominate and which would completely overshadow their own regional body, ASEAN. And although it was not without appeal, the Malaysian proposal for an 'Asians only' regional grouping suffered from a number of problems, not least that the USA remained a crucial economic partner for South-East Asian countries. The strategy adopted by these countries (with the partial exception of Malaysia) was to support the development of APEC, provided it was based on the principles of consensual governance and what became known as open regionalism. This is similar to the approach taken by most other countries in the region, but at odds with the underlying preferences of the USA. Seeing itself as the most open economy, the USA has instead sought a set of arrangements based on specific reciprocity requirements. As we have seen, paralleling these differing national strategies are academic debates about the organisational prerequisites for successful multilateral economic cooperation.

If one judges APEC purely on the basis of the extent to which it has produced concrete economic benefits that would not have come about in its absence, one would have to conclude that, as yet, it has been long on promise and short on results. However, if one accepts that APEC has at least as much to do with broader foreign policy calculations as it does with economic cooperation, then it is readily apparent that its progress is quite remarkable.

---

[1] Grateful acknowledgments to Nancy Viviani and the editors for helpful comments in an earlier draft of this paper.

[2] With the accession of Vietnam, ASEAN now has seven members. For the purposes of this paper, however, the focus is on the previous six members since Vietnam's involvement in the whole process is very recent.

[3] Interestingly, Funabashi (1995) has recently argued that Australia was in substantial measure acting on behalf of the Japanese government. For discussion of Australian reactions to these claims see the various articles in the *Australian*, 5 and 6–7 January 1996.

[4] The participating countries are: Australia, Indonesia, Singapore, Malaysia, Thailand, Brunei, the Philippines, Hong Kong, China, Taiwan, South Korea, Japan, the Soviet Union/Russia, Canada, the USA, Mexico, Chile, Peru and the Pacific Island states. Recently Vietnam has been included as an associate member.

[5] The ASEAN PMC then comprised the six ASEAN states, together with five dialogue partners: Japan, the USA, Canada, Australia and New Zealand.

[6] These include: the Trade and Investment Data group, the Trade Promotion group, the Human Resources Development group, the Energy Cooperation group, the

Marine Resource Conservation group, the Telecommunications group, the Fisheries group, the Tourism group, the Transportation group and the Investment and Industrial Science and Technology group.

7 Appeals to nationalist sentiment do, of course, resonate in most polities. It seems likely, however, that this is particularly the case in Malaysia, given its strong ethnic divisions and the ongoing need to reassure the indigenous Malay majority of its political predominance. Jesudason (1989) provides a valuable overview of the political economy of ethnicity in Malaysia.

8 Or, as Malaysian Prime Minister, Dr Mahathir (1995: 130) said of education: 'We hope Japan will contribute more to the education of Malaysians by increasing the number of places for overseas students at universities and training institutions in Japan. Of course we do not expect Japan to pay for our students' education, but the availability of places is very important to us. Of the 70 000 Malaysians studying abroad, Britain and the USA each account for some 20 000. Only around 2000 (including trainees) are in Japan. This is a very small number. Japan should have more'.

9 In an intriguing parallel to this line of thinking, one of the Australian academic economists most closely associated with APEC, Ross Garnaut (1994), declared that in the event of sustained hostile USA trade measures, even Australia would be attracted to the idea of an EAEC—provided, of course, at least some Western countries could be admitted.

10 On this last point, according to Funabashi (1995: 107) the USA representative to and chair of the APEC Eminent Persons Group, Fred Bergsten, was told by a German GATT negotiator that 'The chief determinant of the successful conclusion of the Uruguay Round was the APEC summit in Seattle; they sent us a clear message. You had an alternative, and we did not'.

11 A parallel but broader observation can be made about the wider corpus of literature on APEC from all sources. By comparison with the domestically oriented comparative political economy literature on South-East Asia (or other developing regions) the international political economy literature on APEC is more narrowly focused in both theoretical and ideological terms. On big domestically-rooted questions, such as industrialisation and democratisation, one finds fundamental differences of interpretation and intellectual approach—differences which divide political scientists and economists, and sub-divide each of these groups. However on big international political economy questions pertaining to APEC, such as the preconditions for multilateral economic cooperation, the divergence is much less. To be sure, there are lively debates, but they do not correspond to the familiar intellectual and ideological cleavages. Indeed, much of the theoretical literature is couched more or less explicitly in terms of game theory, which grew out of mathematics and came to political science via micro-economics. Speculating on the explanations of this intriguing difference between domestic and international

political economy literatures is beyond the scope of this chapter, but in some measure it seems to reflect the fact that, unlike the domestic political economy literature, the international political economy literature is heavily dominated by North American scholarship and the greater intellectual and ideological homogeneity associated with it.

[12] An interesting and clear exception to this proposition has been the strong and direct lobbying efforts in Washington to support the development of APEC by large USA corporations with a major stake in Asia. This support was based on anticipated assistance from' APEC with the opening of Asian markets to foreign participation. How long such firms will be prepared to support APEC if it fails to deliver tangible benefits remains to be seen.

[13] See Higgott (1992) and the various articles in the two special editions of the *Pacific Review* (1994 7 (4) and 1995 8 (1)) examining the impact of epistemic communities in Asia.

[14] The quotes are from the USA representative to the APEC Emminent Persons Group, C. Fred Bergsten (1996: 105) and the Japanese representative, Ippei Yamazawa (Ravenhill 1995b). One wonders whether it is being too cynical to attribute an element of self-interest to such assessments when their authors are likely to be among the direct beneficiaries of increased attention to the institutional development of APEC. The biting critique of Hughes (1991) has some resonance here.

References

Aggarwal, V. (1994) 'Comparing regional cooperation efforts in the Asia-Pacific and North America', in A. Mack and J. Ravenhill (eds), *Pacific Cooperation: Building Economic and Security Regimes in the Asia-Pacific Region*, Sydney: Allen & Unwin.

Ariff, M. (1994) 'Open regionalism à la ASEAN', *Journal of Asian Economics*, 5 (1): 99–117.

ASEAN (1990) Joint Statement of ASEAN Foreign and Economic Ministers, Kuching, Malaysia (February).

Bergsten, C. F. (1994) *APEC: The Bogor Declaration and the Path Ahead*, Washington DC: Institute of International Economics.

Bergsten, C. F. (1996) 'Globalizing free trade', *Foreign Affairs*, 75 (3): 105–20.

Bergsten, C. F. and Noland, M. (1993) *Reconcilable Differences? United States–Japan Economic Conflict*, Washington: Institute for International Economics.

Cohen, S. (1995) 'The variable geometry of Asian trade', in E. Doherty (ed.) *Japanese Investment in Asia: International Production Strategies in a Rapidly Changing World*, San Francisco: Asia Foundation and the Berkeley Roundtable on the International Economy.

Doherty, E. (ed.) (1995) *Japanese Investment in Asia: International Production Strategies in a Rapidly Changing World*, San Francisco: Asia Foundation and the Berkeley Roundtable on the International Economy.

Drysdale, P. (1988) *International Economic Pluralism: Economic Policy in East Asia and the Pacific*, Sydney: Allen & Unwin.

Drysdale, P. and Garnaut R. (1993) 'The Pacific: an application of a general theory of economic integration', in C. F. Bergsten and M. Noland (eds) *Pacific Dynamism and the International Economic System*, Washington DC: Institute for International Economics.

Elek, A. (1992) 'Pacific economic cooperation policy choices for the 1990s', *Asian-Pacific Economic Literature*, 6 (1): 1–15.

Elek, A. (1995) 'APEC beyond Bogor: an open economic association in the Asia-Pacific region', *Asian-Pacific Economic Literature*, 9: (1) 1–16.

Fane, G. (1995) 'APEC: regionalism, globalism, or obfuscation?' *Agenda*, 2 (4): 399–409.

Fishlow, A. and Haggard, S. (1992) *The United States and the Regionalization of the World Economy*, Paris: OECD.

Frankel, J. (1993) 'Is Japan creating a yen bloc in East Asia and the Pacific?', in J. Frankel and M. Kahler (eds) *Regionalism and Rivalry: Japan and the United States in Pacific Asia*, Chicago: University of Chicago Press.

Funabashi, Y. (1995) *Asia Pacific Fusion: Japan's Role in APEC*, Washington DC: Institute for International Economics.

Garnaut, R. (1994) *APEC After Seattle and Uruguay Round*, Kuala Lumpur: PECC X Conference, March.

Garnaut, R. and Drysdale, P. (eds) (1994) *Asia Pacific Regionalism*, Sydney: Harper Educational.

Haggard, D. (1995) *Developing Nations and the Politics of Global Integration*, Washington DC: Brookings Institution.

Haggard, S. (1990) *Pathways from the Periphery: The Politics of Growth in the Newly Industrializing Countries*, Ithaca: Cornell University Press.

Higgott, R. (1992) 'Pacific economic cooperation and Australia: some questions about the role of knowledge and learning', *Australian Journal of International Affairs*, 46 (2): 182–97.

Higgott, R. (1994) 'APEC: a sceptical view', in A. Mack and J. Ravenhill (eds) *Pacific Cooperation: Building Economic and Security Regimes in the Asia-Pacific Region*, Sydney: Allen & Unwin.

Higgott, R. and Stubbs, R. (1995) 'Competing conceptions of economic regionalism: APEC versus EAEC in the Asia Pacific', *Review of International Political Economy*, 2 (3): 516–35.

Hughes, H. (1991) 'Does APEC make sense?', *ASEAN Economic Bulletin*, 8 (2): 125–36.

International Monetary Fund *Direction of Trade Statistics* (various issues), Washington DC: International Monetary Fund.

Jesudason, J. (1989) *Ethnicity and the Economy: The State, Chinese Business, and Multinationals in Malaysia*, Singapore: Oxford University Press.

Kahler, M. (1994) 'Institution-building in the Pacific', in A. Mack and J. Ravenhill (eds) *Pacific Cooperation: Building Economic and Security Regimes in the Asia-Pacific Region*, Sydney: Allen & Unwin.

Katzenstein, P. (forthcoming) 'Japan in Asia: theoretical and comparative perspectives', in P. Katzenstein and T. Shiraishi (eds) *Between Two Worlds: Japan in Asia*, Ithaca: Cornell University Press.

Krugman, P. (1993) 'The uncomfortable truth about NAFTA: it's foreign policy, stupid', *Foreign Affairs*, 72 (5): 13–19.

Lincoln, E. (1990) *Japan's Unequal Trade*, Washington: Institute for International Economics.

MacIntyre, A. (ed.) (1994) *Business and Government in Industrialising Asia*, Ithaca: Cornell University Press.

Mahathir, M. and Ishihara, S. (1995) *The Voice of Asia: Two Leaders Discuss the Coming Century*, Tokyo: Kodansha International.

Manning, R. and Stern, P. (1994) 'The myth of the Pacific community', *Foreign Affairs*, 73 (6): 79–93.

Panagariya, A. (1993) *Should East Asia Go Regional?: No, No, and Maybe*, Policy Research Department, Working Paper no. 1209, World Bank, Washington DC.

Ravenhill, J. (1993) 'The 'Japan problem' in Pacific trade', in R. Higgott, R. Leaver and J. Ravenhill (eds) *Pacific Economic Relations in the 1990s: Cooperation or Conflict?*, Sydney: Allen & Unwin.

Ravenhill, J. (1995a) *Bringing Politics Back In: The Political Economy of APEC*, Seoul: Yonsei University.

Ravenhill, J. (1995b) 'Competing logics of regionalism in the Asia-Pacific', *Journal of European Integration*, 18 (2–3): 179–99.

Ravenhill, J. (1995c) *The Growth of Economic Interdependence in East Asia and Its Effect on the Australia-United States Relationship*, Asia Impact Study Workshop, Australian Centre for American Studies, 11–12 July, Sydney.

Richardson, M. (1995) 'Trade war worries Asia: dispute with Japan tilts U.S. to Europe', *International Herald Tribune*, 23 May: 13.

Shirk, S. (1994) *How China Opened Its Door: The Political Success of the PRC's Foreign Trade and Investment Reforms*, Washington, DC: Brookings Institution.

Smith, C. (1994) 'Man of the moment: they love Malaysia's Mahathir in Japan', *Far Eastern Economic Review*, 24 November: 18.

Soesastro, H. (1994a) *The Institutional Framework for APEC: An Asian Perspective*, Australian, Indonesian, and Japanese Approaches to APEC, conference paper, Australian National University, Canberra (September).

Soesastro, H. (1994b) 'Pacific economic cooperation: the history of an idea', in R. Garnaut and P. Drysdale (eds) *Asia Pacific Regionalism: Readings in International Economic Relations*, Sydney: Harper Educational.

Timberman, D. (ed.) (1992) *The Politics of Economic Reform in Southeast Asia*, Manila: Asian Institute of Management.

USTR (1994) *1994 National Trade Estimate Report on Foreign Trade Barriers*, Washington: USA Trade Representative.

USTR (1995) *1995 National Trade Estimate Report on Foreign Trade Barriers*, Washington: USA Trade Representative.

Woods, L. (1993) *Asia-Pacific Diplomacy: Nongovernmental Organizations and International Relations*, Vancouver: University of British Columbia Press.

Yamazawa, I. (1992) 'On Pacific economic integration', *Economic Journal*, 102, November: 1519–29.

10

Trans-state Developments in South-East Asia: Subregional Growth Zones

James Parsonage

Subregional Growth Zones in Global Context

Economic development in the latter half of the twentieth century has been marked by the expansion of transnational production and financial networks, commonly described as 'globalisation', as technological advances have facilitated the global decomposition of production (Frobel et al. 1980; Hirst and Thompson 1996).[1] Globalisation has also been driven by the increasing prevalence of liberal economic policies involving deregulation of national economies which has stimulated the growth of free trade organisations such as the Asia-Pacific Economic Cooperation forum (APEC) and the World Trade Organisation (WTO). The end of the cold war has accelerated this trend towards a more integrated global economy by incorporating 'socialist' economies more fully within the ambit of global capitalism. In this reorganisation of production on a global scale there are complex variations in the ways that transnational and nationally based corporations undertake their activities (see Gereffi 1992). Fundamentally, though, in all these arrangements the 'national economy' is generally considered to be of diminished conceptual and practical relevance (Ohmae 1995; Reich 1991; Harris 1986).[2]

Paradoxically, globalisation has been accompanied by a regional concentration of trade and investment due to changes in the organisation of production which favour proximity in the production process and the need to address variations in regional markets (UNCTC 1991; Morales and Quandt 1992).[3] These economic dynamics driving regionalism have been reinforced by nation states seeking to enhance national competitiveness through collaboration with neighbouring states. This latter development has been manifested in formal associations such as the North American Free Trade Agreement (NAFTA) and the European Community.

More recently in South-East Asia, especially within the Association of South-East Asian Nations (ASEAN), formal agreements have emerged to facilitate subregional economic networks that combine complementary economic differentials, such as production costs and infrastructure, between neighbouring economies. These subregional arrangements have been the subject of various analyses and descriptions.[4] They have been encapsulated in such geometric metaphors as growth triangles, circles, quadrangles or polygons. In this discussion, the generic term 'Subregional Growth Zones' (SGZs) is used to depict these arrangements for enhancing subregional trade and attracting foreign direct investment (FDI).

SGZs are not entirely unprecedented, having existed for some time in Europe and along the USA–Mexico border (Sree and Siddique 1994: 47; Sklair 1989), in addition to similar proposals in North-East Asia (see Myo et al. 1994). However, in the view of neo-classical economists, their contemporary expansion demonstrates the increasing influence of market forces in organising production according to comparative advantage (see Ohmae 1993 and Lee 1993). In other words, national barriers to the most cost effective organisation of production according to the availability of resources and differences in labour and other costs are being dismantled. Though neo-classical-influenced analyses of SGZs often acknowledge the interplay of political and economic dynamics underlying their emergence, they usually view the role of the state as secondary (Ng and Wong 1991). States are seen as providing a supporting role by implementing policies and providing infrastructure conducive for subregional exchange. This view is consonant with the neo-classical idealisation of globalisation as capital transcending the nation state (Mittelman 1996; Panitch 1994), exemplified by Kenichi Ohmae's (1993: 79) description of SGZs as 'natural economic zones' delineated by economic complementary rather than formal national borders.

In contrast to this approach it will be argued that SGZs are not solely 'market driven' with states merely assisting the inexorable advance of market forces, nor do they represent the victory of market functional 'rationality' over rent-seeking coalitions impeding the transition to a market economy. This will be demonstrated by examining the origins, characteristics and prospects of several South-East Asian SGZs. This will include the Indonesia–Malaysia–Singapore Growth Triangle and the Indonesia–Malaysia–Thailand Growth Triangle hereafter conveniently adopting the generic Growth Zones (GZ), thus respectively abbreviated to IMS-GZ, and IMT-GZ. It will also encompass the East ASEAN Growth Area (EAGA) (see map 10.1) and the Greater Mekong Subregion (GMS). The discussion will highlight the interdependence between economic and political dynamics in the integration of subregional integration of subregional economies within and across national borders (Milne 1993: 293).

Map 10.1 Subregional growth zones in South-East Asia

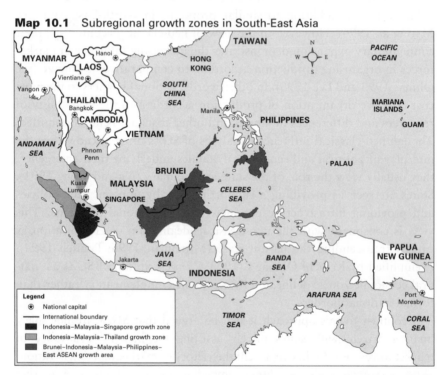

Legend
- ⊛ National capital
- —— International boundary
- ■ Indonesia–Malaysia–Singapore growth zone
- ▨ Indonesia–Malaysia–Thailand growth zone
- ■ Brunei–Indonesia–Malaysia–Philippines–
  East ASEAN growth area

Source: East Asia Analytical Unit (1995): viii.

The need to address the nexus between polity and economy is especially relevant in South-East Asia where political legitimacy is often intimately related to the promotion of economic development. South-East Asian governments have historically viewed the 'market' as a means to attain political goals rather than as an end in itself (Castells 1992), and have extensively intervened in the social and economic spheres to create and maintain nationally-based comparative advantages in human and physical capital (Robison and Rodan 1986; Douglass 1991; Henderson 1993). SGZs involve the continuation and even transformation of state intervention to create the appropriate conditions for new forms of intensified subregional manufacturing and financial networks. Apart from confirming the continued economic significance of the state, SGZs embody the emergence of a new relationship between capital and the state with not only economic but social and political consequences.

Old Wine in New Bottles?

The economic interdependence of Hong Kong and southern China is often regarded as having stimulated the development of subregional economic zones in Asia by demonstrating the benefits of subregional cooperation (Min and Myo 1994: 1). This interdependence stemmed from China's 'open door' policies in the late 1970s which opened specially created 'Special Economic Zones' (SEZs) and coastal cities to foreign investment. These policies enabled Hong Kong-based manufacturing capital to relocate their operations to SEZs in neighbouring Guangdong to take advantage of cheaper land and labour costs. This stimulated the rapid growth of export oriented industrialisation in the SEZs and concomitant expansion of Hong Kong's tertiary sector which handles the bulk of South China's exports.[5] Along with Taiwanese investment in China, this Southern Chinese Subregional Growth Zone or 'Greater China' (Myo et al. 1994) subregion encompassing 120 million people has experienced rapid economic growth with a combined GDP of nearly $400 billion (Mittelman 1995: 289).

However, analysis of South-East Asian SGZs needs to be historically informed as many crossborder linkages in the area, such as those between Johor and Singapore, date from the colonial era (Buchanan 1972; Guinness 1991). Indeed the IMS-GZ invokes a sense of *déjà vu*, alluded to by Singaporean Cabinet Minister Yeo Yong Boon, who believes that little has changed since Singapore's colonial foundation in 1819 as it fundamentally remains a 'value adding switching node' for people, goods, capital, financial risks and information (Yeo 1990). Moreover, in tandem with the relocation of industry from Hong Kong to southern China, an embryonic IMS-GZ was also evident during the late 1970s. This stemmed from Singapore's state-directed industrial upgrading policy which encouraged labour-intensive industries to relocate to the neighbouring Malaysian state of Johor and Indonesian island of Batam (see Rodan 1989; Ho 1979). Both Indonesia and Malaysia responded to this initiative by implementing policies to attract industrial investment, resulting in a 1980 Batam Economic Cooperation Agreement between Singapore and Indonesia and significant investment in Johor by Singaporean-based capital (Parsonage 1992a, 1992b; Regnier 1991: 76).

Despite the apparent similar economic dynamics driving economic interdependence between Hong Kong and Singapore and their respective 'hinterlands' in the late 1970s, Singapore failed to emulate Hong Kong's economic integration with southern China. This was due to resentment over Singapore's role as an intermediary between the region and international capital (Mirza 1986; Indorf 1984), while overall ASEAN economic cooperation was fettered by economic nationalism and direct competition in the same commodity and capital markets (Suriyamongkol 1988; Indorf 1984). The recent emergence of formal SGZs within South-East Asia provides vivid evidence of the extent to which these nationalist concerns have been ameliorated by a conjuncture of national and international political and economic transformations.

Regionalism and the End of the Cold War

The divisive nature of ASEAN economic nationalism mellowed after an economic recession in the early and mid-1980s, linked to a fall in commodity prices and international economic downturn, demonstrated the region's vulnerability (Robison et al. 1987). This led South-East Asian policy-makers, particularly in Malaysia and Indonesia, to liberalise their foreign investment regulations as a means of accelerating industrialisation. The recession also underlined difficulties confronting Singapore's industrial upgrading polices (Rodan 1989), and prompted Singaporean policy-makers to rely less upon market forces to advance regional links.[6] The Singaporean government has since actively encouraged regional linkages to nurture Singapore's envisaged role as a regional support centre for international capital (Singapore 1986).

The convergence in economic policy within ASEAN created a conducive regional political and economic environment for economic regionalism (Hobohm 1989). It also complemented economic restructuring pressures in Japan and the Asian newly industrialising countries (NICs) where the need to relocate labour-intensive processes—due to currency appreciation, increased labour and production costs and the loss of trade privileges—was underpinned by large current account surpluses (Drobnick 1992; Cheng and Haggard 1987). These complementary differentials stimulated a surge in investment from North-East Asia which accelerated the pace of industrialisation in South-East Asia, reflected in manufacturing's share of total exports increasing from 10–30 per cent in Indonesia, 30–45 per cent in Malaysia and 40–55 per cent in Thailand between 1985–89 (Drobnick 1992: 22; Chia 1993: 85; UNCTC 1991). These developments reinforced the positive attitude towards regional economic cooperation among South-East Asian governments.

The emergence of Asian economic regionalism has been slow in comparison with that of regional economic groupings in Europe and America such as NAFTA and the European Community. However, it is now being driven by powerful political and economic dynamics. Apart from the complementary differentials between North-East Asian and South-East Asian economic restructuring policies, regionalism is being supported by ASEAN states due to concern over increased global competition for FDI. Political support for increased regionalism within ASEAN has also been reinforced by ASEAN forfeiting its political utility to the West as a bulwark against communism in South-East Asia due to the unravelling of cold war political and economic structures. This has exposed ASEAN states to criticism over human rights, environmental and labour issues, and raised concern that such criticism could be linked to trade disputes and generate projectionist measures by Western governments. In a more positive sense, the end of cold war divisions reinforced South-East Asian regionalism as the normalisation of relations between China and South-East Asia and the transition from centrally controlled economies towards various forms of 'market socialism' in Vietnam, Cambodia and Laos has created trade and investment opportunities for more mature ASEAN capital.

These geopolitical transformations stimulated a search within ASEAN for new sources of solidarity to replace a now redundant cold war security focus, especially as Vietnam came to be redefined as a 'market opportunity' rather than as a security threat. The main source of solidarity, given the need for continued economic expansion to sustain political legitimacy, has been a common focus on accelerating economic development. It has been this change in ASEAN's world view and common perceived economic challenges that created the political conditions for the creation of the SGZs (see Goh 1991). Crucially, as will be discussed below, these largely external influences have been reinforced by national and regional alliances between political and economic elites with an interest in advancing regional linkages.

History and Status of Subregionalism

A conducive political environment for ASEAN economic cooperation became evident at the 1987 Manila ASEAN Heads of Government Summit which produced several agreements on promoting intra-ASEAN trade and economic cooperation (Hobohm 1989). However, concrete action was largely manifested at the subregional level with Malaysia implementing a 'twinning' policy between the economies of Johor and Singapore in 1988. Thailand announced an intention to foster subregional linkages in 1989 through the creation of a prosperous 'Suwannaphume' or 'Golden Land' by transforming 'Indochina from a battlefield into a trading market' through economic linkages with Indochina and Burma (Sricharatchanya 1989).

Johor's twinning policy was premised on enhancing its longstanding economic links with Singapore which responded by introducing incentives to entice firms to relocate labour-intensive operations to Johor (MIER 1989). This overt political support for crossborder economic cooperation stimulated a substantial increase in capital inflows with Singaporean-based investment alone increasing 200 per cent in 1988 (Parsonage 1992a). However, this influx over-extended Johor's infrastructure and labour supply, leading Singaporean-based capital to approach the Singaporean government for assistance in easing crossborder congestion. Other multinationals diverted investment to other parts of the region, notably Thailand (Low 1994). These developments prompted Singapore policy-makers to seek to keep labour-intensive manufacturing investment within Singapore's immediate environs, thereby ensuring linkages with Singapore's services, by redirecting investment from Johor to Batam (Natatrajan and Tan 1992; Low 1994).[7] Subsequent negotiations between Singapore and Indonesia resulted in Batam's investment regulations being liberalised in 1989, including the private development of industrial estates to alleviate infrastructure shortcomings (Perry 1991).

This improvement in bilateral cooperation between Singapore and Johor/Batam led Deputy Prime Minister Goh Chok Tong to suggest that: 'Singapore, Batam and Johor could form a "triangle of growth" within ASEAN in the decade ahead ... the Indonesian island of Batam and the nearby Malaysian state of Johor could contribute land, gas, water and labour while Singapore could provide management expertise' (*ST* 21 December 1989).

Goh's suggestion appeared to surprise Malaysia,[8] but the idea was endorsed by Malaysia's Prime Minister Mahathir and Indonesia's President Suharto during a bilateral meeting in June 1990. It has since been reinforced by formal agreements between Singapore and Indonesia on joint development of Riau and the signing of a trilateral Memorandum of Understanding in December 1994. In 1995 and 1996 the IMS-GZ was enlarged to include the Malaysian states of Malacca, Pahang and Negri Sembilan and the Indonesian province of West Sumatra (see map 10.2).

Map 10.2   Indonesia–Malaysia–Singapore growth zone

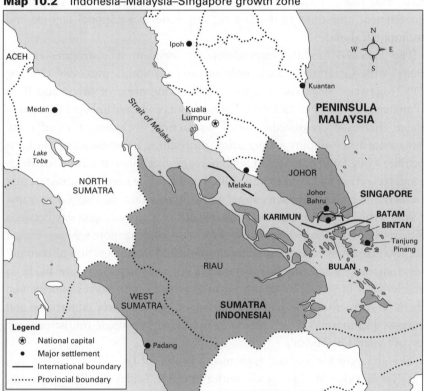

Source: East Asia Analytical Unit (1995): 28.

Assessing the impact of the IMS-GZ is difficult given the advent of wider national liberalisation policies and associated increases in capital inflows. Nonetheless, formal and informal political links between the three governments and state-linked companies have enhanced the area's attractiveness to foreign capital. Johor's industrialisation has been accelerated by an influx of investment in the textile, chemicals, electronics and food manufacturing sectors (Yuhanis et al. 1991; Mubariq 1992). Riau's economy has also been transformed by inflows of investment in electronics, textiles and tourism in Batam and Bintan, heavy industrial projects on Karimun Island and labour migration from mainland Riau to the Riau islands. These developments have also had a positive impact on Singapore's economy by supporting government ambitions to develop the city state as a regional services and technological node.

The IMS-GZ has also engendered an incipient 'transfrontier metropolis' with integrated social, political and economic processes (Herzog 1992) between Singapore and the neighbouring areas of Johor and Riau (Parsonage 1992b; Naidu 1994). Political and economic linkages have facilitated increased crossborder social interaction, including a rapid entry 'smart card' to simplify immigration procedures for commuting business people, while large numbers of Singaporeans visit Johor and Riau for shopping, recreation and even to buy holiday homes. The need for infrastructure to service enhanced flows of capital and labour within the IMS-GZ is transforming the built environment. Projects such as a second causeway between Singapore and Johor, the possible extension of Singapore's Mass Transit Railway to a new 'waterfront' city in Johor and the integration of communications and water supplies between Riau and Singapore, are likely to accelerate subregional integration. These projects have been supplemented by the construction of a spate of airports and 'mega resorts' comprising golf courses and marinas which are being marketed in a single tourist package as an 'Asian Caribbean'.

The apparent success and high media profile of the IMS-GZ acted as a catalyst for other SGZ proposals within ASEAN. Malaysia's Prime Minister Mahathir proposed the 'Northern Growth Triangle' in 1991, encompassing a population of 21 million, by linking Northern Peninsular Malaysia to adjacent provinces in Southern Thailand and Northern Sumatra (EAAU 1995: 52) (see map 10.3). This was formalised as the IMT-GZ during a trilateral conference in July 1993 which commissioned an Asian Development Bank (ADB) feasibility study that subsequently recommended a focus on promoting subregional trade and investment in the primary, industrial and tourist sectors and improvements to labour mobility and infrastructure

(ADB 1994a). Though still in a formative stage, the IMT-GZ has generated significant subregional business and infrastructure proposals with contracts worth US$1.31 billion being signed in 1994 (South China Morning Post (*SCMP* 14 December 1995). Its potential may also be enhanced by infrastructure projects along Thailand's southern seaboard and proposals for a trans-ASEAN gas pipeline and a 'land bridge' from Penang to Songkhla aimed at reducing the need for maritime traffic to use the crowded Malacca Straits (*Star* 16 December 1994; *FEER* 8 June 1995). Manufacturing linkages within the IMT-GZ are likely to be focused on Penang, with its important global electronics production centre, and potential to emulate Singapore's role as a manufacturing growth pole in the IMS-GZ. However, creating production links with Sumatra and Thailand may require considerable enhancement of national and subregional infrastructure (Vatikiotis 1993). Efforts are being made to attract private sector investment to help fund these projects, estimated to cost US$15–20 billion during the next decade, by establishing Special Economic Zones with industrial estates and duty free exemptions along the Thailand–Malaysia border and a joint Malaysian–Indonesian development zone in northern Sumatra (*ADB* 1995).

The IMT-GZ was followed by a 1993 proposal for an East ASEAN-GZ or Growth Area linking Brunei, East Kalimantan and Sarawak (*FEER* 5 August 1993). Assisted by improved Philippine–Malaysian relations over the status of Sabah, this concept was formalised as the Brunei–Indonesia–Malaysia–Philippines East ASEAN Growth Area (EAGA) in March 1994 (*STWE* 29 January 1994) (see map 10.4), and is also referred to as a 'growth polygon'.[9] Like the other ASEAN-SGZs economic cooperation is being reinforced by business councils and trade agreements such as the introduction of tariff free trade between Sulawesi and Sabah (Pushpa and Hamzah 1995; *SCMP* 14 December 1995; EAAU 1995:67).

The EAGA encompasses largely peripheral areas and, unlike the IMS and IMT-GZs, lacks a manufacturing growth pole, prompting the inaugural ministerial meeting to focus on ventures in the primary sector, primarily fishing and logging, and enhancing air and sea communications (EAAU 1995: 69). Joint tourist ventures are also planned, replicating the IMS-GZ's 'Asian Caribbean' approach, including marketing Sarawak and Davao as a single package (*FEER* 10 February 1994). During 1994 and 1995 the EAGA attracted US$200 million in investments including a granite mining venture in Palawan funded by Brunei and Malaysian initiatives for development of a tourist resort in Davao and palm oil plantations in Mindanao (*SCMP* 30 May 1995; 14 December 1995).

Map 10.3   Indonesia–Malaysia–Thailand growth zone

Source: East Asia Analytical Unit (1995): 50.

The establishment of SGZs within maritime ASEAN initially overshadowed developments in mainland South-East Asia. On the mainland, the end of cold war tensions and the end of Vietnam's occupation of Cambodia created the political preconditions for subregional economic cooperation. These developments coincided with Thailand's need for access to regional sources of raw materials and cheap labour and a growing feeling that Thailand could be the growth centre of the region. Thai capital is prominent in the region, being the main source of investment in Laos, with significant financial and technical expertise to support infrastructure and tourist projects in capital starved Laos, Burma and southwest China (Murray 1993: 3). Thailand is also an important trading partner for the countries of the region, sourcing labour and natural resources such as hydroelectric power, timber and natural gas from Laos, Cambodia, Vietnam and Burma and is supporting efforts to establish regional cooperation in the tourism sector.

The increase in mainland South-East Asian economic linkages has generated a proposal for a Greater Mekong Subregion (GMS) involving cooperation among the so-called 'Mekong 6' of Thailand, Burma, Laos, Vietnam, Cambodia and Yunnan in China, all linked by the Mekong river (see map 10.5). The GMS would incorporate areas with a population of 225 million whose current GDP is US$160 billion—which the ADB calculates has the potential to expand to US$681 billion by 2010 (*SCMP* 26 October 1995). The GMS proposal dovetails with initiatives to improve the Mekong river itself for trade, irrigation and hydroelectric power, supported by the ADB and World Bank, both of which have been major proponents of subregional economic growth zones in Asia (Myo 1994). The ADB has provided some US$280 million in loans for infrastructure projects in the GMS for road and rail transportation, river trade, power generation, irrigation telecommunications and tourism (*Asian Business* November 1993; Stuart-Fox 1995; *SCMP* 26 October 1995).

Residual political and military insecurity and infrastructure deficiencies may preclude large-scale subregional manufacturing linkages in the GMS. However, the common need for external economic links to stimulate growth and buttress political regimes after decades of war may well stimulate regional cooperation in infrastructure and agriculture. This potential has been demonstrated by significant increases in regional trade and investment since formerly socialist countries have began the transition to market economies. Vietnam has approved proposals for 1400 projects worth $17.8 billion primarily from the Asian NICs, ASEAN and Japan (*SCMP* 6 December 1995), with Laos attracting US$980 million in FDI since 1989,

Map 10.4   East ASEAN growth area

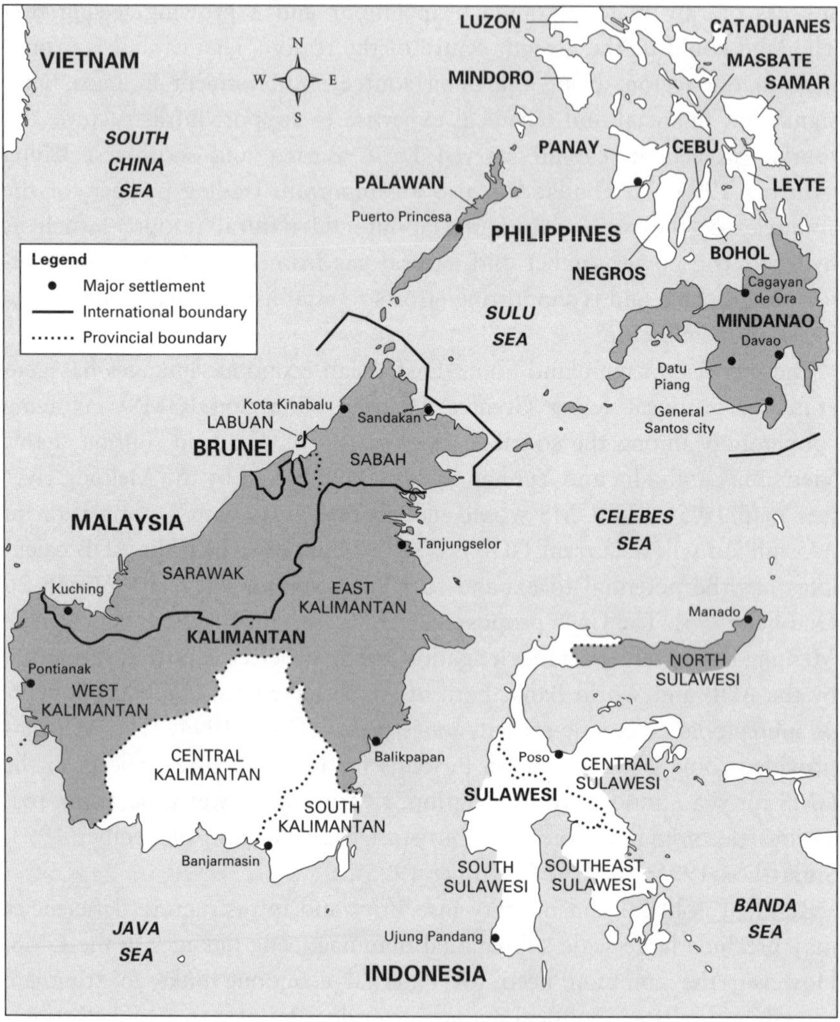

Legend
● Major settlement
— International boundary
······ Provincial boundary

VIETNAM

LUZON
CATADUANES
MASBATE
SAMAR
MINDORO

SOUTH
CHINA
SEA

PANAY
CEBU
LEYTE

PALAWAN
Puerto Princesa

PHILIPPINES

NEGROS
BOHOL
Cagayan
de Ora
MINDANAO
Davao

SULU
SEA

Datu
Piang

General
Santos city

Kota Kinabalu
LABUAN
Sandakan
BRUNEI
SABAH

MALAYSIA

CELEBES
SEA

SARAWAK
Tanjungselor

Kuching

EAST
KALIMANTAN

KALIMANTAN

Manado

NORTH
SULAWESI

Pontianak
WEST
KALIMANTAN

CENTRAL
KALIMANTAN

Balikpapan

Poso

CENTRAL
SULAWESI

SULAWESI

SOUTH
KALIMANTAN

Banjarmasin

SOUTH
SULAWESI

SOUTHEAST
SULAWESI

BANDA
SEA

JAVA
SEA

Ujung Pandang

INDONESIA

Source: East Asia Analytical Unit (1995): 66.

and Cambodia is establishing free trade zones and attracting investment from ASEAN and the Asian NICs.[10] China's recent focus on developing inland provinces has also enhanced the development of economic links between southwest China and mainland South-East Asia and resulted in a significant Chinese economic presence through crossborder transport and hydroelectric projects, labour exports and trade links with Laos and Burma (d'Hooghe 1994).

The preceding empirical description reveals major differences between the various SGZs, especially between the GMS and EAGA which are oriented towards infrastructure development and the primary sectors, and the IMS-GZ, and to a lesser extent IMT-GZ, which are centred on existing manufacturing complexes in Singapore, Johor and Penang. The disparities are even more stark when comparing levels of foreign investment, as segments of the IMS and IMT-GZs have well-established links with international capital while the GMS is struggling to attract investment due to doubts over the profitability of its infrastructure projects (*SCMP* 21 November 1995), and along with the EAGA, is more a statement of intention than reality.

It is noteworthy that the only well-established SGZ, the IMS-GZ, is similar to the Hong Kong–China relationship in being based on an existing 'world city' with pre-existing linkages to both regional and international capital (see Friedmann 1986; 1995). It is likely that the viability of the other SGZs will also be contingent on the ability of the participating states to fashion national and subregional transactional environments conducive for investment from international and local capitals. Indeed, despite their disparate nature, South-East Asian SGZs are distinguished by extensive state intervention premised on creating a conducive subregional environment for investment and intra-regional trade. The nature of this relationship between states, capital and SGZs will now be examined.

Subregionalism and Capital

Like other subregional economic zones in Hong Kong–China, Europe and along the USA–Mexico border, South-East Asian SGZs reflect the spatial consequences of globalisation (Herzog 1992: 519; Sklair 1989; Hart 1995). Their trans-state nature illustrates the extent to which South-East Asian policy-makers have been compelled to respond to the reduced efficacy of nationally oriented incentives as international capital requires a mix of skilled and unskilled labour and developed infrastructure which a single national economy is unlikely to contain. This means comparative advantage

Map 10.5 The Greater Mekong subregion

Source: Asian Development Bank (1994b): 2.

is no longer viewed as an exclusively national feature (Nunez 1990: 40; UNCTC 1991) exemplified by Goh's vision of the IMS-GZ combining Malaysian and Indonesian land and labour with Singapore's management functions. Such visions have been supported by capital from Japan and the Asian NICs (see Clad 1989: 231; Lincoln 1992: 25).

Despite their apparent similarity to other subregional zones, there are important caveats concerning South-East Asian SGZs. First, as mentioned above, only the IMS-GZ has significantly advanced beyond the planning stage and it has unique features being centred on Singapore's existing regional intermediary role and domination of intra-ASEAN trade (Mirza 1988). Second, unlike the Hong Kong–China and USA–Mexico relationships, capital from participating states is not prominent in the industrial sector within the South-East Asian SGZs.

This dependence on FDI means many of the dynamics underlying SGZs are externally derived, reflecting Japanese and Asian NIC restructuring policies over which South-East Asian states can exert little leverage apart from creating a conducive national and regional investment environment. In addition to the provision of new infrastructure, this has required formal policy changes to facilitate subregional exchange through the harmonisation of subregional customs, land and investment regulations. Due to their utility in providing an insulated enclave for experimenting with economic liberalisation policies, it is likely that SGZs will further influence national policies. This has been demonstrated by the extension of investment exemptions granted to Batam in 1989 to other priority zones, while the extension of foreign land leases in Batam from thirty to fifty, and then eighty, years in response to investor demands may presage changes to national land ownership legislation.

The focus on attracting FDI has led some observers to regard SGZs as being primarily driven by foreign capital as opposed to local development impulses (Douglass 1992: 9; Ho 1992: 50). However, SGZs are more than a state sponsored formalisation of existing intra-firm regional divisions of labour established by the spatial pattern of multinational corporation (MNC) investment (see Scott 1987; Henderson 1989). In addition to being influenced by the needs of international capital, states are also subject to the demands of local capital. SGZs may be of little relevance to smaller capitals, but they have potential for emerging regional MNCs which, like capital globally, need to transcend the limits of national economies in order to meet the challenges of intensified global competition (Aggarwar 1985). Crucially, SGZs do not threaten the domestic position of the various major national capitals which have previously impeded regional economic cooperation (Indorf 1984; Clad 1989). Instead, pressure from Singaporean based capital for state assistance in crossborder ventures and the alacrity with which regional capitals have seized on investment opportunities in SGZs suggests a congruence between SGZs and economic needs of local capitals. This has been evident in the IMS-GZ where large and often politically well-connected domestic capitals, such as Singaporean-based Wah Chang, Indonesia's Salim and United Engineers in Malaysia, have been prominent investors in infrastructure and tourism projects. The interests of mature Thai capital, such as Charoen Pokpand which is the largest single investor in China (Viraphol 1994: 7), are also likely to have influenced the Thai government's regional policies.

While SGZs are only one aspect of this regional expansion of local capitals—which are engaged in a variety of trade and investment projects both within SGZs and throughout South-East Asia and China[11]—the *de jure* nature of SGZs confers unique advantages for both local and international capital. State involvement provides capital with a degree of political and economic protection for its regional ventures and can also create opportunities for national capitals. This has been demonstrated in the Batam Industrial Park project, a joint venture between Singaporean government-linked companies (GLCs) and Indonesian capital, where Singaporean companies were awarded 50 per cent of the contracts during the construction phase (*Business Times* 13 March 1993). State promotion of the regional expansion of international capital can also benefit local capitals by creating the possibility for local capital to establish or enhance linkages with MNCs, evident in Singapore's role as a regional centre for financial and manufacturing services (see Panitch 1994: 66).

This congruence between SGZs and the interests of major local capitals reflects the priority South-East Asian governments are according economic regionalisation in national development policies. South-East Asian governments are actively supporting the regional expansion of their respective national capitals by economic diplomacy, establishing SGZs and even direct economic involvement using state-linked capital as either a joint venture partner with private capital or as an independent investor. This trend is exemplified by Singapore's policy of using GLCs, which represent the majority of large local capitals, to spearhead the development of an external economy through regional infrastructure, industrial and tourist ventures (Ministry of Trade and Industry 1993; Parsonage 1994). The Malaysian government has also adopted a high profile regionalisation drive by sending trade missions to China and Indochina to assist Malaysian capital which is currently prominent in resource extraction and development of tourist resorts and industrial estates in the states of Indochina, China and the Philippines. The formation of SGZs is being driven by this regionalist trend which is fostering a diverse range of intra-regional economic cooperation, such as proposals for Kazanah Malaysia, a government investment company that manages the investment of Malaysia's national reserves, to cooperate with Singapore's Government Investment Corporation in regional joint ventures (*Asian Business* October 1994) and supporting private sector funded bilateral ventures.

Implications for Labour

SGZs are likely to have a varied impact on the different segments of labour, primarily female rural-based unskilled labour from peripheral areas and comparatively skilled labour from the developed cores such as Penang and Singapore, which they bring together in subregional manufacturing and service networks.[12] Though intra-regional labour migration has been long extant within South-East Asia the *de jure* nature of Growth Triangle arrangements involves a formalisation and transformation of labour flows. The emergence of formal labour policies within SGZs allows labour migration to be contained within parameters set by the respective national governments in order to maximise the returns for capital and limit labour's organisational capacity. This is evident in Batam where the right to reside, usually in workers' dormitories, is restricted to labour recruited on two-year contracts, in order to pre-empt labour problems such as job-hopping (Perry 1991: 148). While these regulations may not be totally effective, evident in the expansion of shanty towns on Batam, they indicate SGZs may utilise harsher labour regulations than those currently extant nationwide.

The regularisation of labour migration also enables participating states to pre-empt intra-regional friction over illegal immigration and guest labour from undermining subregional linkages. This is especially relevant in the more developed states such as Singapore, Thailand and Malaysia where there is concern that immigrant labour threatens social stability and inhibits industrial upgrading by contradicting incentives for capital to shift to higher value added processes less dependent on cheap labour. The subregional division of labour within SGZs resolves this contradiction between the economic need for, and socio-political aversion to, migrant labour by facilitating access to unskilled labour *in situ* thereby insulating the more developed areas from the social and political problems governments associate with migrant labour.

SGZs may also have implications for labour in the more developed economies such as Singapore and Thailand where there is concern that subregional production networks may result in industrial 'hollowing out' and lead to increased unemployment for low-skilled workers. There is evidence that MNCs relocating their labour-intensive operations to Johor and Batam resulted in over 9000 job losses in Singapore during 1994 (*Asia Inc.* June 1995). While this accords with Singapore's upgrading ambitions and foreign labour is usually laid off before local labour, Singaporean leaders continue to draw attention to the need for Singaporean labour to remain competitive. The enhanced ability of capital to relocate to contiguous sites within SGZs is therefore likely to have a similar effect as labour imports in capping wage increase for low-skilled labour in Singapore and other manufacturing cores such as Penang.

The consequences for managerial and technical labour are likely to be more benign as the establishment and maintenance of subregional manufacturing and financial networks within SGZs is accelerating the emergence of a mobile and affluent middle class in the region. These functionaries require housing, education and other allowances for residing in, or commuting to, peripheral areas, as demonstrated by Singaporean government policies to encourage Singaporean executives to take up foreign postings (Parsonage 1994). The necessity for SGZs to combine these different segments of labour means they may bring disparate social classes into close contact as in the IMS-GZ where businessmen and tourists are accommodated within a sanitised 'pleasure periphery' of resorts and golf courses in Riau and Johor adjacent to low-wage labour housed in dormitories or shanty towns (Douglass 1992; McGee and Macleod 1992).[13] Such concentrated class disparities, which often parallel ethnic divisions, have the potential to arouse class-based resentment (Douglass 1992; Parsonage 1992a). However, this seems to have been anticipated by government polices that pre-empt labour militancy, such as the use of rural female labour on short-term labour contracts which can be easily repatriated (McGee and Macleod 1992: 55).

State and Capital: A New Relationship?

National and subnational state institutions have been prominent actors in establishing and maintaining SGZs including involvement in projects outside their own national borders. This growing internationalisation of the economic role of South-East Asian states has been most evident in the IMS-GZ which is overseen by trilateral ministerial meetings in addition to numerous bilateral agreements and coordination boards (Naidu 1994). The role of state institutions formerly premised on national economic development has also been transformed as they become involved in trans-border developments. Singapore's Economic Development Board (EDB), formerly dedicated to enticing investment in Singapore, has assumed more of an external role in establishing the Riau leg of the IMS-GZ. Subsidiaries of the Jurong Town Corporation and GLCs are involved in numerous joint ventures in Asia as well as inside the IMS-GZ where GLCs are engaged in infrastructure, tourism and heavy industry projects. Capital linked to the Malaysian state, such as Renong and the Johor State Economic Development Corporation (JSEDC), are also prominent actors in the IMS-GZ. In addition to developing industrial estates the JSEDC has significant local and regional business ventures and is developing an industrial park in Padang, West Sumatra, in concert with the local government (*Business Times* 6 March 1996). Thailand's Board of Investment is also taking on the added role of promoting investment by Thais in the region.

The extensive nature of state intervention in South-East Asian SGZs illustrates the necessity for the continuation of functions traditionally provided by the state, such as provision of infrastructure and maintenance of a legal and social environment conducive for accumulation, to be maintained as capital transcends formal national borders (Murray 1971; Panitch 1994: 65). This is particularly relevant in subregional complexes such as South-East Asian SGZs where capital needs to be reassured given the history of intra-regional hostility and disparities in legal systems. This has required significant levels of state involvement in creating the appropriate conditions for subregional accumulation and resolving the adverse political, social and economic consequences of increased regional interaction.

The efficacy of such intervention has been demonstrated by cooperation between Indonesian capital and Singaporean GLCs in transforming Riau's lack of infrastructure and labour into a subregional comparative advantage by importing labour to specially constructed industrial estates contiguous to Singapore's developed infrastructure and international linkages (Parsonage 1994). This creation of transnational comparative advantage challenges the neo-classical dictum that comparative advantage is best exploited by allowing market forces free rein. For, in contrast to the neo-classical contention that state intervention distorts market forces, intervention by the Indonesian and Singaporean governments in Riau has pre-empted market forces. It is unlikely the current levels of investment in Riau would have occurred if these governments had not intervened to enhance Riau's attractiveness (Pangestu 1991: 106).

Though there are significant national variations in the economic role of the state, the widespread nature of state participation in SGZs reflects indistinct boundaries between elite and national interests and close relations between the state and capital in South-East Asia (Clad 1989, Yoshihara 1988).

Contrary to Yoshihara (1988), the networks between political elites and international and local capital have not contradicted the development of capitalism. Instead, they have enhanced the responsiveness of the state to the needs of capital, evident in the *de jure* nature of SGZs, which has been instrumental in insulating foreign capital from informal barriers to investment such as corruption and alleviating infrastructure and labour shortfalls.

This alliance between state and capital can be viewed as a regional variant of what Leslie Sklair (1991) terms a 'transnational capitalist class' (TCC) involving close cooperation between state apparatuses, local and external capital which is assisting the expansion and deepening of capitalism in the region (see also Becker et al. 1987). This regional TCC has a common interest in fostering SGZs as they confer both political and economic benefits, such as enhancing national development policies and providing profitable opportunities for private and state-linked capitals. This class has been instrumental in fashioning the subregional policies and infrastructure required by SGZs illustrated by the cooperation between Singaporean government institutions and Indonesian capital in developing Riau. Examples include Ho Kwon Ping of the Singaporean-based Wah Chang International which has joined Singaporean GLCs in tourism projects in Riau. In addition to having been a member of several Singaporean government committees, Ho is managing the transformation of the Public Utilities Board into a privatised entity, Singapore Power, which is already engaged in regional ventures including the possibility of supplying electricity to Riau (*Business Times* 7 February 1996).[14]

International capital's need for an intermediary in regional ventures has been especially opportune. Singapore's policy-makers are endeavouring to exploit the island's well-established links with MNCs and regional governments to advance the export of local services such as the construction and management of industrial estates (Ministry of Trade and Industry 1986: 12). This role is not limited to states, as it also involves mature regional capitals, for example, Indonesia's Salim Group has partnered the Singaporean state in developing industrial parks in Riau, China and Vietnam, and others are developing industrial estates in South-East Asia.

This alliance between regional states and capitals is likely to further blur the distinction between 'national interest' and the self-interest of political and economic elites. This may be seen in the involvement of a son of the Indonesian President and a brother of Dr B. J. Habibie, the Indonesian Minister for Research and Technology, in the Batam Industrial Park project (*FEER* 18 October 1990). The emergence of a regional TCC has political and social, as well as economic, implications. Mutually reinforcing economic interests among regional political and economic elites are likely to buttress regionalism by ameliorating nationalist impediments to regionalism. It may also buttress the authoritarian nature of many of the region's regimes by complementing an emerging regional political consciousness derived from a neo-Confucian culturalist explanation for Asian economic 'success'. This posits that strong state direction of corporatist relations between the state, labour and national capitals not only ensures economic development but is consonant with a 'traditional' deference to an enlightened governing authority (Clegg and Redding 1990).

Domestic Significance of Subregional Growth Zones

From their inception SGZs have provided a platform for domestic politics. Goh Chok Tong's association with the IMS-GZ, which served to symbolise Singapore's recovery from the mid-1980s recession, also enhanced his status prior to becoming Singapore's Prime Minister. The IMS-GZ also diverted attention from a number of internal security arrests which had attracted adverse national and international reaction (Bowring 1990; Asia Watch 1990). Johor's rapid growth associated with the IMS-GZ and twinning policy benefited the political profile of Chief Minister Muhyiddin Yassin who was elected an UMNO Vice President in 1993. Support for the IMT-GZ can also be linked to the fact that it includes the home states of the Malaysian Prime Minister and his deputy and Chuan Leekpai, Thailand's Prime Minister from 1992 to 1995. The strong support for the IMS-GZ shown by Dr Habibie, who has advocated the construction of bridges linking Batam to the neighbouring islands of Barelang and Rempang, and even envisioned a future link to Singapore, may also be linked to his political ambitions. Habibie is also Chairman of the Batam Island Development Authority and its participation in the IMS-GZ could enhance any Presidential ambitions he holds (*AWSJ* 29 April 1993).

In most cases, SGZs assume a prominent place in broader national development strategies. The IMS-GZ has been promoted as an essential aspect of Singapore's 'external economy' (Parsonage 1994), and conducive for economic growth and balanced regional development in Malaysia (Malaysia 1991a: 51). Both the IMT-GZ and EAGA complement the Sixth Malaysian Plan's strategy of developing electronic industries in the northern region (Penang, Kedah, Perak, Perlis) and resource-based industries in Sabah and Sarawak, as well as enhancing Labuan's potential as a regional financial centre (Malaysia 1991b). The Sultan of Brunei has described the EAGA as the 'cornerstone of Brunei's economic development' (EAAU 1995: 71), and Brunei is attempting to establish itself as the main entry point for the EAGA by enhancing its transport and communication sectors and developing a free trade zone near Muara port (Pushpa and Hamzah 1995: 119). The EAGA also reflects President Ramos' policy of enhancing economic development in the Philippines through regional economic diplomacy and investment which has resulted in significant increases in the Philippines' share of intra-ASEAN trade (*SCMP* 28 September 1995).

The nexus between national agendas and emerging SGZs is less formalised but potentially no less important. The GMS is congruent with the Thai government's 'Suwannaphume' initiative and aspiration to establish Thailand as a regional 'core' economy and intermediary between Indochina and the global economy. Laos views the GMS as a means of achieving its ambitions to act as a regional service and transport node and source of hydroelectric power (Stuart-Fox 1995), while subregional economic cooperation provides a conduit to the global economy for pariah states such as Burma and unstable states like Kampuchea.

Impediments to Economic Regionalism

Notwithstanding their apparent economic and political benefits, GZs do not necessarily insulate national and regional economics from wider changes in the global political economy. Regional production networks in Europe and North America and increased automation in the process may be reducing capital's need for global offshore sourcing of 'cheap labour' (Oman 1994). While South-East Asian SGZs stand to benefit from the need of North-East Asia capital for cheaper production costs, this may result in an over-reliance upon North-East Asian capital, constraining South-East Asia's options for industrial upgrading. Subregional arrangements based on city states, such as the IMS-GZ which is focused on Singapore, may also be vulnerable as they depend upon larger states for the maintenance of international regimes conducive for subregional economic exchange. They are unlikely to tolerate a 'free ride' on such matters if subregional groupings threaten their own national economies (Hirst and Thompson 1994).

While one may not agree with the contention that the NICs are carving out a regional empire (Carlsson and Shaw 1988: 8), a division of labour implies dependence and therefore a political relationship (Marx in Gilpin 1987: 50). This means subregional cooperation ostensibly based on natural economic complementarities may be undermined by national rivalries. Ideally, differentials in economic development among participants should mean SGZs target a diverse range of FDI and economic niches, thereby reducing direct competition among the participating states and subregions. This has been demonstrated by MNCs such as Philips, Seagate Technology and Sumitomo enhancing their use of Singapore as a regional management and technology centre while either relocating or introducing new labour-intensive operations in Johor and Batam (*STWE* 1 June 1991; 25 May 1991). Thailand's regional financial aspirations and focus on developing higher value-added industries also avoids direct competition with its GMS partners whose main requirements are labour-intensive manufacturing, development of infrastructure and establishment of external economic linkages.

Nonetheless, complementary economic differentials do not necessarily eliminate competition as borders are still crucial in defining and encapsulating cost differentials that are crucial for national competitiveness.[15] In addition to a lack of economic complementarity in many South-East Asian SGZs, there is considerable intra-regional competition over developing regional financial centres, and air and sea transport nodes. This has been evident in discord over the ADB study on the IMT-GZ and between participants in the IMS-GZ. Indeed, previous development of both Johor and Batam was motivated by the desire to reduce dependence upon Singaporean services, and participation in the IMS-GZ is not inconsistent with this aim as both areas are developing extensive links with foreign capital and enhancing their infrastructure.

The persistence of nationalist dynamics became obvious in the controversy over the nature of the IMS-GZ following Singapore's high profile release of an insensitive portrayal of the IMS-GZ as a static, vertical division of labour based upon Indonesian and Malaysian land and labour and Singapore's more advanced economic sectors. This made it seem that the high value operations would be located in Singapore and resulted in the perception that the SGZ was premised on furthering Singapore's status as a regional services centre. This has been contradicted by Johor targeting investment in high technology and services. In a similar vein Indonesia has threatened to 'close' the Riau Islands to low technology operations and has evicted textile manufacturers from Batam on the grounds that they contradict Indonesia's aim to nurture high technology manufacturing (*STWE* 6 August 1994). These tensions have been aggravated by infrastructure, labour supply and bureaucratic problems in Johor and Batam (SMA 1992). They have also adversely affected investor sentiment towards the IMS-GZ compared to alternative sites for investment in Vietnam and China, resulting in delays in completing industrial estates on both Batam and Bintan (Economist Intelligence Unit 4/92; 1/93).

Subregional linkages may also exacerbate existing political, economic and social tensions. Despite the allure of Thai investment, there is concern in Burma, Cambodia, and Laos over potential Thai economic hegemony. This has aggravated suspicions that Thailand is maintaining links with anti-government groups within these countries (Murray 1993; *FEER* 18 April 1996: 23). The perceived Singaporean domination of the IMS-GZ and imported inflation resulting from Singaporeans visiting Johor have aggravated a wide range of disputes between Singapore and Malaysia (Parsonage 1992). The prominence of overseas Chinese capital in ventures within SGZs could also reignite latent ethnic hostility in the region. Opponents of the IMS-GZ in Indonesia and Malaysia have attempted to exploit resentment against perceived Chinese economic dominance, especially as Singapore is regarded as a Chinese city state, and there is resentment over the presence of Chinese labour and traders in Burma.[16] Conflict may also be inherent in the process of economic integration between areas with wide socio-economic differences. Increased crossborder interaction has been perceived as responsible for undesirable social effects, such as increased prostitution, the prevalence of AIDS and the practice of businessmen maintaining mistresses or bigamous marriages during their crossborder visits.

Though ostensibly premised on fostering national development, SGZs may contradict national integrity. This fear initially beleaguered Malaysia's participation in the IMS-GZ due to concern that Johor's orientation towards Singapore could undermine Federal-State relations. Similarly, there is concern in Laos that linkages with China, Vietnam and Thailand could diminish its control over border regions (Stuart-Fox 1995: 180), while the IMT-GZ may create the opportunity for southern Thailand's independent-minded Malay-Muslims to enhance their identity through strengthened ties with Malaysia (Vatikiotis 1996). A concern over equity can also arise nationally as SGZs may privilege already-developed areas or be perceived as central exploitation of peripheral areas. This has been evident in Riau where BIDA and central government directives have marginalised the local government and, in one instance, local residents angry over compensation for land acquisition forced workers to leave a golf course development (Mubariq 1992; *Straits Times* 22 May 1993).

ASEAN and Subregional Growth Zones

Despite the common ASEAN interest in fostering economic cooperation, the impact of SGZs is not unambiguous. Competition between SGZs may contradict wider ASEAN integration. The close triangular relationship between Singapore, Malaysia and Indonesia underlying the IMS-GZ and Thailand's emphasis on its purported natural economic and cultural affinity with its Indochinese neighbours, if successful, could presage a schism between the maritime and mainland sections of ASEAN. Overlaying these issues is uncertainty over the ambitions of China as a military and economic superpower and concern that it is not only participating in the GMS and other regional ventures for economic reasons but also to enhance its military profile through improved crossborder access and a naval presence in Burma and the Andaman sea (*FEER* 22 December 1994: 23).

However, the fractious issues among participants in SGZs are likely to be ameliorated by the increasing *de jure* nature of SGZs. This has been demonstrated in the IMS-GZ where the signing of a trilateral Memorandum of Understanding in 1994 reduced discord among the participating governments. Kuala Lumpur has now granted Johor considerable freedom in its dealings with Singapore which in turn has been more tactful in its public approach towards the IMS-GZ. The potential for SGZs to enhance national security reinforces their potential as a force for increased regionalism. Both the EAGA and IMT-GZ involve the development of peripheral areas which have previously been marked by secessionist unrest. Furthermore, subregional cooperation may facilitate access to essential commodities. More generally, the extent of subregional cooperation involved in SGZs should enhance regional security amid regional and global geopolitical transformations by fostering regional interdependence and give each state a self-interest in a peaceful regional environment.

In summary, SGZs are a complex dialectic of both cooperation and competition whose success is contingent upon the management of a wide range of issues which may not necessarily be directly related to SGZs, such as economic rivalry, territorial disputes, discord over refugees and sharing of natural resources. Despite their potential contradictions, SGZs have enhanced ASEAN economic cooperation, formally recognised at the 1992 ASEAN Heads of Government Summit in Singapore, which resulted in a Framework Agreement on Enhancing ASEAN Economic Cooperation permitting two or more ASEAN members to proceed with subregional development. SGZs have also been perceived as providing a base for wider regional cooperation such as AFTA, also agreed during the Singapore Summit, and a means of reinforcing an ASEAN identity amid wider regional groupings such as APEC (Imada and Naya 1992; Weatherbee 1995).

Subregional cooperation in the form of SGZs can also provide the means of facilitating the eventual ascension of Laos, Cambodia and Burma into full ASEAN membership (Sree and Siddique 1994), already known as the 'ASEAN 10', which would create a powerful trading entity with the potential to counterbalance the region's dependence upon external capital. The ability of SGZs to facilitate wider regional economic cooperation has been demonstrated in the expansion of the IMS-GZ to include West Sumatra, and potentially Malacca and Negri Sembilan. Along with the IMT-GZ this would mean a large portion of the western coast of Peninsula Malaysia, Singapore and Sumatra participating in SGZs. Paradoxically, this would probably dilute the original focus of the SGZs, especially the Singapore, Riau and Johor nexus, as SGZs become incorporated within a more encompassing international form of economic cooperation and integration within ASEAN.

Conclusion

Though historical circumstances and specific national characteristics mean that there are significant differences between the various SGZs, some generalisations can be made. The emergence of SGZs is not surprising given the conducive political and economic environment for economic cooperation created by the end of cold war divisions, complementary economic differentials between North-East Asian and South-East Asian restructuring policies and regionalist trends in the world economy. SGZs can therefore be partially explained as a response by South-East Asian states to changed regional and global political and economic circumstances (Parsonage 1992). This means SGZs must be considered as symptomatic of wider international and regional dynamics and should not be analysed in isolation from these influences.

However, SGZs have particular significance as the leading edge of a South-East Asian statist form of regionalism in response to the challenges and opportunities of economic globalisation (see Mittelman 1995). Certainly, as this chapter has shown, the significance of states as economic actors has not been transcended. State involvement has been crucial in providing the physical, legal and human prerequisites for subregional transnational environments.

Indeed states are important actors in resolving the contradictions inherent in the transnationalisation process. For though national boundaries are becoming increasingly irrelevant for flows of capital, states also have an interest in preserving national borders as a means of containing highly valued services and industrial niches in order to enhance national economic competitiveness. National interests are, therefore, a prime aspect in the establishment of SGZs and will also determine their future course. Capital is also dependent on the state to maintain the conditions for economic regionalism and manage potentially fractious political and social consequences.

States also have a direct economic interest in regionalism and are acting as entrepreneurs themselves through the auspices of state-linked capitals which are prominent investors, often in concert with private local capital, in the IMS and IMT-GZs. SGZs further both the economic and political interest of this alliance between state and capital, who constitute an incipient 'transnational capitalist class' with a strong stake in fostering economic regionalism. It is this conjuncture of national, regional and global political and economic dynamics which highlights the need for a social theory of political economy approach in analysing SGZs. Apart from exposing the false dichotomy between state and market, such an approach demonstrates that, far from being 'natural economic zones', SGZs are politically constructed and controlled subregional environments reflecting dynamic and changing patterns of social and economic power.

---

[1] For contrasting views of contemporary globalisation see Hirst and Thompson (1992; 1996) who argue that globalisation is an ideal type description that is unlikely to eventuate and Sklair's (1991) conception of globalisation as a world system based on transnational practices and institutions.

[2] Some analysts regard the state as a barrier to economic growth, best encapsulated by Ohmae's (1993: 79) concept of a 'borderless world' where 'dysfunctional nation states' cease to exist as a meaningful economic entity.

[3] 'Flexible production' practices based on shorter product cycles and the need to address regional variations in market are a main cause of contemporary geographical concentration of production (see Oman 1994: 86–8).

[4] For overviews of South-East Asian Growth Zones see Lee (1991), Toh and Low (1993), Myo et al. (1994) and EAAU (1995). Growth Zones have also been incorporated into the analysis of Asian 'extended metropolitan regions' whose self-sufficiency in food and energy and access to surplus rural labour facilitates rapid growth based on regionally based urbanisation processes (McGee and Greenberg 1992). The popularity of the term 'Growth Triangle' dates from its use by Goh Chok Tong, then a Deputy Prime Minister of Singapore, in describing his vision of the IMS-GZ in December 1989.

[5] Hong Kong-based capital employs over 2 million workers in China and just 700 000 in Hong Kong (Drobnick 1992: 38).

[6] During the late 1970s Singapore preferred the private sector to establish regional linkages (Wilson 1972: 34, 109; Regnier 1991: 77) and did not actively pursue Indonesian suggestions for bilateral government cooperation in developing Batam in the late 1970s (Reza 1994).

[7] This option was based on existing linkages with Batam and a Singaporean-government commissioned study that recommended government support for enhanced regional linkages (Low 1994).

[8] Ironically Goh's IMS-GZ announcement may have been intended to assuage Malaysian resentment over Singapore redirecting investment to Batam (McGee and

Macleod 199: 11). Goh (1991) later stated that his IMS-GZ proposal was only 'an opinion of what was happening' instead of a proposal for economic cooperation.

9 The East ASEAN Growth Area incorporates the Philippine island of Mindanao, Sabah and Labuan in East Malaysia and the Indonesian provinces of East and West Kalimantan, Maluku, central, north and southeast Sulawesi and may be extended to include trade and investment links with Darwin in northern Australia.

10 Official statistics underestimate the amount of crossborder economic activity due to poor statistics collection, smuggling and corruption.

11 For example, Singaporean electronic firms have a significant presence in Penang, Indonesian aircraft are being marketed in Malaysia and Malaysia's 'Proton' car is being assembled in the Philippines, Vietnam and Indonesia. Indonesian-owned firms are also prominent in Singapore, while networks of overseas Chinese capital also have a long history of operating across formal political boundaries within the region.

12 Both Singapore and Penang offer relatively cheap skilled labour compared to Europe and the USA (Henderson 1989), while the inclusion of West Sumatra in the IMS-GZ provides additional sources of cheap unskilled labour (EAAU 1995: 47).

13 It should be noted that labour in Riau and Johor is well paid by national standards due to the relatively high cost of living in both areas and the higher wages paid by MNCs.

14 Elements of this regional TCC have also cooperated in acquiring equity in foreign conglomerates outside the region.

15 Both Malaysia and Singapore have filed complaints against each other with the WTO, Singapore over Malaysian tariffs that protect Johor's petrochemical industry while Malaysia has complained over Singaporean tariffs on its steel exports.

16 However, both the Malaysia and Indonesia governments have been stressing the value of local Chinese capital and UMNO de-emphasised distinctions between indigenous and Chinese communities during the 1995 federal elections.

References

Aggarwar, Raj (1985) 'Emerging Third World multinationals: a case study of the foreign operations of Singapore firms', *Contemporary Southeast Asia*, 7 (3) December: 193–208.

Applebaum, R. P. and Henderson, J. (ed.) (1992) *States and Development in the Asian-Pacific Rim*, London: Sage.

Asian Development Bank (1994a) *Indonesia–Malaysia–Thailand Growth Triangle Development Project, Final Report*, June, Manila: Asian Development Bank.

Asian Development Bank (1994b) *Economic Cooperation in the Greater Mekong Subregion: Toward Implementation*, Manila: Asian Development Bank.

Asian Development Bank (1995) 'Asia's newest growth triangle beckons private investors', *ADB Review*, Manila, Asian Development Bank, June: 3–7.

Asia Watch (1990) *Silencing All Critics: Human Rights Violations in Singapore*, New York: Asia Watch.

Becker D. G., Frieden, J., Schatz, S. and Sklair, R. (eds) (1987) *Postimperialism: International Capital and Development in the Late Twentieth Century*, Boulder: Lynne Rienner.

Bowring, Phillip (1990) 'Change in emphasis', *Far Eastern Economic Review*, 28 June.

Buchanan, Iain (1972) *Singapore in Southeast Asia*, London: Bell.

Carlsson, J. and Shaw, T. M. (1988) *Newly Industrialising Countries and the Political Economy of South-South Relations*, London: Macmillan.

Castells, Manuel (1992) 'Four Asian tigers with a dragon head: a comparative analysis of the state, economy and society in the Asian Pacific rim', in R. P. Applebaum and J. Henderson, (eds) *States and Development in the Asian Pacific Rim*, London: Sage.

Chan, Steve (ed.) (1995) *Foreign Direct Investment in a Changing Political Economy*, London: Macmillan.

Cheng Tun-jen and Haggard, Stephan (1987) *Newly Industrializing Asia in Transition*, Berkeley: Institute of International Studies, University of California.

Chia Siow Yue (1993) 'Foreign direct investment in ASEAN economies', *Asia Development Review*, 11 (1): 60–102.

Clad, J. (1989) *Behind the Myth: Business, Money and Power in Southeast Asia*, London: Unwin Hyman.

Clegg, S. R. and Redding, S. G. (1990) *Capitalism in Contrasting Cultures*, Berlin: De Gruyter.

d'Hooghe, Ingrid (1994) 'Regional economic integration in Yunnan', in David S. G. Goodman and Gerald Segal, *China Deconstructs: Politics, Trade and Regionalism*, London: Routledge.

Douglass, Mike (1991) 'Transnational capital and the social construction of comparative advantage in Southeast Asia', *Southeast Asian Journal of Social Science*, 10 (1) and (2): 14–43.

Douglass, Mike (1992), 'Global interdependence and urbanization: planning for the Bangkok Mega-Urban Region', paper presented at the International Conference on Managing the Mega-Urban Regions of ASEAN Countries: Policy Challenges and Responses, 30 November–3 December, Asian Institute of Technology, Bangkok.

Drobnick, Richard (1992) 'Economic integration in the Pacific Region', *OECD Technical Paper No 65*, Paris.

East Asia Analytical Unit (EAAU) (1995) *Growth Triangles of South East Asia*, Parkes, ACT: Department of Foreign Affairs and Trade, Commonwealth of Australia.

Economist Intelligence Unit (1992) *Singapore Country Report*, no. 4, London: Economist Intelligence Unit.

Economic Intelligence Unit (1993) *Singapore Country Report*, no. 1, London: Economist Intelligence Unit.

Friedmann, J. (1995) 'Where we stand: a decade of world city research', in Paul Knox and Peter J. Taylor (eds) *World Cities in a World System*, Cambridge: Cambridge University Press.

Friedmann, J. (1986) 'The world city hypothesis', *Development and Change*, 17 (1): 69–83.

Frobel, Folker, Heinrichs, Jurgen and Kreyo, Otto (1980) *The New International Division of Labour*, Cambridge: Cambridge University Press.

Gereffi, Gary (1992) 'New realities of industrial development in East Asia and Latin America: global, regional and national trends', in R. P. Applebaum and J. Henderson (eds) *States and Development in the Asian Pacific Rim*, London: Sage.

Gilpin, Robert (1987) *The Political Economy of International Relations*, Princeton University Press.

Goh Chok Tong (1991) A speech on 'ASEAN Competitiveness', CSIS Asia Society Conference, Bali 4 March, in *Speeches*, Ministry of Information and the Arts, 15 (2) Singapore.

Goodman David S. G. and Segal, Gerald (1994) *China Deconstructs: Politics, Trade and Regionalism*, London: Routledge.

Guinness, Patrick (1992) *On the Margin of Capitalism: People and Development in Mukim Plentong, Johor, Malaysia*. Singapore: Oxford University Press.

Harris, Nigel (1986) *The End of the Third World*, Harmondsworth: Penguin Books.

Hart, Jeffrey (1995) 'Maquiladorization as a global process', in Steve Chan (ed.) *Foreign Direct Investment in a Changing Political Economy*, pp. 25–38, London: Macmillan.

Henderson, Jeffrey (1989) *The Globalisation of High Technology Production*, London: Routledge.

Henderson, Jeffrey (1993) 'Against the economic orthodoxy: on the making of the East Asian miracle', *Economy and Society*, 22 (2): 200–17.

Herzog, Lawrence A. (1992) 'Cross-national urban structure in the era of global cities: the US–Mexico transfrontier metropolis', *Urban Studies*, 28 (4).

Hewison, K., Robison, R. and Rodan, G. (eds) (1993) *Southeast Asia in the 1990s: Authoritarianism, Democracy and Capitalism*, Sydney: Allen & Unwin.

Hirst, Paul and Thompson, Grahame (1992) 'The problem of globalisation: international economic relations, national economic management and the formation of trading blocs', *Economy and Society*, 21 (4): 357–96.

Hirst, Paul and Thompson, Grahame (1994) 'Globalisation and the future of the nation state', *mimeo*.

Hirst, Paul and Thompson, Grahame (1996) *Globalisation in Question*, London: Blackwell.

Ho Kwon Ping (1979) 'Birth of the second generation', *Far Eastern Economic Review*, 18 May.

Ho Kwon Ping (1992) 'Singapore 1991', in *Singapore: the Year in Review*, Singapore: Times Academic Press.

Hobohm Sarwar (1989) *ASEAN in the 1990s: Growing Together*, London: Economist Intelligence Unit special report no. 1131.

Imada, Pearl and Naya, Seiji (1992) *AFTA: The Way Ahead*, Singpore: ASEAN Economic Research Unit, Institute of Southeast Asian Studies.

Indorf, Hans H. (1984) *Impediments to Regionalism in SE Asia*, Singapore: Institute of Southeast Asian Studies.

Institute of Southeast Asian Studies, *Southeast Asian Affairs, Yearbook*, various issues, ISEAS, Singapore.

Knox, Paul and Taylor, Peter J. (eds) (1995) *World Cities in a World System*, Cambridge: Cambridge University Press.

Kumar Sree (1994) 'Johor-Singapore-Riau Growth Triangle: a model of subregional economic cooperation', in Myo thant, Min tang and Hiroshi Kakazu (eds) *Growth Triangles in Asia: A New Approach to Economic Cooperation*, Hong Kong: Oxford University Press.

Lee Tsao Yuan (ed.) (1991) *Growth Triangle: The Johor-Singapore-Riau Experience*, Singapore: Institute of Southeast Asian Studies.

Lee Tsao Yuan (1993) 'Subregional economic zones in the Asia-Pacific: an overview', in Toh Mun Heng and Linda Low (eds) (1993) *Regional Cooperation and Growth Triangles in ASEAN*, Singapore: Times Academic Press.

Lim, Linda Y.C. and Pang Eng Fong (1994) 'The Southeast Asian economies: resilient growth and expanding linkages', in *Southeast Asian Affairs 1994*, Singapore: Institute of Southeast Asian Studies.

Low, Linda (1994) 'Government approaches to Sijori', paper delivered at the Joint ISEAS–Pacific Forum CSIS Conference on Economic Interdependence and Challenges to the Nation State: The emergence of Natural Economic Territories in the Asia-Pacific, 27–28 October, Singapore.

Malaysia (1991a) *The Second Outline Perspective Plan 1991–2000*, Kuala Lumpur, Government Press.

Malaysia (1991b) *The Sixth Malaysia Plan 1991–1995*, Kuala Lumpur: Government Press.

Malaysian Institute of Economic Research (MIER)(1989) *Johor Economic Plan 1990–2005 Final Report*, Kuala Lumpur.

McCloud, Donald G. (1986) *System and Process in Southeast Asia: The Evolution of a Region*, Boulder: Westview.

McGee, T. G. and Greenberg, Charles (1992) 'The emergence of extended metro-politan regions in ASEAN', *ASEAN Economic Bulletin*, 9 (1): 22–44.

McGee, T. G. and Macleod, Scott (1992) 'Emerging extended metropolitan regions in the Asia-Pacific urban system: a case study of the Singapore–Johor–Riau Growth Triangle', paper presented at the Workshop on Asia-Pacific Urban Systems: Towards the 21st Century, Chinese University of Hong Kong, February.

Miliband, R. and Panitch, L. (eds) (1994) *Socialist Register 1994: Between Globalism and Nationalism*, London: Merlin.

Milne, R. S. (1993) 'Singapore's Growth Triangle', *Round Table*, 291–304 July.

Ministry of Trade and Industry, Economic Committee (1986) *The Singapore Economy: New Directions, Singapore*: Singapore National Printers.

Mirza, Hafiz (1986) *Multinationals and the Growth of the Singapore Economy*, New York: St Martin's Press.

Mirza, Hafiz (1988) 'Peripheral intermediation : Singapore and the emerging inter-national economic order', in J. G. Taylor and A. Turton (eds) *Sociology of Developing Societies: Southeast Asia*, New York: Monthly Review Press.

Mittelman, James H. (1995) 'Rethinking the international division of labour in the context of globalisation', *Third World Quarterly*, 16 (2): 273–95.

Morales, Rebecca and Quandt, Carlos (1992) 'The new regionalism and regional collaborative competition', *International Journal of Urban and Regional Research*, 16 (3): 462–75.

Mubariq, Ahmad (1992) 'Economic cooperation in the Southern Growth Triangle: an Indonesian perspective', paper presented at a Conference on Regional Cooperation and Growth Triangles in ASEAN, 23–24 April, National University of Singapore.

Murray, David (1993) 'From golden triangle to economic hexagon—recent devel-opment proposals for regional economic linkages in mainland Southeast Asia', IOCPS Occasional Paper no. 34. University of West Australia.

Murray, Robin (1971) 'The internationalization of capital and the nation state', *New Left Review* (67): 84–109.

Myo thant, Min tang and Hiroshi Kakazu (eds) (1994) *Growth Triangles in Asia: A New Approach to Economic Cooperation*, Hong Kong: Oxford University Press.

Naidu G. (1994) 'Johor–Singapore–Riau Growth Triangle: progress and prospects', in Myo thant, Min tang and Hiroshi Kakazu (eds) *Growth Triangles in Asia: A New Approach to Economic Cooperation*, pp. 218–42, Hong Kong: Oxford University Press).

Natarajan S. and Tan Juay Miang (1992) *The Impact of MNC Investments in Malaysia, Singapore and Thailand*, Singapore: ASEAN Economic Research Unit, Institute of Southeast Asian Studies.

Ng Chee Yuen and Wong Poh Kam (1991) 'The growth triangle a market driven response?', *Asia Club Papers* (2): 123–52, Tokyo: Asia Club.

Nunez Wilson Peres (1990) 'From globalisation to regionalization: the Mexican case', OECD *Technical Paper* no. 24, Paris.

Ohmae, Kenichi (1993) 'The rise of the region state', *Foreign Affairs*, Spring 78–87.

Ohmae, Kenichi (1995) *The End of the Nation State–The Rise of Regional Economies*, New York: Free Press.

Oman, Charles (1994) *Globalisation and Regionalisation: The Challenge for Developing Countries*, Paris: OECD.

Pangestu, Mari (1991) 'An Indonesian perspective', in Lee Tsao Yuan (ed.) *Growth Triangle: The Johor–Singapore–Riau Experience*, Singapore: Institute of Southeast Asian Studies.

Panitch, Leo (1994) 'Globalisation and the state', in R. Miliband and L. Panitch (eds) *Socialist Register 1994: Between Globalism and Nationalism*, London: Merlin.

Parsonage, J. (1992a) 'South-East Asia's growth triangle': a subregional response to global transformation', *International Journal of Urban and Regional Research*, 16 (2): 307–17.

Parsonage, J. (1992b) 'Southeast Asia's growth triangle—an extended metropolitan region?', paper presented at the International Conference on Managing the Mega-Urban Regions of ASEAN Countries: Policy Challenges and Responses, Asian Institute of Technology, 30 November–3 December, Bangkok.

Parsonage, J. (1994) *The State and Globalisation: Singapore's Growth Triangle Strategy*, Asia Research Centre Working Paper no. 23, Perth: Asia Research Centre, Murdoch University.

Perry, Martin (1991) 'The Singapore growth triangle: state, capital and labour at a new frontier in the world economy', *Singapore Journal of Tropical Geography*, 12 (2): 139–51.

Pushpa, Thambipillai and Hamzah, Sulaiman (1995) 'Brunei Darussalam after a decade of independence', in *Southeast Asian Affairs 1995*, Singapore: Institute of Southeast Asian Studies, Singapore.

Regnier, Philippe (1991) *Singapore: City State in Southeast Asia*, Honolulu: University of Hawaii Press.

Reich, Robert B. (1991) *The Work of Nations*, London: Simon & Schuster.

Reza, Yamora Sirejer (1994) 'Integrating business opportunities in SIJORI: from the perspectives of some theoretical concepts', paper delivered at the Joint ISEAS-Pacific Forum CSIS Conference on Economic Interdependence and Challenges to the Nation State: The Emergence of Natural Economic Territories in the Asia-Pacific, 27–28 October, Singapore.

Robison, Richard and Rodan, Garry (1986) 'In defence of state intervention', *Far Eastern Economic Review*, 23 October: 54–6.

Robison, R., Hewison, K. and Higgott R. (eds) (1987) *South East Asia in the 1980s*, Sydney: Allen & Unwin.

Rodan, Garry (1989) *The Political Economy of Singapore's Industrialisation: National State and International Capital*, Basingstoke: Macmillan.

Scott, A. (1987) 'The semiconductor industry in Southeast Asia: organisation, location and the international division of labour', *Regional Studies*, 21: 143–60.

Singapore, Ministry of Trade and Industry, Economic Committee (1986) *The Singapore Economy: New Directions*, Singapore: Singapore National Printers.

Sklair, Leslie (1989) *Assembling for Development: The Maquila Industry in Mexico and the US*, London: Unwin Hyman.

Sklair, Leslie (1991) *Sociology of the Global System*, Hertfordshire: Harvester.

SMA (1992) *Singapore Manufacturer's Association Survey report on the JSR GT*, Singapore: SMA Research Director.

Sree, Kumar and Siddique, Sharon (1994) 'Beyond economic reality: new thoughts on the growth triangle', in *Southeast Affairs 1994*, Singapore: Institute of Southeast Asian Studies.

Sricharatchanya, Paisai (1989) 'The Golden Land', *Far Eastern Economic Review*, 23 February: 11–12.

Stuart-Fox, Martin (1995) 'Laos: towards subregional integration', in *Southeast Asian Affairs 1995*, Singapore: Institute of Southeast Asian Studies.

Suriyamongkol, M. L. (1988) *The Politics of ASEAN Economic Cooperation*, Singapore: Oxford University Press.

Taylor, J. G. and Turton, A. (eds) (1988) *Sociology of Developing Societies: Southeast Asia*, New York: Monthly Review Press.

Toh Mun Heng and Low, Linda (eds) (1993) *Regional Cooperation and Growth Triangles in ASEAN*, Singapore: Times Academic Press.

United Nations (UNCTC) (1991*) World Investment Report 1991: the Triad in Foreign Direct Investment*, New York: United Nations Centre on Transnational Corporations.

Vatikiotos, Michael (1993) 'Three's company', *Far Eastern Economic Review*, 5 August.

Vatikiotos, Michael (1996) 'Ties of faith', *Far Eastern Economic Review*, 11 April.

Viraphol, Sarasin (1994) 'Hong Kong and the "golden triangle"', lecture given at the University of Hong Kong, 29 October.

Weatherbee, Donald E. (1995) 'Southeast Asia at mid-decade: independence through interdependence', in *Southeast Asian Affairs 1995*, Singapore: Institute of Southeast Asian Studies.

Wilson, Dick (1972) *The Future Role of Singapore*, London: Oxford University Press.

Yeo Yong Boon (1990) Speech at the Global Strategy conference, 5 June, in *Speeches*, Singapore Ministry for Information and the Arts, 14 (4).

Yoshihara, Kunio (1988) *The Rise of Ersatz Capitalism*, Singapore: Oxford University Press.

Yuhanis, Kamil, Pangestu, Mari and Fredericks, Christina (1991) 'A Malaysian perspective', in Lee Tsao Yuan (ed.) *Growth Triangle: The Johor–Singapore–Riau Experience*, Singapore: Institute of Southeast Asian Studies.

Newspapers and Magazines

ADB Review

Asia Inc.

Asian Business

Asian Wall Street Journal (AWSJ)

Economist

Far Eastern Economic Review (FEER)

Far Eastern Economic Review, Asia Yearbook, various issues.

Singapore Business

South China Morning Post (SCMP)

Star

Straits Times (ST)

Straits Times Weekly Edition (STWE)

Appendix 1
Indonesia: Basic Social and Economic Data

Population (1994)	192.2 million
Population density (1993)	98.5 persons per sq. km
Average annual population growth (%, 1980–92)	1.8
Urban population (%, 1994)	32
Total labour force (1994)	81.2 million
Major ethnic groups	Javanese, Sumatran
Capital city	Jakarta
Land area	1 919 317 sq. km
Official language	Bahasa Indonesia
Other major languages	Javanese, Sundanese, other local languages
Administrative division	Provincial divisions based on large islands or groups of islands

Education

Primary enrolments (% of age group, 1991)	116
Secondary enrolments (% of age group, 1991)	45
University enrolments (% of age group, 1991)	10
Adult literacy rate (%, 1993)	81.5
Educational expenditure (% of GNP, 1992)	1.4

Health

Life expectancy (1992)	60
Infant deaths per 1000 births (1992)	66
Persons per hospital bed (1991)	1 645
Persons per physician (1991)	10 991

Economy

GNP at market prices (1992)	US$108 909.7 million
GNP per capita at market prices (1993)	US$654
GDP growth (%, 1994)	7.3

Trade

Exports, value (1993)	US$36 607 million
Imports, value (1993)	US$31 487 million
Main exports (1993)	Manufactures (45.6%), mineral fuels (28.4%), food & live animals (8%)
Main imports (1993)	Machinery & transport equipment (38.6%), manufactures (19%), chemicals (12.8%)
Foreign debt (1994)	US$90 467 million
Foreign reserves (1994)	US$13 321 million
Energy consumption per capita (1992)	309 kg oil equiv.

Communications

Circulation of daily newspapers per 1000 people (1992)	24
Number of television receivers per 1000 people (1993)	62
Telephones per 100 people (1993)	0.9
Number of mobile telephone subscribers (1993)	53 438

Transport

Passenger cars in use (1993)	1.7 million
Religions	Predominantly Muslim

Sources: Economics and Development Resource Center, Asian Development Bank (1995) *Key Indicators of Developing Asian and Pacific Countries 1995*, vol. 26, Manila: Oxford University Press; Asian Development Bank (1995) *Asian Development Outlook 1995 and 1996*, Manila: Oxford University Press; Far Eastern Economic Review (1995) *Asia 1995 Yearbook: A Review of the Events of 1994*, Hong Kong: Review Publishing Company; Europa Publications (1995) *The Far East and Australasia 1995*, London: Europa Publications; UNESCO (1995) *Statistical Yearbook 1995*, Paris: UNESCO; United Nations (1995) *Statistical Yearbook Fortieth Issue*, New York: United Nations.

Appendix 2
Malaysia: Basic Social and Economic Data

Population (1994)	19.3 million
Population density (1993)	57.8 persons per sq. km
Average annual population growth (%, 1980–92)	2.5
Urban population (%, 1994)	44
Total labour force (1994)	7 846 000
Major ethnic groups	Malays and other indigenous groups, Chinese, Indians
Capital city	Kuala Lumpur
Land area	329 758 sq km
Official language	Bahasa Malaysia
Other major languages	Malay, Chinese, English, Tamil, Itan Dusan, Bajau
Administrative division	14 states
Education	
Primary enrolments (% of age group, 1991)	93
Secondary enrolments (% of age group, 1991)	58
University enrolments (% of age group, 1991)	7
Adult literacy rate (%, 1993)	78.5
Educational expenditure (% of GNP, 1992)	4.8
Health	
Life expectancy (1992)	71
Infant deaths per 1000 births (1992)	14
Persons per hospital bed (1992)	481
Persons per physician (1992)	2411
Economy	
GNP at market prices (1994)	US$69 302 million
GNP per capita at market prices (1993)	US$3260
GDP growth (%, 1994)	8.7
Trade	
Exports, value (1994)	US$61 634 million
Imports, value (1994)	US$62 529 million
Main exports (1994)	Machinery & transport equipment (53.5%), manufactures (18.6%), crude materials (7.5%), mineral fuels (7.2%)
Main imports (1994)	Machinery & transport equipment (60.1%), manufactures (19.3%), chemicals (6.8%)
Foreign debt (1994)	US$23 335 million
Foreign reserves (1994)	US$30 235 million
Energy consumption per capita (1992)	1388 kg oil equiv
Communications	
Circulation of daily newspapers per 1000 people (1992)	117
Number of television receivers per 1000 people (1993)	151
Telephones per 100 people (1993)	12.6
Number of mobile telephone subscribers (1993)	309 030
Transport	
Passenger cars in use (1993)	2.29 million
Religions	Islam, Buddhism, Hinduism

Sources: Economics and Development Resource Center, Asian Development Bank (1995) *Key Indicators of Developing Asian and Pacific Countries 1995*, vol. 26, Manila: Oxford University Press; Asian Development Bank (1995) *Asian Development Outlook 1995 and 1996*, Manila: Oxford University Press; Far Eastern Economic Review (1995) *Asia 1995 Yearbook: A Review of the Events of 1994*, Hong Kong: Review Publishing Company; Europa Publications (1995) *The Far East and Australasia 1995*, London: Europa Publications; Economist Intelligence Unit (1995) *Country Profile: Malaysia & Brunei 1995–96*, London: Economist Intelligence Unit; UNESCO (1995) *Statistical Yearbook 1995*, Paris: UNESCO; United Nations (1995) *Statistical Yearbook Fortieth Issue*, New York: United Nations.

Appendix 3
Philippines: Basic Social and Economic Data

Population (1994)	68.6 million
Population density (1992)	218 persons per sq. km
Average annual population growth (%, 1980–92)	2.4
Urban population (%, 1994)	44
Total labour force (1994)	27 483 000
Major ethnic groups	Indo-Malay, Mestizo, Chinese
Capital city	Manila
Land area	300 000 sq. km
Official languages	Tagalop and English
Other major languages	Spanish, local dialects
Administrative division	14 regions

Education

Primary enrolments (% of age group, 1991)	110
Secondary enrolments (% of age group, 1991)	74
University enrolments (% of age group, 1991)	28
Adult literacy rate (%, 1993)	93.5
Educational expenditure (% of GNP, 1991)	2.6

Health

Life expectancy (1992)	65
Infant deaths per 1000 births (1992)	40
Persons per hospital bed (1992)	756
Persons per physician (1991)	6648

Economy

GNP at market prices (1994)	US$66 920 million
GNP per capita at market prices (1994)	US$975
GDP growth (%, 1994)	4.3

Trade

Exports, value (1994)	US$13 483 million
Imports, value (1994)	US$22 638 million
Main exports (1993)	Electrical equipment & components (31.2%), garments (20%), food (8.8%)
Main imports (1993)	Machinery & transport equipment (33.5%), mineral fuels (11.5%), base metals (8.9%)
Foreign debt (1994)	US$35 269 million
Foreign reserves (1994)	US$6850 million
Energy consumption per capita (1992)	267 kg oil equiv.

Communications

Circulation of daily newspapers per 1000 people (1992)	50
Number of television receivers per 1000 people (1993)	47
Telephones per 100 people (1993)	1.3
Number of mobile telephone subscribers (1993)	76 880

Transport

Passenger cars in use (1993)	0.53 million
Religions	Christian, Muslim

Sources: Economics and Development Resource Center, Asian Development Bank (1995) *Key Indicators of Developing Asian and Pacific Countries 1995*, vol. 26, Manila: Oxford University Press; Asian Development Bank (1995) *Asian Development Outlook 1995 and 1996*, Manila: Oxford University Press; Far Eastern Economic Review (1995) *Asia 1995 Yearbook: A Review of the Events of 1994*, Hong Kong: Review Publishing Company; Europa Publications (1995) *The Far East and Australasia 1995*, London: Europa Publications; Economist Intelligence Unit (1995) *Country Profile: Philippines 1995–96*, London: Economist Intelligence Unit; UNESCO (1995) *Statistical Yearbook 1995*, Paris: UNESCO; United Nations (1995) *Statistical Yearbook Fortieth Issue*, New York: United Nations.

Appendix 4
Singapore: Basic Social and Economic Data

Population (1994)	2.96 million
Population density (1993)	4480.5 persons per sq. km
Average annual population growth (%, 1980–92)	1.8
Urban population (%, 1994)	100
Total labour force (1994)	1 693 100
Major ethnic groups	Chinese, Malay, Indian
Capital city	Singapore
Land area	641 sq. km
Official languages	English, Chinese, Malay, Tamil
Administrative division	Island city state
Education	
Primary enrolments (% of age group, 1991)	108
Secondary enrolments (% of age group, 1991)	70
University enrolments (total number, 1993)	73 650
Adult literacy rate (%, 1993)	89
Educational expenditure (% of GNP, 1988)	3.0
Health	
Life expectancy (1992)	75
Infant deaths per 1000 births (1992)	5
Persons per hospital bed (1992)	290
Persons per physician (1992)	742
Economy	
GNP at market prices (1994)	US$74 361 million
GNP per capita at market prices (1994)	US$25 379
GDP growth (%, 1994)	10.1
Trade	
Exports, value (1994)	US$104 500 million
Imports, value (1994)	US$111 000 million
Main exports (1994)	Machinery & transport equipment (63.9%), manufactures (13.6%), mineral fuels & bunkers (9.6%)
Main imports (1994)	Machinery & transport equipment (56.5%), manufactures (20.3%), mineral fuels & bunkers (8.8%)
Foreign debt (1994)	US$5514 million
Foreign reserves (1994)	US$58 177 million
Energy consumption per capita (1992)	5952 kg oil equiv.
Communications	
Circulation of daily newspapers per 1000 people (1992)	336
Number of television receivers per 1000 people (1993)	381
Telephones per 100 people (1993)	43.4
Number of mobile telephone subscribers (1993)	179 000
Transport	
Passenger cars in use (1993)	0.32 million
Religions	Buddhism, Islam, Hinduism, Christian

Sources: Economics and Development Resource Center, Asian Development Bank (1995) *Key Indicators of Developing Asian and Pacific Countries 1995*, vol. 26, Manila: Oxford University Press; Asian Development Bank (1995) *Asian Development Outlook 1995 and 1996*, Manila: Oxford University Press; Far Eastern Economic Review (1995) *Asia 1995 Yearbook: A Review of the Events of 1994*, Hong Kong: Review Publishing Company; Europa Publications (1995) *The Far East and Australasia 1995*, London: Europa Publications; Economist Intelligence Unit (1995) *Country Profile: Singapore 1995–96*, London: Economist Intelligence Unit; UNESCO (1995) *Statistical Yearbook 1995*, Paris: UNESCO; United Nations (1995) *Statistical Yearbook Fortieth Issue*, New York: United Nations.

Appendix 5
Vietnam: Basic Social and Economic Data

Population (1993)	70.98 million
Population density (1993)	214.4 persons per sq. km
Annual population growth (%, 1992)	2.41
Urban population (%, 1994)	20
Total labour force (1993)	32.7 million
Major ethnic groups	Vietnamese, Chinese, various hill groups
Capital city	Hanoi
Land area	330 363 sq. km
Official language	Vietnamese
Other major languages	English, French, Russian
Administrative division	43 provinces
Education,	
Primary enrolments (1993/94)	9 725 095
Secondary enrolments (1993/94)	3 815 852
University enrolments (1993/94)	118 589
Adult literacy rate (%, 1993)	88
Health	
Life expectancy (1987–92)	67
Infant deaths per 1000 births (1987–92)	36
Persons per hospital bed (1992)	351
Persons per physician (1992)	2533
Economy	
GNP at market prices (1993)	US$12 121 million
GNP per capita at current prices (1993)	US$170
GDP growth (%, 1994)	8.8
Trade	
Exports, value (1994)	US$3800 million
Imports, value (1994)	US$5000 million
Main exports (1991)	Mineral fuels (33.6%), food & live animals (32.4), manufactures (13%), crude materials (12.7%)
Main imports (1991)	Machinery & transport equipment (24.5%), manufactures (23%), chemicals (15.3%)
Foreign debt (1994)	US$24 700 million
Foreign reserves (1988)	US$14.6 million
Energy consumption per capita (1992)	107 kg coal equiv.
Communications	
Circulation of daily newspapers per 1000 people (1992)	8
Number of television receivers per 1,000 people (1993)	42
Telephones per 100 people (1993)	0.4
Number of mobile telephone subscribers (1993)	4060
Transport	
Roads, paved (1992)	12 000 km
Religions	Buddhism, Confucianism, Christian

Sources: Economics and Development Resource Center, Asian Development Bank (1995) *Key Indicators of Developing Asian and Pacific Countries 1995*, vol. 26, Manila: Oxford University Press; Asian Development Bank (1995) *Asian Development Outlook 1995 and 1996*, Manila: Oxford University Press; Far Eastern Economic Review (1995) *Asia 1995 Yearbook: A Review of the Events of 1994*, Hong Kong: Review Publishing Company; Europa Publications (1995) *The Far East and Australasia 1995*, London: Europa Publications; Economist Intelligence Unit (1995) *Country Profile: Indochina, Vietnam, Laos, Cambodia 199--95*, London: Economist Intelligence Unit; UNESCO (1995) *Statistical Yearbook 1995*, Paris: UNESCO; United Nations (1995) *Statistical Yearbook Fortieth Issue*, New York: United Nations; World Bank (1995) *World Development Report 1995: Workers in an Integrating World*, Washington: World Bank.

Appendix 6
Thailand: Basic Social and Economic Data

Population (1994)	59.4 million
Population density (1993)	112 persons per sq. km
Average annual population growth (%, 1990–94)	1.5
Urban population (%, 1994)	35.0
Total labour force (1993)	32.8 million
Major ethnic groups	Thai, Lao, Malay
Capital city	Bangkok
Land area	513 115 sq. km
Official language	Central Thai
Other major languages	Lao, other Thai dialects
Administrative division	72 provinces
Education	
Primary enrolments (% of age group, 1991)	113
Secondary enrolments (% of age group, 1991)	33
University enrolments (% of age group, 1991)	16
Adult literacy rate (%, 1993)	93.5
Educational expenditure (% of GNP, 1991)	3.6
Health	
Life expectancy (1992)	69.5
Infant deaths per 1000 births (1992)	26
Persons per hospital bed (1991)	606
Persons per physician (1991)	4425
Economy	
GNP at market prices (1993)	US$122 billion
GNP per capita at market prices (1993)	US$2083
GDP growth (%, 1994)	8.5
Trade	
Exports, value (1994)	US$44 577 million
Imports, value (1994)	US$53 781 million
Main exports (1994)	Manufactures (36.6%), machinery & transport equipment (32.9%), food & live animals (20.8%)
Main imports (1994)	Machinery & transport equipment (47.4%), manufactures (22.2%), mineral fuels (6.6%)
Foreign debt (1994)	US$55 billion
Foreign reserves (1994)	US$30.3 billion
Energy consumption per capita (1992)	614 kg oil equiv.
Communications	
Circulation of daily newspapers per 1000 people (1992)	85
Number of television receivers per 1000 people (1993)	113
Telephones per 100 people (1993)	3.7
Number of mobile telephone subscribers (1993)	436 000
Transport	
Passenger cars in use (1993)	1.6 million
Religions	Buddhism, Islam

Sources: Economics and Development Resource Center, Asian Development Bank (1995) *Key Indicators of Developing Asian and Pacific Countries 1995*, vol. 26, Manila: Oxford University Press; Asian Development Bank (1995) *Asian Development Outlook 1995 and 1996*, Manila: Oxford University Press; Far Eastern Economic Review (1995) *Asia 1995 Yearbook: A Review of the Events of 1994*, Hong Kong: Review Publishing Company; UNESCO (1995) *Statistical Yearbook 1995*, Paris: UNESCO; United Nations (1995) *Statistical Yearbook Fortieth Issue*, New York: United Nations.

Index

General Agreement of Tariffs and Trade
(GATT) 43, 236
    and South-East Asia 239
    Uruguay Round 226, 234, 241
globalisation 248, 249, 261, 274
    impact on labour regimes 212
Goh Chok Tong, Prime Minister 167, 168,
    169, 171, 263
    association with IMS-GZ 269
    recommends growth triangle 254, 263
governance, and new institutional political
    economy 11–12
Greater Mekong Subregion (GMS) 249, 259
    map of 262
    regional cooperation in infrastructure and
        agriculture 259, 261
growth theory, and East Asian economies
    7–8, 95–6
Growth Triangle see IMS-GZ; IMT-GZ

Habibie, B. J., 'engineers' and economics pol-
    icy struggles 51–3
Heavy Industry Corporation of Malaysia
    (HICOM) 132
Hong Kong
    economic inter-dependence, Singapore
        comparison 252
    special economic zone in southern China
        251, 261, 263

import-substitution industrialisation
    Latin America 219
    Malaysia 124, 134
    Philippines 16, 68–9, 71, 75, 76, 85,
        219–20
    Singapore 150, 151
    Thailand 18, 102–4
IMS-GZ 24, 161, 162, 249, 273
    and ethnic hostilities 272
    and nationalist dynamics 271–2
    and 'transfrontier metropolis' 256
    capital investment 263–4
    comparison with other subregional zones
        263
    enhancement of national security 273
    formal agreements 254–5, 273, 273–4
    Goh's proposal 254, 263
    historical origins 251–2
    impact of 256
    importance for Singapore and Malaysia
        269
    intra-regional competition 271
    labour issues 265–6
    map of 255
    perceived Singapore domination of 272
IMT-GZ
    domestic support for 269
    formation 256
    importance for Malaysia 269
    infrastructure proposals 256–7
    intra-regional competition 271

map 258
Indian workers, Malaysia 19, 122, 123,
    129, 131
Indonesia
    and World Bank 33, 37, 39, 42, 51
    automobile industry 3 7–8, 55–7
    banking industry
        bad and doubtful debts 40, 44–5,
            50–1
        deregulation 36
        regulatory framework controls
            49–50
    bourgeoisie 30, 32
    capitalism, 1949–86 2, 32–3
    cartels and licenses 43, 44–5
    company bankruptcies 40, 48
    conglomerates 31, 33, 35, 36, 37, 39,
        40, 42, 57
        and anti-Chinese sentiment 42
        and state bank credit 40
    corporate empires 16, 33, 37, 39, 46
    corruption and collusion disclosed 12,
        43–4, 48–9
    debt-rescheduling 32–3
    deregulation 15, 30, 35–6, 53, 55
    domestic trade monopolies 44–5
    economic nationalism 29, 33
    economic populism 29
    electricity generation and distribution
        40
    export manufacturing 34
    export oriented industrialisation 211,
        212
    family interests 16, 40, 44–5, 50
        see also Soeharto family groups
    fertiliser pellet monopoly 37
    financial sector reform 34–5
    fiscal and monetary discipline 47–8
    food industry 37, 44–5
    foreign debt 43
    foreign investment 34, 39, 48, 56, 58,
        252
    forestry industry 38–9
    GDP from labour 206
    governance issues 12
    Growth Triangle see IMS-GZ; IMT-GZ
    industry licenses 38, 39, 40
    industry policy conflicts 51–7
    infrastructure investment 38, 39
    interest rates 48
    labour control 212
    labour organisations 206, 207
    liberalism 30, 42, 43, 46–7
    manufacturing 34
    markets and the concentration of econom-
        ic and political power 16, 35–41
    mega-project investments 38–9
    monopolies 36, 37–40, 44–5
    oil prices 33, 34
    petrochemical industry 40, 41, 54–5
    plywood industry 38, 44

natural economic zones   249, 275
neo-classical economics
   and political economy   9–11
   South-East Asia   6, 8–9
neo-liberalism, South-East Asia   21, 205
neo-Weberian views on rent-seeking   65–6
new institutional political economy   9,
   11–13
newly industrialising countries (NICs), Asian
   136, 137, 252, 261
North American Free Trade Association
   (NAFTA)   23, 226, 230, 238, 248
   significance of   240
North-East Asia
   and EAEC   233
   export oriented industrialisation   1
   investment in South-East Asia   253
   market institutions   13
   newly industrialising countries (NICs)   1
NTUC *see* National Trades Union Congress
numerical flexibility   215

oligarchy, Philippines   72–4, 75, 79, 81, 83,
   85, 86
open regionalism   236, 242
Organisation for Economic Cooperation and
   Development, views on Singapore   171
organised labour
   South-East Asia   206–11, 221
   suppression, Thailand   207, 212
   weakness of   209

Pacific Basin Economic Council (PBEC)   227
Pacific Economic Cooperation Council
   (PECC)   22, 227
Pacific Trade and Development Conference
   (PAFTAD)   227
participative flexibility   216
Permodalan Nasional Berhad (PNB)   131–2,
   135
Pertamina (oil company)   33, 41
Phibun government   100, 101
Philippines
   agricultural 'green revolution'   73–4
   agricultural labour force   73
   and World Bank   71, 85
   annual average growth rate   64
   authoritarianism   80–1
   bourgeoisie   17, 75, 78
   capitalist development   78
   corporate land holdings, size limitations
      73
   crisis management   84–6
   cronyism   17, 69, 77–8, 83
   currency exchange controls   68–9
   developmental sequencing   218–20
   economic liberalisation   17, 66, 67, 83,
      84
   elite democracy   79–80, 81
   employment structure   73, 76, 77
   export agriculture   67, 68

export manufacturing   70, 76, 84
export oriented industrialisation   211,
   212
export restructuring   69–70, 71
foreign debt   17, 71–2, 85
foreign investment   71–2
garment industry expansion   16, 70, 78
GDP structure   68, 70
historical background   67–70
import-substitution industrialisation   16,
   68–9, 71, 75, 76, 85, 219–20
industrialisation   75–7
labour as major export   70
labour control   207, 212
labour force   76
labour market deregulation   218
labour organisations   206, 207
land and labour   72–4
Local Government Code   82, 83
manufacturing   17, 68, 69
martial law introduced   81, 84
middle classes   77–9
NGO-PO sector and local government
   82, 86
oligarchy and class power   2, 17, 72–4,
   75, 79, 81, 83, 85, 86
political decentralisation   81–2, 83
political economy   64–86
political landscape   79–80
poor economic performance and rent-
   seeking   65–6
private sector investors   85–6
privatisation, opposition to   213–14
rent-seeking   17, 64–7, 72, 79, 80, 83,
   85
rice agriculture and class relations   72,
   73–4
social structure and development   72
sugar industry   69, 71, 72, 74
'technocrats' policies   84–5
telecommunications industry   85
ties with USA   68, 69, 70, 71
unionism and strikes   75–6, 206, 207,
   213–14, 218–19
'weak state'   79, 84, 85
see also Aquino regime; Marcos regime;
   Ramos regime
pluralist and Marxist-derived political economy   9, 13–15
post-Fordist flexible production systems   21,
   22, 205, 211, 213, 220
Prajogo Pangestu   40, 41
predatory bureaucratism, Indonesia   30
pribumi business, resentment of Chinese   42
privatisation   2, 213
   Indonesia   45–7
   Malaysia   133–4
   Philippines   213–14
   Singapore   160
   Thailand   213
public choice theory   10

pre-industrialisation   149–50
privatisation   160
public policy programs and subsidies
   168–9
regional broadcasting centre   165
Regional Business Forum   159
regional economic integration   160,
   161–6
regional links   252
'Second Industrial Revolution'   155–6,
   157, 165
self-government   150
skill development levy   222
small and medium enterprises master plan
   159
social divisions   166–70
social mobility   148–9, 171
Strategic Economic Plan   159
structural change theories   165
subregionalism   254–6
   see also IMS-GZ
tariff levels   239
technology-deepening   155, 220–1
trade unions   152, 207, 208, 209, 211
unemployment   153, 156
workforce productivity   156
Singapore Inc.   161
skill development levy   222
social classes *see class headings under specific*
   countries; class relations
Soeharto, President, endorses growth triangle
   254
Soeharto, Tommy   44, 51, 56
Soeharto family groups   57
   and Korean automobile company   38,
      55–6
   corporate empires   33, 37, 39, 46
   state lottery monopoly   45
Soeharto government
   changes   32–4
   political uncertainty after   31, 58
   support for Habibie   52
Soekarno, Guided Economy   32
South-East Asia
   and APEC   237–42
   broad state-capital alliances   24–5
   comparison with East Asia   7, 8
   cooperation between different states   24
   economic liberalisation   22, 213–14
   export oriented industrialisation   22,
      211–13
   industrial and social indicators   210
   investment from North-East Asia   253
   labour and industrial restructuring
      21–2, 205–22
   labour market   206–11
      deregulation   22, 218
   labour movement, suppression and stag-
      nation   207, 208–11
   labour systems changes   205, 206, 220

labour under early export oriented industrialisation   211–13
manufacturing and industrial employment
   209–10
modernisation theories   3–4
neo-classical economics debate   6, 8–9
neo-liberalism   21, 205
political economy approaches   9–15
post-Fordist flexible production systems
   21, 22, 205, 211, 213, 220
state role in economic growth   1–2
tariff reductions   239
trade pattern directions   231
trade unions and organised labour   206–9
trans-state developments   248–75
working class and democratic reforms
   21–2
   see also specific countries
South-East Asian countries
   and APEC   230
   liberal capitalism, problems with   2–3
   primary commodity production   1
   private banking systems   2
   privatisation and market deregulation   2
South-East Asian governments
   and APEC   23
   polity and economy   251
Special Economic Zones (SEZs)   251, 257
state autonomy   79
state capacity   79
static flexibility, and labour   214–15, 216
subregional growth zones   24–5, 248–75
   and ASEAN   272–4
   and broader national development strate-
      gies   269–70
   and capital   24, 261–5, 268, 274
   and intra-regional competition   271, 274
   and multinational corporation investment
      263, 271
   dependence on FDI   249, 253, 263
   domestic significance   269–70
   fears over national integrity   272
   global context   248–51
   history and status   254–61
   impediments to economic regionalism
      270–2
   international capital role   268
   labour implications   265–6
   map of   250
   neo-classical influences   249
   origins   251
   provision of state-linked capital   264
   state participation in   266–9, 274–5
   structural differences between types
      261, 263
   within maritime ASEAN   259, 273
   see also East ASEAN Growth Area;
      Greater Mekong Subregion; IMS-GZ;
      IMT-GZ
Sumitro, Professor   42–3